DYNAMICS OF REACTIVE SYSTEMS PART II: *MODELING AND HETEROGENEOUS COMBUSTION*

Edited by
J. R. Bowen
University of Washington
Seattle, Washington

J.-C. Leyer
Universite de Poitiers
Poitiers, France

R. I. Soloukhin
Institute of Heat and Mass Transfer
BSSR Academy of Sciences
Minsk, USSR

Volume 105
PROGRESS IN
ASTRONAUTICS AND AERONAUTICS

Martin Summerfield, Series Editor-in-Chief
Princeton Combustion Research Laboratories, Inc.
Monmouth Junction, New Jersey

Technical papers presented from the Tenth International Colloquium on Dynamics of Explosions and Reactive Systems, Berkeley, California, August 1985, and subsequently revised for this volume.

Published by the American Institute of Aeronautics and Astronautics, Inc.
1633 Broadway, New York, NY 10019

American Institute of Aeronautics and Astronautics, Inc.
New York, New York

Library of Congress Cataloging in Publication Data

International Colloquium on Dynamics of Explosions and
   Reactive Systems (10th:1985:Berkeley, Calif.)
   Dynamics of Reactive Systems. Part II.
   Modeling and Heterogeneous Combustion.
   (Progress in astronautics and aeronautics; v.105: Part II)
   "Technical papers presented from the Tenth
International Colloquium on Dynamics of Explosions and
Reactive Systems, Berkeley, California, August 1985."
   Companion volume to: Dynamics of explosions.
   Includes index.
   1. Modeling—Congresses. 2. Heterogeneous Combustion—
Congresses. I. Bowen, J.R. (J. Raymond) II. Leyer, J.-C. III.
Soloukhin, Rem Ivanovich. IV. Title. V. Series.
ISBN 0-930403-14-2

Table of Contents

Table of Contents for Volume 105: Part I

Chapter I   Flame Ignition and Propagation

Chapter II   Diffusion and Premixed Flames   **113**

Table of Contents for Companion Volume 106

Preface

Companion volumes, *Dynamics of Explosions* and *Dynamics of Reactive Systems,* present revised and edited versions of 90 papers given at the Tenth International Colloquium on the Dynamics of Explosions and Reactive Systems held in Berkeley, California, in August 1985.

The colloquia originated in 1966 as a result of the wide held belief among leading researchers that revolutionary advances in the understanding of detonation wave structure warranted a forum for the discussion of important findings in the gasdynamics of flow associated with exothermic processes—the essential feature of detonation waves—and other, associated phenomena.

Dynamics of Reactive Systems principally concerns the interrelationship between the rate processes of energy disposition in a compressible medium and the concurrent nonsteady flow as it occurs typically in explosion phenomena. *Dynamics of Reactive Systems* (Volume 105:Parts I and II) spans a broader area, encompassing the processes of coupling the dynamics of fluid flow and molecular transformations in reactive media, and occurring in any combustion system. The colloquium, then, in addition to embracing the usual topics of explosions, detonations, shock phenomena, and reactive flow, included papers that deal primarily with the gasdynamic aspect of nonsteady flow in combustion systems, the fluid mechanics aspects of combustion, with particular emphasis on the effects of turbulence, and diagnostic techniques used to study combustion phenomena.

In this volume, *Dynamics of Reactive Systems, Part II:Modeling and Heterogeneous Combustion*, the papers have been arranged into chapters on models, simulations, and experiments in turbulent reacting flows; heterogeneous combustion; and combustion modeling and kinetics. While the brevity of this preface does not permit the editors to do justice to all papers, we offer the following highlights of especially noteworthy contributions.

In Chapter I, Models, Simulations, and Experiments in Turbulent Reacting Flows, *Raiszadeh and Dwyer* report a sensitivity analysis for the k-ϵ model in the case of a turbulent round-jet diffusion flame. The sensitivity study indicated the dependence of the diffusion-flame solution on the semiempirical model constants and the degree of coupling between the flame and the turbulent motion. The random-vortex method, the subject of the plenary lectures, was used by *Ng and Ghoniem* to model a confined shear layer at high Reynolds numbers and by *Hasegawa et al.* to model premixed flames stabilized by a cylinder. *Dibble et al.* use diatomic carbon fluorescence to mark the flame-front position in turbulent non-premixed jet flames with carbonaceous fuels. This technique takes advantage of the fact that diatomic carbon is a short-lived transient produced and consumed in a very thin zone near the fuel-rich side of the flame.

In Chapter II, Heterogeneous Combustion, *Nicholls, Lee, Wolanski*, and their respective coworkers present results of their recent investigations on the combustion of dusts with or without gaseous fuels. Dust or dust hybrid flames are markedly different from gas flames: flame thickness is about fivefold that of gas flame at similar conditions, and its surface is about twice that of a gas flame. *Lee and coworkers* have found that the character of the near-limit flame in hybrid mixtures and the mode of propagation changes as the concentrations of dust and gaseous fuel are varied. Application of perturbation analysis to droplet combustion is another subject covered in this chapter. *Higuera and Linan* consider the effects of surface tension, phase change, and internal fluid motion on the stability of a droplet vaporizing in a hot atmosphere. *Rangel and Fernandez-Pello* report an analysis of thermal ignition in the gas phase surrounding an evaporating droplet.

In Chapter III, Combustion Modeling and Kinetics, *Pitz and Westbrook* report a numerical analysis of interactions between a laminar flame and end gas auto-ignition at approximately 30 atmospheres. The flame is found to have little influence on the predicted rates of auto-ignition, but the resulting high rates of energy release produce strong acoustic waves. *Creighton and Oppenheim* present numerical solutions of detailed chemical kinetics in the concentration-temperature phase plane, and observe that after very short times the kinetics are such that the radical concentrations form a pool, determined by a pseudosteady state, that shifts with changes in temperature. For the cases examined, heat release is proportional to the radical-pool concentrations. *Frenklach et al.* propose a systematic method for the development of reduced reaction mechanisms for dynamic modeling. They postulate that, if a reduced reaction mechanism accurately describes the temporal behavior of both thermal and chain reaction processes characteristic of a more complete mechanism, then the reduced mechanism will describe those chemical processes in a chemically reacting flow with approximately the same degree of accuracy. Tests of this postulate are presented, and extensions to flames considered.

In Part I of this volume, *Dymanics of Reactive Systems: Flames* and *Configurations*, the papers have been arranged into chapters on flame ignition and propagation; diffusion and premixed flames; flame instabilities and acoustic interactions; and practical combustion configurations.

The companion volume, *Dynamics of Explosions,* presents papers on flame acceleration and transition to detonation; initiation and transmission of detonations; detonation structure and limit propagation; detonation kinetics, structure and boundary effects; explosions, shock reflections, and blast waves; heterogeneous detonations and explosions; and condensed phase shocks and detonations (Volume 106 in the AIAA *Progress in Astronautics and Aeronautics* series).

Both volumes, we trust, will help satisfy the need first articulated in 1966 and will continue the tradition of augmenting our understanding of the dynamics of explosions and rective systems begun the following year in Brussels with the first colloquium. Subsequent colloquia have been held on a biennial basis since then (1969 in Novosibirsk, 1971 in Marseilles, 1973 in La Jolla, 1975 in Bourges, 1977 in Stockholm, 1979 in Gottingen, 1981 in Minsk, 1983 in Poitiers, and 1985 in Berkeley). The colloquium has now achieved the status of a prime international meeting on these topics, and attracts contributions from scientists and engineers throughout the world.

The proceedings of the first six colloquia have appeared as part of the journal, *Acta Astronautica*, or its predecessor, *Astronautica Acta*. With the publication of the Seventh Colloquium, the proceedings now appear as part of the AIAA *Progress in Astronautics and Aeronautics* series.

Acknowledgments

The Tenth Colloquium was held under the auspices of the College of Engineering, University of California-Berkeley, August 4-9, 1985. Arrangements in Berkeley were made by Dr. A. C. Fernandez-Pello. The publication of the proceedings has been made possible by grants from the National Science Foundation (USA) and the Army Research Office (USA).

Preparations for the Eleventh Colloquium are underway. The meeting is scheduled to take place in August 1987 at the Warsaw University of Technology, Poland.

J. Ray Bowen
J.-C. Leyer
R. I. Soloukhin
February 1986

Chapter I. Models, Simulations, and Experiments in Turbulent Reacting Flows

Sensitivity Analysis of Turbulent Variable Density Round Jet and Diffusion Flame Flows

F. Raiszadek* and Harry A. Dwyer†

University of California at Davis, Davis, California

Abstract

The turbulent round jet diffusion flame has been systematically studied with the use of sensitivity analysis applied to the k-ε turbulence model. Sensitivity analysis exhibits clearly how the diffusion flame solution depends on the semiempirical constants of the turbulence model and gives a clear picture of the interaction between the flame and the turbulence. The results have shown that the turbulence model is less sensitive for the H_2 diffusion flame than the incompressible jet because of the flame sheet and the relatively heavy density of the outer fluid (air). The sensitivity coefficients are surprisingly small and have a dominant pattern imposed on them by the local reacting flow. An analysis of these results leads to the conclusion that possibly the mean velocity profiles are insensitive to organized turbulent structure. This result could be extremely important if proven to be true.

Introduction

A detailed study has been made of the statistical turbulent structures of variable density round jets and diffusion flames. The method employed to understand these flows has been sensitivity analysis and it has been applied to the k-ε turbulence model. This model has been chosen because of the success achieved with it by Bilger (1976) and Jones and Whilelaw (1982), and also because of the fact that our basic knowledge of turbulent reacting

---

Presented at the 10th ICDERS, Berkeley, California, August 4-9, 1985.
*Postdoctoral Research Associate.
†Professor.

flows is not detailed enough to distinguish the fine points of the more complicated turbulence models.

Over the past few years, the present authors have used sensitivity analysis to understand the turbulent flow structure of both wall boundary layers (Dwyer and Peterson, 1980) and free jets (Raiszadek and Dwyer, 1983). Sensitivity analysis has yielded a unique and systematic understanding of how the model depends on its parameters in a given flow. For example, the incompressible round jet studies (Raiszadek and Dwyer, 1983) have shown that the jet flow is dominated by its conservation property and that there is no unique way of changing the k-ε model's constant to give a correct spreading rate. In the present study, this work has been extended to variable density and chemically reacting flows. It will be shown that these flows are also strongly influenced by conservation of jet momentum and that the reacting flow has an additional strong constraint imposed on it by the flame. Both of these factors tend to limit the importance of the turbulence model and to cause these simple flows to not be prime candidates for determining turbulence model constants. It is the opinion of the present authors that other flows will have to be utilized and that these other flows will have to have the important feature of being sensitive to the turbulence model constants.

Basic Analysis

In the present study, the hydrogen diffusion flame, originally studied by Kent and Bilger (1976), has been used as the flow to be investigated as well as the nonreacting variable density round jet. The turbulence model used has been the k-ε model (Jones and Whilelaw, 1982) and the approach of a conserved scalar in local equilibrium has been utilized. The basic boundary-layer equations, which are given in their mass weighted form are

Continuity:

$$\frac{\partial}{\partial x}(\bar{\rho}\tilde{u}) + \frac{\partial}{\partial y}(\bar{\rho}\tilde{v}) + \frac{\bar{\rho}\tilde{v}}{y} = 0$$

Momentum:

$$\frac{\partial}{\partial x}(\bar{\rho}\tilde{u}\tilde{u}) + \frac{\partial}{\partial y}(\bar{\rho}\tilde{u}\tilde{v}) + \bar{\rho}\frac{\tilde{u}\tilde{v}}{y} = \frac{1}{y}\frac{\partial}{\partial y}[y(\mu_1 + \mu_t)\frac{\partial \tilde{u}}{\partial y}]$$

Specie:

$$\frac{\partial}{\partial x}(\overline{\rho}\tilde{u}\tilde{Y}_i) + \frac{\partial}{\partial y}(\overline{\rho}\tilde{v}\tilde{Y}_i) + \frac{\overline{\rho}\tilde{v}\tilde{Y}_i}{y} = \frac{1}{y}\frac{\partial}{\partial y}[y(\mu_1 + \frac{\mu_t}{\sigma_c})\frac{\partial \tilde{Y}_i}{\partial y}]$$

k:

$$\frac{\partial}{\partial x}(\overline{\rho}\tilde{u}\tilde{k}) + \frac{\partial}{\partial y}(\overline{\rho}\tilde{v}\tilde{k}) + \frac{\overline{\rho}\tilde{v}\tilde{k}}{y}$$

$$= \frac{1}{y}\frac{\partial}{\partial y}[y(\mu_1 + \frac{\mu_t}{\sigma_k})\frac{\partial k}{\partial y}] + \mu_t(\frac{\partial \tilde{u}}{\partial y})^2 - \overline{\rho}\tilde{\epsilon}$$

$\overline{\epsilon}$:

$$\frac{\partial}{\partial x}(\overline{\rho}\tilde{u}\tilde{\epsilon}) \frac{\partial}{\partial y}(\overline{\rho}\tilde{v}\tilde{\epsilon}) + \frac{\overline{\rho}\tilde{v}\tilde{\epsilon}}{y}$$

$$= \frac{1}{y}\frac{\partial}{\partial y}[y(\mu_1 + \frac{\mu_t}{\sigma_\epsilon})\frac{\partial \epsilon}{\partial y}] + C_1\frac{\tilde{\epsilon}}{\tilde{k}}\mu_t(\frac{\partial \tilde{u}}{\partial y})^2 - \frac{C_2\overline{\rho}\tilde{\epsilon}^2}{\tilde{k}}$$

Turbulent stress:

$$\mu_t = C_\mu \frac{\tilde{k}^2}{\tilde{\epsilon}}$$

where $\tilde{u}$ and $\tilde{v}$ are the mean primary velocities, $\tilde{k}$ and $\tilde{\epsilon}$ the flow, C_μ, C_1 and C_2 the major kinetic energy and dissipation rate empirical constants of the model, $\tilde{Y}_i$ the species mass fraction, $\overline{\rho}$ the local density of the fluid, and x and y the respective coordinates of u and v.
    The density was determined from the mixture equation of state and the temperature from an equilibrium correlation with the hydrogen element mass fraction (Kent and Bilger, 1976). In order for the temperature–hydrogen element correlation to be used, the Lewis number was assumed to be unity. With this simplification, only the hydrogen element mass fraction Y_H has to be calculated to determine both the temperature and density in the jet. The turbulent Schmidt and Prandtl numbers chosen have the same values as those used by Kent and Bilger(1976), Pr=0.7. Thus, the sensitivity study centers very closely around their model. As can be seen from the above problem description the present work is directed to understanding

the presently used k-ε formulation, and it will not be concerned with other models. The results presented do not include the influence of the fluctuation in conserved species mass fraction, since this correction for the diffusion flame is not explicitly part of the k-ε model. Also, the influence of the temperature fluctuations on the mean velocity profiles are rather small for the jet diffusion flame.

The procedure for the application of sensitivity analysis to the above system of equations is to differentiate these equations with respect to the parameter to be studied. The resulting equation for a sensitivity variable such as

$$\bar{U} = \frac{\partial \tilde{u}}{\partial \mu_t} \quad \text{is}$$

Momentum sensitivity:

$$\frac{\partial}{\partial x}(\bar{R}\tilde{u}^2 + 2\bar{\rho}\tilde{u}\bar{U}) + \frac{\partial}{\partial y}(\bar{R}\tilde{u}\tilde{v} + \bar{\rho}\tilde{u}\bar{V} + \bar{\rho}\bar{U}\tilde{v}) + \frac{\bar{R}\tilde{u}\tilde{v} + \bar{\rho}\bar{U}\tilde{v} + \bar{\rho}\tilde{u}\bar{V}}{y}$$

$$= \frac{1}{y}\frac{\partial}{\partial y}(y\frac{\partial\tilde{u}}{\partial y}) + \frac{1}{y}\frac{\partial}{\partial y}[y(\mu_1 + \mu_t)\frac{\partial\bar{U}}{\partial y}]$$

where the remaining sensitivity variables are defined as

$$\bar{E} = \frac{\partial\tilde{\varepsilon}}{\partial\mu_t} \qquad \bar{V} = \frac{\partial\tilde{v}}{\partial\mu_t} \qquad \bar{R} = \frac{\partial\tilde{\rho}}{\partial\mu_t} \qquad \bar{K} = \frac{\partial\tilde{k}}{\partial\mu_t}$$

and where these variables are obtained from the differentiated form of the other equations (Jones and Whilelaw, 1982).

For most applications, it is convenient to present the sensitivity results in terms of normalized sensitivity variables that can be interpreted directly in terms of percentage changes. A typical "percent" sensitivity variable is

$$U\mu_t = \frac{\partial\tilde{u}}{\partial\mu_t}\frac{\mu_t}{\tilde{u}}$$

and the majority of the results will be presented in this form except in those cases where the normalization variables are zero. It should also be pointed out that the mathematical nature of the sensitivity equations is

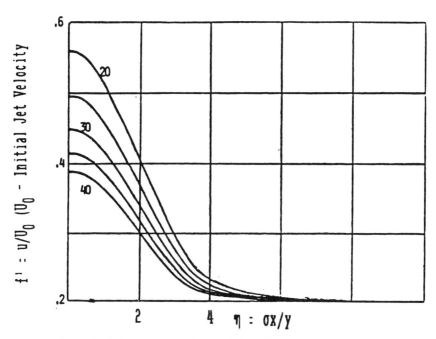

Fig. 1 Primary velocity profiles for diffusion flame.

identical to the original equations and that the same numerical method can be used for both the physical transport and the sensitivity equations. The physical equations were solved with an implicit marching scheme that employed iteration for the nonlinear turbulent transport, while the linear sensitivity equations were solved with the same implicit method but without iteration. Also, it should be mentioned that the calculations were carried out in a coordinate system which expanded with the jet. This coordinate system is

$$\xi = x, \qquad n = \sigma y/x$$

where σ is a stretching parameter. The value used for σ in the present paper was approximately .15 and this insured that the jet flow was always inside the computational boundaries.

Results

In the present paper the results of the study will be presented by comparing the variable density round jet with the hydrogen diffusion flame. With this approach, it is much easier to understand how the diffusion flame is

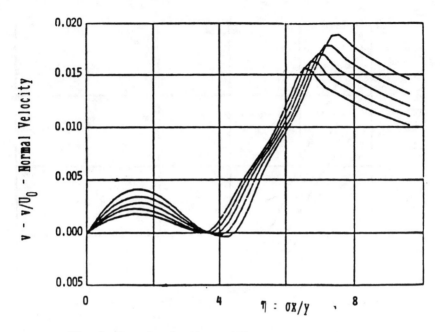

Fig. 2   Normal velocity profiles for diffusion flame.

different from other flows and it is also much easier to interpret the sensitivity results. The presentation will begin with the mean velocities, temperature, and density distributions for the diffusion flame shown in Figs. 1-4 for the region between 20 and 40 diam (X/D) from the jet exit. The axial or primary mean velocity distribution is quite similar to those of the variable density jet without chemical reaction, but the velocity normal to the jet axis and the density distribution are radically different. It should be noted that both velocity components have been normalized with the initial jet velocity U_0, and the velocity distribution was assumed to be flat at the jet exit.

Shown in Figs. 5-7 are the velocities and density distributions for an isothermal round jet at X/D = 40 with density ratios between the outer and inner flows of 1,2,4, and 8 (Note -- the numbers on the curve have the following correspondence: 1) $\rightarrow$ ρ_C/ρ_J = 1.0, 2) $\rightarrow$ ρ_C/ρ_J = 2.0, 3) $\rightarrow$ ρ_C/ρ_J = 4.0, and 4) $\rightarrow$ ρ_C/ρ_J = 8.0.) One of the keys to understanding the differences between these flows is by comparing the normal velocities shown in Figs. 2-6. The normal velocity for the diffusion flame exhibits a central region where the fluid particles are moving

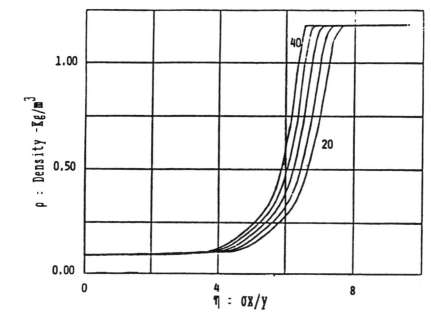

Fig. 3   Density profiles for diffusion flame.

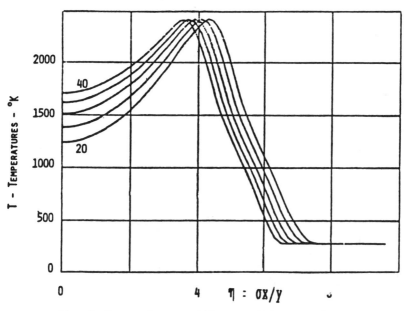

Fig. 4   Temperature profiles for diffusion flame.

10   F. RAISZADEK AND H.A. DWYER

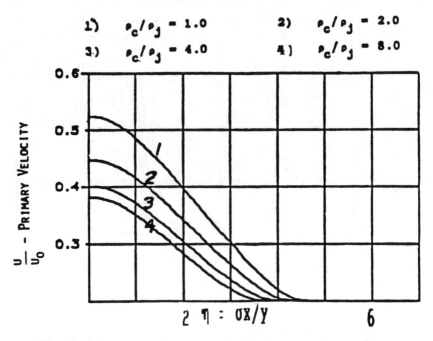

Fig. 5 Primary velocity profiles for variable-density flow.

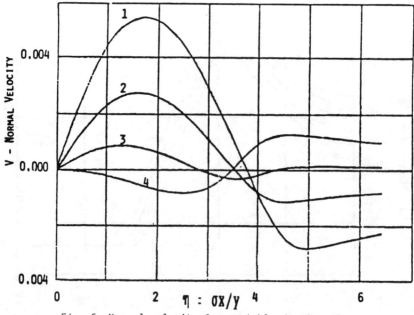

Fig. 6 Normal velocity for variable-density flow.

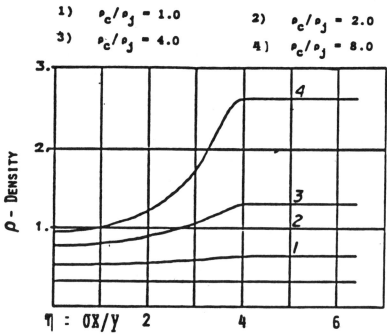

Fig. 7 Density distribution for variable-density flow.

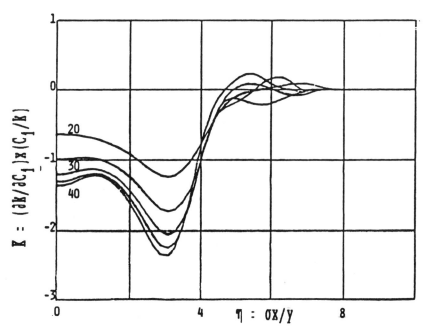

Fig. 8 Diffusion flame kinetic energy sensitivities.

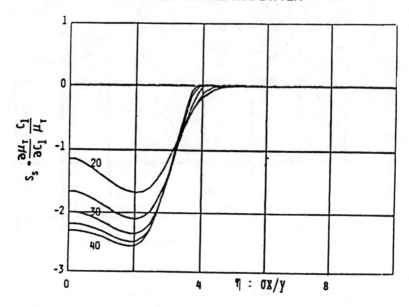

Fig. 9 Diffusion flame turbulent viscosity sensitivities.

outward due to primary velocity decay (Fig.2) and an outer region where the normal velocity is very large because of the expansion of the gas due to the flame. The variable density round jet (Fig. 6) has a very different distribution that is very sensitive to the jet density ratio and the Schmidt number. With a Schmidt number of 0.7, the mass diffusion from the jet outer region dominates the velocity decay for large density ratios and the normal velocity reverses sign. It will be shown in the following paragraphs that the above normal velocity behavior has a significant influence on how the flow depends on the turbulence model and its parameters.

The present combustion model is sometimes referred to as the "flame sheet" model since the reaction zone is reduced to a thin sheet and the limiting process is turbulent diffusion. It will be shown that this description is also useful to describe the lack of turbulent interaction between the flow on the fuel and oxidyzer sides of the flame. The sensitivity results clearly exhibit that the mixing and entrainment processes are less dependent on the semiempirical constants of the k-ε model. In order to illustrate this point, the sensitivity variable for the parameter C_1 in the ε equation will be presented for both the variable density

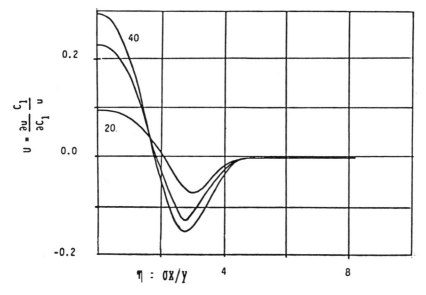

Fig. 10 Diffusion flame primary velocity sensitivities.

and diffusion flame flows. This sensitivity variable has been chosen because a small change in its value has the largest influence on the results and thus the greatest sensitivity.

The sensitivity variables for the kinetic energy k, the turbulent viscosity μ_t, and the mean primary velocity u are shown in Figs. 8-10 for the turbulent diffusion flame and for the parameter C_1 in the ϵ equation. In general these variables have a qualitative similarity with the variable density flows, but their quantitative values are much different. In Figures (11) through (13), the equivalent sensitivities for the variable density round jet are shown and the magnitude of the sensitivity variables are almost an order of magnitude larger. For the variable density flow, there is good coupling between the outer and inner regions of the jet flow, while the expansive character of the flame sheet has essentially decoupled the two regions. The sensitivity variables of the diffusion flame are at least a factor of three smaller than the 8/1 variable density flow.

The significance of these results is that a statistical model of a turbulent diffusion flame is constrained to a great extent by the flame and that the diffusion flame flow is a mild test of the turbulence behavior. The flame is a very dominant kinematic

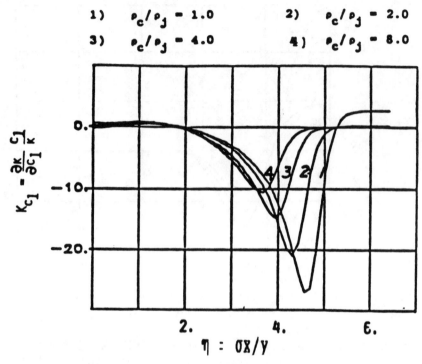

1) $\rho_c/\rho_J = 1.0$ 2) $\rho_c/\rho_J = 2.0$

3) $\rho_c/\rho_J = 4.0$ 4) $\rho_c/\rho_J = 8.0$

Fig. 11 Kinetic energy sensitivities for variable-density flow.

constraint, independent of whether the flow is laminar or turbulent, and the behavior of the turbulence model, although important, is not as important as in other flows such as turbulent separation. The diffusion flame in the flow does not cause a delicate balance of the various terms in the transport equation, but generates a very expansive mean flow that dominates the flow characteristics. The large value of normal velocity V essentially decouples the flow outside of the flame from that on the inside, and a decrease in the turbulence model sensitivity is thus achieved.

The sensitivity variables themselves tell an interesting story about the influence of variable density and combustion on jet flow. With variable density by itself, it is seen from Figs. 11–13 that increasing the density of the outer coflowing stream decreases the sensitivity coefficients. From the physical point of view, this makes sense, since the inner flow has relatively less momentum and kinetic energy. This behavior can also be seen by observing the shift in the maximum

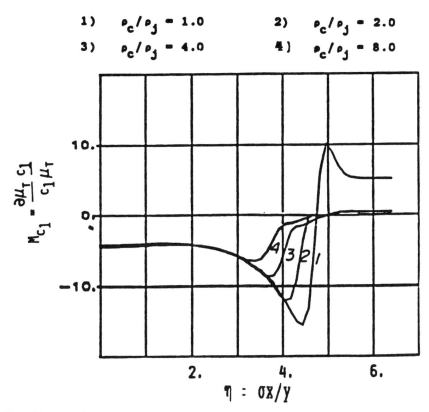

1) $\rho_c/\rho_j = 1.0$ 2) $\rho_c/\rho_j = 2.0$

3) $\rho_c/\rho_j = 4.0$ 4) $\rho_c/\rho_j = 8.0$

Fig. 12 Turbulent viscosity sensitivities for variable-density flow.

sensitivities inward, away from the heavy density boundary
fluid.

For the diffusion flame flow, the sensitivity
variables are both lower and shifted inward away from jet
boundaries. This behavior is exactly the opposite of the
incompressible jet flow, where the sensitivity variables
maximize in the jet boundary region. The decreased
sensitivities are actually fortunate for the turbulence
modeling of diffusion flames, since there is better
confidence in the prediction in the mean properties of the
flow (The opposite would be true if the case of a heavier
inner fluid was considered.) The same conclusion cannot
be made concerning the prediction of nonequilibrium
chemical effects in a flame. It is very probable that
nonequilibrium influences will depend on the fine
structure of the turbulent flow. Such influences may be
lost in present turbulence models due to the domainance of
flame expansion and density differences.

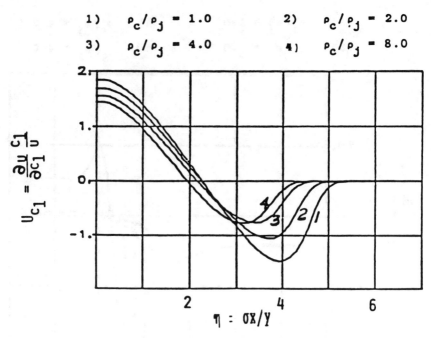

1) $\rho_c/\rho_j = 1.0$ 2) $\rho_c/\rho_j = 2.0$

3) $\rho_c/\rho_j = 4.0$ 4) $\rho_c/\rho_j = 8.0$

Fig. 13 Primary velocity sensitivities for variable-density flow.

Conclusion

It has been shown in the present research study that sensitivity analysis can be employed to understand the structure of statistical turbulence models in complex flow systems. For the general category of variable density flows, it has been demonstrated that increased outer flow density decreases the influence of the inner jet, and thus the influence of the turbulence model.

The influence of turbulence and the turbulence model is further decreased for diffusion flames because of the kinematic constraint imposed by the expansion caused by heat release. The expansion and corresponding large increase in jet normal mean velocity essentially decouples the "inner" and "outer" parts of the diffusion flame. Without the flame separating these "inner" and "outer" parts, there is much better communication and transport of mass, momentum, and energy.

A positive benefit of this study is that, with the constraint of a flame, it is possible to have more confidence in the prediction of the mean flow properties.

However, it should be remembered that it is still unclear what detailed role the variable density k-ε turbulence model has played, since the flow is less sensitive to its characteristics.

Acknowledgment

The authors would like to thank the U.S. Department of Energy, Division of Chemical Science, for its support and aid.

References

Bilger, R. W., Progress Energy Combust. Sci. 1, 87, 1976.

Jones, W. P., and Whilelaw, J. H., "Calculation Methods for Reacting Turbulent Flows: A Review," Combustion and Flame Vol. 48, pp.1. 1982.

Dwyer, H. A., and Peterson, T., "A Study of Turbulent Flow with Sensitivity Analysis," AIAA 13th Fluid and Plasma Dynamics Conference, Snowmass, Colorado, July, 1980, Also AIAA J.

Raiszadek, F., and Dwyer, H. A., "A Study with Sensitivity Analysis of the k-ε Turbulence Model Applied to Jet Flows," AIAA Paper 83-0285, 21st Airspace Sciences Meeting, Reno, 1983: Also submitted to AIAA Journal.

Kent, J. H., and Bilger, R. W., "The Prediction of Turbulent Diffusion Flames Fields and Nitric Oxide Formation," 16th Sym (Int) Combustion (1976), pp. 1643-1655.

Raiszadek, F., "The Sensitivity Analysis of k-ε Turbulence Modeling Applied to Jet Flows," Ph.D. Dissertation University of California, Davis, June, 1984.

Numerical Simulation of a Confined Shear Layer

Kenneth K. Ng* and Ahmed F. Ghoniem†

Massachusetts Institute of Technology, Cambridge, Massachusetts

Abstract

Vortex simulation of the Navier-Stokes equations is used to study a confined shear layer at high Reynolds number. The vorticity field is discretized among Lagrangian vortex elements of finite cores and the velocity field is evaluated by direct summation over the modified Biot-Savart law, augmented by vortex images due to solid boundaries. Average velocity profiles and streamwise fluctuations agree with experimental data at high Reynolds numbers, while cross-stream fluctuations are closer to experimental data at low Reynolds numbers since the simulation is limited to two dimensions. The vorticity field is a synthesis of eddies that merge by successive pairing, accomplished without noticeable growth in the total area occupied by vorticity. However, the area of each eddy increases by circumferentially entraining irrotational fluid. The momentum thickness of a naturally excited layer grows linearly. A substantial increase in the growth rate of the layer, accompanied with an early and, at times, multiple merging of small eddies, is achieved by superimposing a periodic perturbation at the pairing frequency of the naturally excited layer on the approaching streams.

Introduction

The deterministic dynamics of turbulent shear flow, which was established during the last two decades through extensive and elaborate experimental research (viz. the reviews by Cantwell, 1981 and Ho and Huerre, 1984), was largely complemented by numerical studies based on the

---
Presented at the 10th ICDERS, Berkeley, California, August 4-9, 1985. Copyright © 1986 by the American Institute of Aeronautics and Astronautics, Inc. All rights reserved.

*Graduate Research Assistant, Department of Mechanical Engineering.

†Associate Professor, Department of Mechanical Engineering.

18

fundamental form of the Navier-Stokes equations. The agreement between the numerical results and the experimental data provided evidence that direct numerical simulation of turbulence is possible. In the early research on separating flows, two-dimensional, spanwise large vortical structures were shown to dominate the dynamics of flowfields characterized by high rates of shear. Roshko (1976) analyzed long-time behavior of these vortical structures and confirmed that their interactions and growth, as well as their geometrical configurations, are quite repeatable. The spatial growth of a shear layer has been attributed to either the pairing between large vortex eddies or the entrainment of irrotational fluid into these structures.

As experiments have shown that, at high Reynolds numbers, two-dimensional shear layers develop streamwise vortex structures, a two-dimensional simulation may not predict accurately the properties of the flow. In this work, a vortex simulation of the most generic form of a shear flow, a confined shear layer developing downstream of a splitter plate, is used to gain some insight into the mechanism of growth and entrainment within the layer and to quantify the deviation of the real flow from the assumption of two-dimensionality. Harmonic modulation is then used to discern the fundamental ingredients of the dynamics of the flow by analysis of the response of the layer to external excitations.

Numerical investigations of shear layers can be divided into two categories: temporally and spatially developing layers. For the first, a coordinate system moving with the average streamwise velocity is used and periodic boundary conditions are imposed on the two sides of the domain to limit the computational effort. Riley and Metcalfe (1980) applied a pseudospectral method to solve the governing equations, while Corcos and Sherman (1984) obtained a solution with a finite difference method. In both studies, results were obtained at low Reynolds numbers, limited by the properties of the corresponding numerical schemes. The results demonstrated the importance of initial excitations on the development of the large-scale structures and showed that the presence of subharmonic perturbations is necessary for the process of pairing.

Aref and Siggia (1980) used a vortex-in-cell method with a large number of vortex elements to resolve the small scales at higher Reynolds numbers. Their analysis reproduced the roll-up of the initial vorticity layer, followed by the pairing of these eddies into larger structures. Since the Navier-Stokes equations are not invariant to a transformation in which the coordinate system moves with a variable speed, it is not obvious that these results apply to the spatially developing layer; and the conclusions of these studies should be checked against computations for a spatially developing flow.

Ashurst (1979) demonstrated that vortex methods could predict the formation and pairing of the large structures and that velocity statistics could be calculated from the vorticity field for a free, spatially developing shear layer. His interpretation of the results showed that pairing is the primary growth mechanism. With a similar Lagrangian analysis of a free shear layer and using passive markers which were engulfed by the large vortex structures, Inoue (1985) showed that entrainment is the main mechanism for the growth of the layer. Mansour (1985) developed a computationally efficient hybrid vortex-in-cell finite difference method to analyze the same flow. As his results were limited to low Reynolds numbers, the growth mechanism could not be analyzed.

In this work, a confined spatially developing shear layer is simulated using an accurate Lagrangian vortex scheme. Confined shear layers are considered since they are more likely to be encountered in experimental apparatus and in practical applications such as mixing and combustion devices. The vortex method used in this paper is derived as a numerical approximation of the Navier-Stokes equations. Means of improving the accuracy, through different choices of the core function of vortex elements and the time-integration method, are introduced within the general construction of the scheme. The application of the method to a confined shear flow is described next, with emphasis on the treatment of the boundaries. The effect of varying the numerical parameters is then discussed to show internal consistency, and further refinement is used to determine the invariability of the results.

The model results are compared with experimental data to assess the accuracy of the predictions and the adequacy of the assumptions of the model, in particular the two-dimensionality. The results of the computations, in terms of the vorticity field, are used to discriminate between the effect of pairing and entrainment on the growth of the layer and to ascertain the role of each mechanism in the mixing of the initially separated streams. Finally, the vorticity field is computed under conditions of periodic external perturbations of small amplitude and at the preferred frequencies of the naturally excited layer.

Numerical Scheme: Vortex Methods

The vorticity transport equations for an unsteady, two-dimensional, viscous, incompressible flow are:

$$\frac{\partial \omega}{\partial t} + \mathbf{u} \cdot \nabla \omega = \frac{1}{R} \Delta \omega \tag{1}$$

$$\nabla \times \mathbf{u} = \omega \tag{2}$$

$$\nabla \cdot \mathbf{u} = 0 \tag{3}$$

where ω is the vorticity, $\mathbf{u}=(u,v)$ the two-dimensional velocity vector, $\mathbf{x}=(x,y)$, x and y are the streamwise and the cross-stream directions, respectively, t the time, R the Reynolds number, and

$$\nabla = (\frac{\partial}{\partial x}, \frac{\partial}{\partial y}) \qquad \Delta = \frac{\partial^2}{\partial x^2} + \frac{\partial^2}{\partial y^2}$$

All quantities are made nondimensional with respect to the appropriate combination of the streamwise velocity jump across the layer, $\Delta U = U1 - U2$, the main channel height H, and $R = \Delta U H / \nu$, where ν is the kinematic viscosity.

Equation (3) is satisfied by the introduction of a stream function ψ such that

$$u = \frac{\partial \psi}{\partial y} \qquad v = - \frac{\partial \psi}{\partial x} \tag{4}$$

Substitution of Eq. (4) into Eq. (2) leads to Poisson equation,

$$\Delta \psi = - \omega \tag{5}$$

The solution of Eq. (5) in a domain without boundaries is given by

$$\psi = \int G(\mathbf{x} - \mathbf{x}') \, \omega(\mathbf{x}') \, d\mathbf{x}' \tag{6}$$

where

$$G(\mathbf{x}) = -(1/2\pi) \ln r$$

while $r^2 = (x^2 + y^2)$, $d\mathbf{x} = dx \, dy$ and the integration is performed over the area where $|\omega| > 0$. The velocity distribution is recovered by substitution of Eq. (6) into Eq. (4), so that

$$\mathbf{u} = \int K(\mathbf{x} - \mathbf{x}') \, \omega(\mathbf{x}') \, d\mathbf{x}' \tag{7}$$

where

$$K(\mathbf{x}) = -\frac{1}{2\pi} \frac{(y,-x)}{r^2}$$

and K is the integral kernel of the Poisson equation (5).

In vortex methods, the vorticity distribution is expressed in terms of a collection of vortex elements of finite cores (viz., e.g., Chorin, 1973; Hald, 1979; and Leonard, 1980). Thus,

$$\omega(\mathbf{x}) = \sum_{i=1}^{N} \Gamma_i \, f_\delta(\mathbf{x}-\mathbf{\chi}_i) \qquad (8)$$

where Γ_i is the circulation of the vortex element i, whose center is at $\mathbf{x}=\mathbf{\chi}_i$, and f_δ is the core function or the distribution of vorticity associated with each element. δ is the core radius such that

$$f_\delta(\mathbf{x}) = \frac{1}{\delta^2} \, f(\frac{r}{\delta}) \qquad (9)$$

The approximate velocity distribution is obtained upon substitution of Eq. (8) into Eq. (7),

$$\mathbf{u}_\delta(\mathbf{x}) = \sum_{i=1}^{N} \Gamma_i \, K_\delta(\mathbf{x}-\mathbf{\chi}_i) \qquad (10)$$

where

$$K_\delta(\mathbf{x}) = K(\mathbf{x}) \, \kappa(r/\delta) \quad \text{and} \quad \kappa(r) = 2\pi \int_0^r r \, f(r) dr$$

κ represents the total circulation of the element i within radius r. Changes in the vorticity distribution within the core due to the strain field are neglected, while the circulation carried by each element is constant with time, in accordance with Kelvin's circulation theorem.

The vorticity transport equation, Eq. (1), is solved in two fractional steps, namely,

$$\frac{\partial \omega}{\partial t} + \mathbf{u} \cdot \nabla \omega = 0 \qquad (11)$$

$$\frac{\partial \omega}{\partial t} = \frac{1}{R} \, \Delta \omega \qquad (12)$$

Equation (11) expresses the fact that, in an inviscid flow,
vorticity remains constant along a particle path. This
result can be used in the numerical solution after
transformation of Eq. (11) in Lagrangian form.
If $\chi_i=\chi(X_i,t)$ is the trajectory of a vortex element that
starts at X_i with a total circulation Γ_i, then Eq. (11)
implies that

$$\frac{d\chi_i}{dt} = u_\delta(\chi_i) \tag{13}$$

with $\omega(\chi_i)=\omega(X_i)$. Vortex elements move as rigid particles
under the influence of other elements, governed by the
modified Biot-Savart law [Eq. (10)].
     The solution of Eq. (12), governing the diffusive
transport of vorticity, can be obtained in a Lagrangian
frame of reference from the recognition that the Green
function of the one-dimensional form of Eq. (12),

$$G(y,\Delta t) = \sqrt{\frac{R}{4\pi \Delta t}} \exp(-\frac{R}{4\Delta t} y^2)$$

where y is either x or y and Δt the time step, is identical
to the probability density function of a Gaussian random
variable η with a zero mean and a standard deviation σ,

$$P(\eta,\Delta t) = \sqrt{\frac{1}{2\pi\sigma^2}} \exp(-\frac{1}{2\sigma^2} \eta^2)$$

if $\sigma=\sqrt{2\Delta t/R}$ (for more detail, see Chorin, 1973; Ghoniem and
Sherman, 1985; and Hald, 1984). Thus, diffusive transport
of vortex elements can be simulated stochastically by the
addition to their motion of an extra displacement, drawn
from a Gaussian population with a zero mean and a standard
deviation σ. In two-dimensions, two independent random
displacements are implemented in the two perpendicular
directions. The random walk algorithm is compatible with
vortex schemes because of its grid-free Lagrangian form and
can be applied near solid boundaries to satisfy the no-slip
condition without loss of resolution.
     The total displacement of each vortex element is,
thus,

$$\chi_i(X_i,t+\Delta t) = \chi_i(X_i,t) + \sum_k u_\delta(\chi_{ik}) \Delta t + \eta_i \tag{14}$$

where $\sum_k$ is a kth order time-integration scheme used to

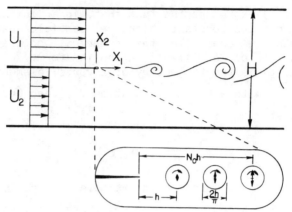

Fig. 1   Schematic diagram of the shear layer model, detailing
the generation of vortices.

integrate Eq. (13) and $\mathbf{n}_i=(\eta_x,\eta_y)_i$ a two-dimensional
Gaussian random number used to simulate the solution of Eq.
(12).   (For more detail on the construction of the
algorithm, Ghoniem and Gagnon (1986) may be consulted.)
The scheme was extended to turbulent combustion modeling by
Ghoniem et al. (1982).   Their model was applied to predict
the formation and inflammation of a turbulent jet by
Ghoniem et al. (1985), and to a turbulent flame stabilized
behind a rearward-facing step by Hsiao et al. (1985). In
the next section, this scheme is used to solve for the flow
in a confined shear layer.

Physical Model: Confined Shear Layer

     The flow under investigation is a confined shear layer
formed downstream of a splitter plate.   The velocity
discontinuity between two coflowing streams, initially
separated by the plate, causes a high rate of shear that
continuously produces vorticity at the trailing edge of the
plate to satisfy the Kutta condition. The channel height
is $H=1$ and the splitter plate is placed halfway between the
two channel walls, with its trailing edge at (0,0).
Upstream of $x=0$, $u=U1=1$ for $y>0$ and $u=U2=\alpha$ for $y<0$. The
geometry of the physical plane is shown in Fig. 1.   At each
time step Δt, the total circulation $\Delta\Gamma=-\Delta U\Delta\ell$, where $\Delta\ell=Um\Delta t$
and $Um=(U1+U2)/2$, is generated and assigned to N_o
vortex elements of equal strength $\Gamma=\Delta\Gamma/N_o$.   These elements
are placed apart by a distance $h=\Delta\ell/N_o$ downsteam of the

splitter plate. Boundary-layer buildup along the two
surfaces of the splitter plate is neglected in the current
calculation.

The potential boundary condition along the channel
walls and the splitter plate is satisfied by mapping the
physical z domain onto one side of the ζ plane and using
image vortices on the other side to enforce a zero normal
velocity along the boundary. This is achieved by applying
the Schwarz-Christoffel theorem to transform the channel
onto the upper half of the ζ plane. The differential
equation governing this transformation is

$$\frac{d\zeta}{dz} = \pi \frac{(\zeta^2-1)}{\zeta}$$

while its integration yields

$$\zeta = \pm \sqrt{\exp(2\pi z)+1}$$

where the plus sign is used for y $<$ 0.0 and the minus is
used for y $\geq$ 0.0. In the ζ plane, the two incoming streams
are represented by two sources, placed at the corresponding
transformations of $(-\infty,+0)$ and $(-\infty,-0)$, with strengths 0.5
and $\alpha/2$, respectively. A correction factor is necessary to
cancel an erroneous self-induced velocity due to the
curvature of the boundaries. It can be shown (e.g.,
Clements, 1973) that the complex velocity in the z plane
should be modified by adding the following complex
velocity,

$$w = \frac{\iota\Gamma}{4\pi} \frac{d^2\zeta/dz^2}{d\zeta\alpha/dz} = -\frac{\iota\Gamma}{4\pi}\frac{dF}{d\zeta} \text{ , with } F(\zeta) = \frac{d\zeta}{dz} \qquad (15)$$

where w=u-ιv and $\iota=\sqrt{-1}$. F(ζ) is the differential
expression obtained from the Schwarz-Christoffel theorem.
For the geometry considered here,

$$w = -\frac{\iota\Gamma}{4} (\frac{\zeta^2+1}{\zeta^2})$$

The importance of this term was realized in the
computations when, if neglected, the deflection of the
shear layer centerline was predicted with a large error.

The viscous boundary condition along the walls is
neglected in this model to reduce the amount of
computational labor. Thus, the results correspond to a
high Reynolds number flow in which the developing boundary
layer on the channel walls are thin and are not expected to

U2/U1=0.333 Re=******   time step=0.10   s=0.002358

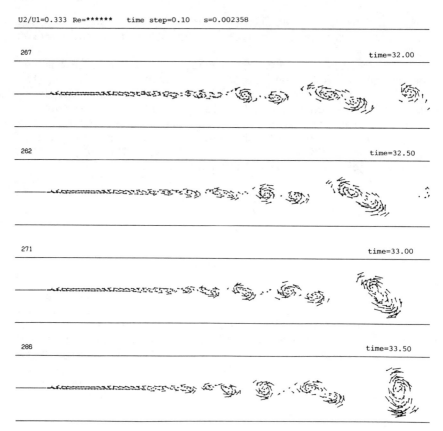

Fig. 2   Vorticity field (inviscid, α=0.333, N_o=3, x_{max}=6,
Heun's   method)   depicting   the   onset   of   Kelvin-Helmholtz
instability in a spatially growing shear layer, the generation
of large eddies,and their pairing interactions.

interact with the interior of   the flow within the range of
interest.

Results

Results of a typical simulation, presented in terms of
the location and velocity   of   vortex   elements used in the
computations, are shown in Figs. 2 and 3 for an inviscid
flow and a viscous flow at R=4000, respectively. In both
cases, the large scale eddy structure of a shear layer,
with the associated pairing dynamics, is clearly depicted.
Each vortex element is represented by a small circle and
its velocity is depicted by a line vector starting at the
center of the circle.   In these plots, the velocity of a

U2/U1=0.333   Re=4000.   time step=0.10   s=0.002358

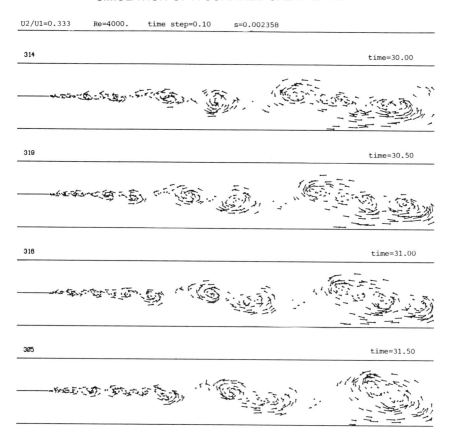

Fig. 3   Vorticity field (R=4000, α=0.333, N_o=3, x_{max}=6, Heun's method) showing the early onset of linear growth due to the simulation of diffusion. An uncommon merging of three eddies is occurring at the right.

vortex is measured relative to an observer moving with the mean velocity of the flow Um.   Fig. 4 shows the variation of the total number of vortex elements within the computational domain with time for the case of Fig. 3, indicating that the field reaches a stationary state after 200 time steps for Um=0.667. The oscillations in the total number of elements is due to the temporal variation of the flux of vorticity at the exit boundary.

Effect of Numerical Parameters

        The numerical parameters used in this simulation are: the order of the time-integration scheme, the time step Δt, the strength of each vortex element or the number of

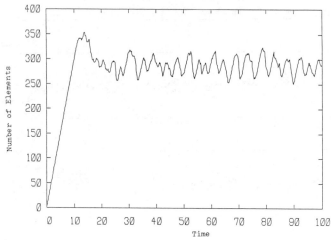

Fig. 4   Number of vortex elements in the computational domain
vs time for the case of Fig. 3 shows that a fully developed
stationary shear layer is achieved after time=20. Oscillations
are due to deletion of eddies at the exit.

vortices generated each time step N_o, the length of the
computational domain x_{max} that defines the point at which
vortices are deleted, the form of the core function, and
the core radius δ. In the following, the dependence of the
accuracy of the computations on some of these parameters is
discussed.

A minimum of N_o=3 was found necessary to describe the
large scale structure accurately. Results of this run are
shown in Fig. 3. Increasing N_o to 6 and 9, however,
improves the resolution of the eddy structure as shown in
Figs. 5 and 6, and produces more accurate predictions for
the statistics of the fluctuations, without affecting the
time-averaged velocity profiles. Table 1 depicts a
comparison between the numerical results of the velocity
statistics for the three values of N_o with the experimental
data of several investigations.

Three different core functions were used to test the
effect of discretizing the vorticity among elements of
different structure on the accuracy of the algorithm.
These are: the Chorin core, characterized by

$$f(r) = 1/(2\pi r) \quad \text{and} \quad \kappa(r) = r \quad \text{for} \quad r \leq 1 \qquad (16)$$
$$\quad\quad = 0 \qquad\qquad\qquad\quad = 1 \quad \text{for} \quad r > 1$$

U2/U1=0.333 Re=4000. time step=0.10

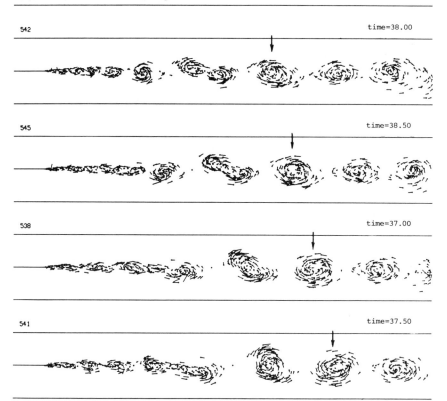

Fig. 5   Vorticity field (R=4000, α=0.333, N_o=6, x_{max}=6, Heun's method) depicting events of pairing and entrainment in neighboring eddies.   Resolution of the field increases as N_o increases.

the Rankine core, for which:

$$f(r) = 1/\pi \qquad \text{and} \quad \kappa(r) = r^2 \quad \text{for } r \leq 1 \qquad (17)$$
$$ = 0 \qquad\qquad\qquad\quad = 1 \quad \text{for } r > 1$$

and the Gaussian core that has

$$f(r) = (1/\pi)\,e^{-r^2} \qquad \text{and} \quad \kappa(r) = 1 - e^{-r^2} \qquad (18)$$

In all cases, δ=h/π, where h is the initial spacing between vortices. Numerical results for the three cores were almost identical, in agreement with the theoretical results

of Hald (1985) who showed that the spatial accuracy of the
three schemes is the same.   This may be explained by the
following observation: variation   within   a   core   that
confines most of the modification of the velocity field of
a point vortex to within a radius δ only, such as the three
cores that were used here, cannot change the accuracy
appreciably when the distance between vortex elements is
larger than the core diameter. It should be noted that the
choice of δ did   not   allow   any   overlap of vortex cores,
since vorticity is confined within a circle of diameter 2δ,
with the exception of   the Gaussian core, while the
separation between vortices is h=πδ.   The fact that
nonoverlapping cores produce accurate results is rather
surprising,since most theoretical   studies (e.g., Beale and
Majda, 1982) indicate that δ should be larger than h.
However, the vorticity field in this simulation changes in
time and may not conform identically with the assumptions
of the theory.

A time step $\Delta t=0.1$ was used in all the computations,
in conjunction with an Euler first-order integration scheme
or a Heun second-order integration scheme in Eq. (14). The
latter scheme was used by Ghoniem and Sethian (1985) in
their study of a recirculating flow. Higher-order time-
integration schemes were not employed since a vortex
approximation using any of the core functions described
above is not expected to be more than second-order
accurate. Results for the first-order scheme are shown in
Fig. 7. When compared with the results of Fig. 3, they
indicate that the numerical diffusion associated with the
first order scheme acts to smear out the vortex structures
and, concomitantly, reduce the values of the fluctuations.
The diffusive nature of the time-integration error was
observed and quantified by Nakamura et al. (1982). They
found that numerical diffusion associated with low-order
integration schemes acts as molecular diffusion and reduces
the effective Reynolds number of the simulation.

The numerical exit boundary condition, in which
vorticity is deleted beyond a downstream section x_{max}, had
an insignificant effect on the calculations beyond one
channel height upstream of x_{max}. This is confirmed by the
results in Table 1 for $x_{max}=6$ and 7. The growth of the
momentum thickness θ, defined as

$$\theta(x) = \frac{1}{\overline{\Delta U}^2} \int_{-0.5}^{+0.5} (\overline{U}_1 - \overline{u})(\overline{u} - \overline{U}_2)\,dy \qquad (19)$$

U2/U1=0.333   Re=4000. time step=0.10   s=0.000786

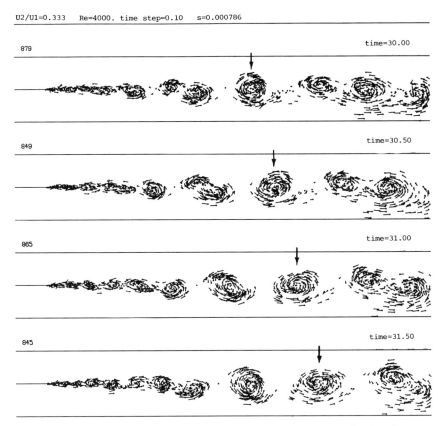

Fig. 6   Vorticity field (R=4000, α=0.333, N_o=9, x_{max}=6, Heun's method) depicting definite growth due to entrainment, but no additional mixing by the pairing mechanism alone.

was particularly sensitive to the value of x_{max}. In Eq. (19), $\overline{U}1(x)=\overline{u}(x,0.0)$, $\overline{U}2(x)=\overline{u}(x,-0.0)$, and $\overline{\Delta U}(x)=\overline{U}1(x)-\overline{U}2(x)$, since both U1 and U2 vary downstream due to flow confinement. The computed value of θ grows linearly with x until x_{max}^* where the growth rate changes due to the removal of vorticity at the exit boundary. This criterion was used to determine the range of the computational domain that is independent of the exit boundary condition. The value of x_{max}^* is approximately $(x_{max}-1)$, depending on the magnitudes of the incoming velocities. A run for α=0.333, using x_{max}=11,

Table 1 Comparison between numerical and experimental results

Quantity[a]	$R=4000, \alpha=0.333, x_{max}=6$			
	$N_o=3$	$N_o=6$	$N_o=9$	Exp.[b]
$|\bar{v}/\bar{U}1|_{max}$	1.1×10^{-2}	1.1×10^{-2}	9.0×10^{-3}	$10^{-3} - 10^{-2}$
$(\sqrt{\overline{v'^2}}/\overline{\Delta U}_{max})$	0.188	0.175	0.165	0.19
$(\sqrt{\overline{v'^2}}/\overline{\Delta U})$	0.235	0.220	0.217	0.12
$-(\overline{u'v'}/\overline{\Delta U \Delta U})_m$	1.1×10^{-2}	9.6×10^{-3}	1.0×10^{-2}	1.3×10^{-2}
q^2_{max}	0.045	0.039	0.037	0.035
$d\theta/dx$	0.017	0.018	0.018	Fig. 9

Quantity[a]	$R=4000, \alpha=0.6, N_o=3$		First Order		Exp.[c]
	$x_{max}=6$	$x_{max}=7$	$N_o=3$	$N_o=6$	
$|\bar{v}/\bar{U}1|_{max}$	5.9×10^{-3}	5.9×10^{-3}	2.0×10^{-2}	1.4×10^{-2}	**
$(\sqrt{\overline{u'^2}}/\overline{\Delta U}_{max})$	0.197	0.196	0.148	0.120	0.10
$(\sqrt{\overline{v'^2}}/\overline{\Delta U})_{max}$	0.230	0.231	0.161	0.138	0.16
$-(\overline{u'v'}/\overline{\Delta U \Delta U})_m$	9.4×10^{-3}	1.1×10^{-2}	5.4×10^{-3}	5.1×10^{-3}	8.0×10^{-3}
q^2_{max}	0.046	0.046	0.024	0.017	**
$d\theta/dx$	0.010	0.010	****	****	0.011

[a] A long overbar indicates spatial averaging for $1 \leq x \leq 5$).
[b] Experimental data from Spencer and Jones (1971).
[c] Experimental data from Browand and Weidman (1976) and Winant and Browand (1974).

showed no significant deviation from the results of the shorter computational window of $x_{max}=6$.

Velocity Statistics

The time-averaged statistics of the velocity field were obtained in terms of the average streamwise velocity $\bar{u}$, the root mean square streamwise fluctuations $\sqrt{\overline{u'^2}}$, the root mean square cross-stream

U2/U1=0.333   Re=4000. time step=0.10

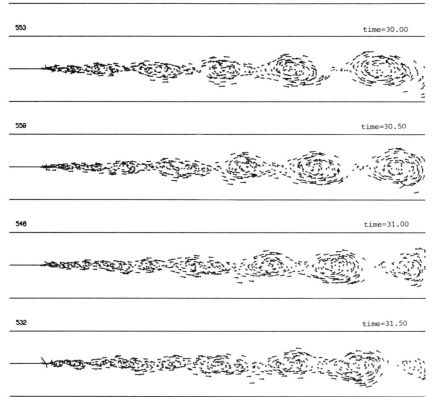

Fig. 7   Vorticity field (R=4000, α=0.333, N_o=6, x_{max}=6, Euler's method) showing the effect of time-integration diffusion. Eddy structures are smeared and the pairing mechanism is inhibited, but growth due to entrainment continues.

fluctuations $\sqrt{\overline{v'^2}}$, and the average streamwise cross-stream correlation $-\overline{u'v'}$. Plots in Figs. 9-12 were produced using a sample of 200 time steps, for t=20-40. An increase in the sample size to 800 did not produce any significant difference in the results. All comparisons were made using the experimental data of Spencer and Jones (1971), unless otherwise stated.

In Fig. 8, the average streamwise velocity is plotted against the similarity variable y^* defined by

$$y^* = (y-y^0)/(x-x^0)$$

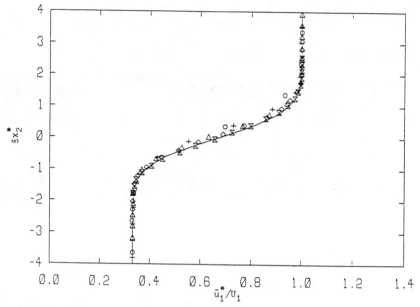

Fig. 8   Average streamwise velocity profiles for the case of
Fig. 5 and averaging time of 20<t<39.9, (o x=1; + x=2; Δ x=3; <>
x=4; and X: x=5) compared with Gortler's error function_____.

where x^0 is the virtual origin of the layer, defined by the
intersection of the centerline of the channel and the layer
centerline where u=Um, and y^0 is the y coordinate of the
centerline of the layer.   The similarity variable is used
to collapse the data at different x stations onto one
curve. It is derived using the Gortler error function
profile, which predicts that, based on the mixing length
theory for an unconfined flow, the average streamwise
velocity distribution is

$$\bar{u}_1 = U2 + (\Delta U/2) [1+erf(\beta y_2^*)]$$

where β is the growth parameter (Brown and Roshko, 1974)
and erf is the error function.

     In the plots of the computational results, $\bar{u}$, U2 and
ΔU are replaced by $\bar{u}^*$, $\bar{U}2(x)$, and $\overline{\Delta U}(x)$,
respectively. The values of x^0, y^0 and β are obtained from
the experimental data of Spencer and Jones (1971) by

interpolation to α=0.33 and 0.6. Figure 8 shows the
profiles of $\bar{u}^*$ for x=1,2,3,4, and 5 for a velocity ratio
α=0.33.   The computational results fit the experimental

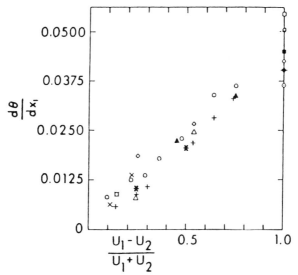

Fig. 9   Comparison of the dependence of the spreading rate on a velocity parameter is shown.   Present   results (*) are plotted along with experimental data gathered by Ho and Huerre (1984).

data with very   good   accuracy   for   1<x<6, where the shear layer exhibits self-similar behavior.   For x<1, the early transitional effects due to the onset of the instability in the separating vortex   sheet   dominate,   while for x>6, the wall   effects   become   more   important.   The self-similar behavior of the shear layer   in this range was also checked by calculation   of   the   rate   of   growth   in   the momentum thickness, as defined in Eq.   (19).   The linear growth was very accurately predicted for α=0.33   and 0.6, with slopes that fall well within the experimental results collected in the review of Ho and   Huerre   (1984)   shown in Fig. 9.   The numerical values of   the   growth   rate   for   both cases are given in Table 1.

Figure   9   indicates   that   the   rate   of   growth   at different values of α is   the same when scaled with respect to the shear factor $\Delta U/U_m$.   The universal scaling can also be   obtained   when   comparing   the   maximum   average normal velocity for different values of α.   Results in Table 1 indicate that $(|v|_{max}/\Delta U)=0.015$ for α=0.33 and 0.6.

Profiles for all the velocity statistics were obtained for the two cases of α=0.33 and 0.6 and were found to be similar if they were scaled by the appropriate power of ΔU.

The profiles of $\sqrt{\overline{u'^2}}$ are plotted in Fig. 10 for x=1-5.   They exhibit   the   trend of self-similarity and the

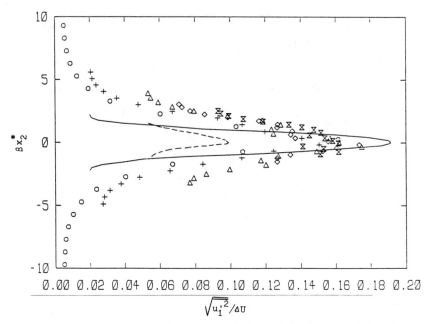

Fig. 10 Average streamwise fluctuating velocity profiles for the case of Fig. 8 compared with the data of Spencer and Jones (1971)_____ and Browand and Weidman (1976)_ _ _ _ _ .

centrally bulged shape of free shear layers and also depict a reasonable agreement with the experimental data of Spencer and Jones (1971). More precisely, they fall closer to the experimental results of Spencer and Jones for a high Reynolds number flow than those of Browand and Weidman (1976) for a lower Reynolds number flow. As indicated in Table 1, these profiles are most sensitive to the numerical parameters, in particular to the time-integration scheme. The numerical results of Ashurst (1979) and Mansour (1985) compared better with the results of Browand and Weidman (1976) since the first used extra diffusion in the vortex aging process and the second employed a combined vortex-in-cell with a finite difference scheme, which is expected to have more numerical diffusion than a Lagrangian vortex scheme. Ashurst (1979) obtained better agreement with the data of Spencer and Jones when he used a higher Reynolds number, in agreement with our results for N_o=6 in Table 1.

Numerical results for $\sqrt{\overline{v'^2}}$ are depicted in Fig. 11 for the same downstream sections x=1,2,3,4, and 5. Here, the numerical results are closer to the moderate Reynolds number data of Browand and Weidman (1976) than with the

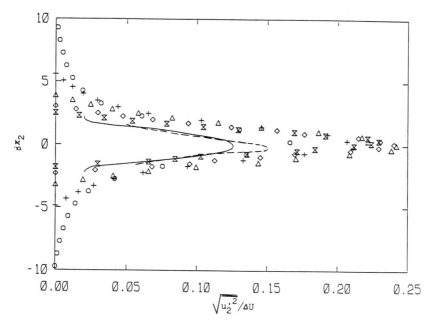

Fig. 11 Average cross-stream fluctuating velocity profiles for
the case of Fig. 8 compared with the data of Spencer and Jones
(1971)_____ and Browand and Weidman (1976)‒ ‒ ‒ ‒ ‒..

high Reynolds number data of Spencer and Jones (1971). At
high Reynolds numbers, three-dimensional effects become
more important and cause this deviation.   It has been shown
(Roshko, 1981) that at high Reynolds numbers, the flow
acquires   more   structure,   associated   with   stronger
fluctuations in the   spanwise   direction.   This effect is
neglected in the computations reported here. Part of the
kinetic energy predicted by the numerical solution to exist
in the y direction would have been
transferred to the spanwise-direction, resulting in
smaller values for $\sqrt{\overline{v'^2}}$. This hypothesis is
supported by the fact that the total kinetic energy
associated with fluctuation, defined by $q^2 = (\overline{u'^2} +$
$\overline{v'^2} + \overline{w'^2})/2\overline{\Delta U}^2$, is in better agreement
with the experimental results, both for the high and the
low Reynolds number cases, as shown in Table 1. If this
hypothesis were true, turbulence would have been almost
isotropic, a conclusion that has been confirmed before in
turbulent shear layers.   The results of Ashurst (1979)
confirm this result, while the results of Inoue (1985) show

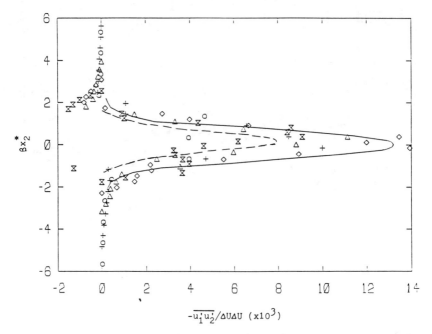

Fig. 12 Average Reynolds stress profiles for the case of Fig. 8
compared with the data of Spencer and Jones (1971)_____ and
Browand and Weidman (1976)_ _ _ _ _.

a lower value for the cross-stream fluctuations produced by
the lower order numerical integration scheme.
       At low and moderate values of the Reynolds number,
$\sqrt{\overline{u'^2}}$ is smaller than $\sqrt{\overline{v'^2}}$ in both the
experimental data and the computational results. On the
other hand, at high Reynolds numbers, the experimental data
display   the   opposite   trend,   while   two-dimensional
computational   results   do   not.   This   is   another
manifestation of the three-dimensionality of the flow at
high Reynolds numbers.   A transitional value of the
Reynolds number must exist where the two fluctuating
components are equal.   The experimental data of Keller
(1982) for a confined shear layer approaches this limit.
       The fluctuation correlations, $-\overline{u'v'}$, are presented in
Fig. 12. The profiles do not exhibit true self-similarity
and negative values appear on the high velocity side of the
layer beyond x=3.   The profiles at x=2-4, when
appropriately scaled, fall close to the
profiles of Spencer and Jones (1971). However, the value
of $-\overline{u'v'}$ changes sign during pairing, as confirmed

by the computations of Aref and Siggia (1980). Negative
values may persist after averaging at a section where the
dynamics of the flow is dominated by these interactions.
This change of sign, coupled with the increasing effect of
the walls, causes a deviation from self-similarity for the
correlation profiles, contrary to the autocorrelation of
u'^2 and v'^2 where the time-dependent values do not change
sign and self-similarity survives for longer distances
downstream.

Negative fluctuation correlations were observed before
in forced shear layers, both in the experiments of Browand
and Ho (1983) and the numerical simulations of Riley and
Metcalfe (1980). They were related to eddy nutation, i.e.,
periodic changes in the relative orientation of the major
axis of an elliptically shaped eddy with respect to the
streamwise direction of the flow. Ho and Huerre (1984)
observed that pairing interactions are accompanied by
similar changes in configuration and the results in Fig. 5
indicate that the line of centers of the two pairing eddies
turn through a 90 deg angle while they are merging.
Moreover, a newly composed eddy, such as that marked by
arrows in Fig. 5, continues to rotate producing a similar
effect. If pairing occurs frequently at a particular
location x, as in the case of a forced layer, negative
correlations arise more often and will dominate if the
correlations are averaged over a long period of time. Weak
forcing may have resulted from perturbations associated
with numerical integration errors and the stochastic
solution of the diffusion equation in the numerical
simulation.

Negative correlations indicate that momentum transfer
is in the direction opposite to the gradient of the average
velocity, i.e., counter-gradient diffusion, and that energy
is transferred back to the main flow from the large
structures. The full implication of this phenomenon
requires further study, in particular of forced shear
layers.

Growth: Pairing and Entrainment

The transition of a spatially growing inviscid shear
layer from a thin vortex sheet extending downstream of the
splitter plate to a set of well-defined large-scale eddies
that interact predominantly by pairing is illustrated in
Fig. 2. Different stages of development can be identified,
namely: roll-up, growth of vortex structures by pairing,
and growth of the area of a large vortex eddy by entraining
irrotational fluid. Each stage will be described in some
detail.

A vortex sheet is linearly unstable to perturbations of wavelengths longer than its thickness via the Kelvin-Helmholtz instability mechanism. This instability causes perturbations to grow exponentially downstream at a rate inversely proportional to the wavelength of the perturbation. In the inviscid computations, perturbations are introduced by the white noise associated with the round-off and time-integration errors. The displacement of one vortex element to a location $|y|>0$ forces a single row of vortex elements to form a thick layer of vorticity consisting of two rows of vortices, with pairs of vortex elements aligned on top of each other. A pair of vortex elements, whose line of centers is arranged normal to the streamwise flow direction, represents the element of a rolled-up eddy, i.e., an elementary eddy. The vorticity initially contained in the velocity discontinuity is now distributed among a set of elementary eddies, arranged in a row downstream. This picture of the initial stages of development of an expanding shear layer was previously seen with high resolution in the vortex simulation of Aref and Siggia (1980) and the experimental observations of Roberts et al. (1982).

Beyond this stage, small eddies grow by successive pairing. During this interaction, two neighboring eddies wrap around each other by their mutual velocity field and deform by the concomitant strain field until their separate identities are lost when their common axis is about 90 deg of the primary flow direction. Pairing doubles the size of each individual eddy and the distance between neighboring eddies. Since in an inviscid flow the area occupied by vorticity is constant according to the Kelvin-Helmholtz theorem, pairing redistributes the vorticity in the normal direction to the flow and produces void zones of irrotational flow between eddies. Results in Figs. 5 and 6, where the resolutions are high enough to enable some quantitative judgment, indicate that the area of a new eddy is very close to the summed area of the two original eddies. Thus, pairing does not increase the total area occupied by vorticity. The same conclusion can be inferred from the computations of Christiansen (1973) for the merging of two finite-area vortices. Successive merging of vortices continues downstream, as seen in Fig. 2 for the inviscid case and Figs. 5 and 6 for the viscous case. This mechanism was presented by Winant and Browand (1974), who suggested that eddy pairing is the primary mechanism of growth of the shear layer in its early stages. However, more evidence to the contrary is shown below.

Eddies grow by pairing to a size at which their strain field is strong enough to tear vorticity from neighboring

smaller eddies and form the links that are seen in Fig. 5,
the so-called braids. The presence of small eddies between
large eddies can be explained by the fact that some eddies
experience less pairings than others.   Braids are also
formed during the process of vortex pairing from the most
elongated part of each eddy (see Christiansen, 1973, Fig.
6). These braids may be entrained and ingested into the
large eddies at later stages, while engulfing some of the
surrounding irrotational fluid during the process. The
engulfment of extra fluid enlarges the area of the original
eddy. The eddies marked by arrows in Figs. 5 and 6 are
clear examples of growth by entrainment and confirm that
this mechanism is responsible for the increase in area of
each individual eddy.   This mechanism was substantiated by
the results of Hernan and Jimenez (1982), obtained by
computer analysis of Brown and Roshko's (1974) experimental
movies. It was also confirmed by the numerical results of
Inoue (1985), in which marker particles were traced while
they moved into the vortex structures.
     In all cases of viscous flow simulations, presented in
Figs. 3, 5, and 6, the process of pairing starts earlier

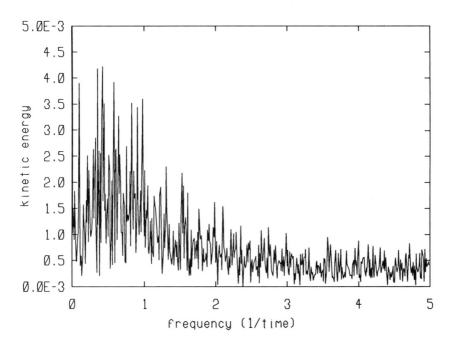

Fig. 13 Spectrum of the kinetic energy of fluctuations,
calculated for the flow in Fig. 3, using a sample of 1000 time
steps at x=5.0 and y=0.0.

than in the inviscid case.   In the former case, the
perturbations initiating the instability are associated
with the random walk simulation of diffusion.   Thus, these
layers can be considered as excited flows, although not
modulated since the statistical excitations have no biased
frequency. Downstream of the onset of instability, the
subsequent processes strongly resemble those in the
inviscid case.   Moreover, the calculations of velocity
statistics that were described in the previous section show
that the profiles for the inviscid case will match those of
the viscous case if they are shifted downstream about two
channel heights.   Thus, both pairing and entrainment are
inviscid mechanisms. Figure 5 shows an uncommon event in
unforced mixing layers, the merging of three eddies
simultaneously.

In summary, numerical results indicate that the area
of the flow occupied by vorticity increases primarily by
the entrainment mechanism, during which large vortex eddies
"devour" the braids of vorticity that separate from the
pairing eddies, along with some irrotational fluid in their
surroundings. Moreover, the strain field of large vortex
eddies can tear the smaller eddies trapped in between and,
consequently, consume their vorticity with some of the
irrotational fluid. The formation of strong vortex eddies
is caused by the pairing of smaller eddies that appear as a
result of the roll-up of the original vortex sheet, while
the presence of small eddies among large eddies can be
explained by the random selection of eddy pairs during the
merging process (Ho and Huerre, 1984).

Harmonic Modulation

Periodic excitation, or forcing, at a well-defined
frequency has been used in analytical studies and
experimental investigations to elucidate the dynamics of
the vorticity field and to manipulate the growth rate of
the shear layer (Oster and Wygnanski, 1982; and Fiedler and
Mensing, 1985).   In the numerical simulations,
monochromatic perturbations were superimposed on the
approaching streams at an amplitude of 10% of the mean
velocity of each stream.   The frequency of excitation was
chosen as one of the preferred frequencies of the layer,
obtained by a spectral analysis of the kinetic energy of
fluctuations of the natural layer at a set of points
downstream of the splitter plate.

The energy spectrum of the fluctuating component,
constructed using a sample of 1000 time steps of velocity
data computed at $x=5.0$ and $y=0.0$, is shown in Fig. 13.   The

U2/U1=0.333   Re=4000.   time step=0.10   s=0.001179   f=1.000   a=0.100

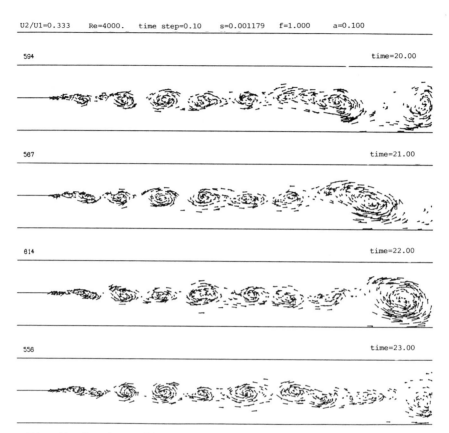

Fig. 14 Vorticity field of a shear layer, computed at the same
conditions as in Fig. 3 with forcing at frequency $\Omega 1$=1.0,
showing the early roll-up of the vorticity into a set of eddies
of the size of E1 and the frequency $\Omega 1$.

energy peaks around two frequencies, $\Omega 1$=1.0 and $\Omega 2$=0.5.
When x is changed, the peak frequencies shift
slightly to $\Omega 1'$ and $\Omega 2'$, increasing upstream and
decreasing downstream, with $\Omega 1'$=2$\Omega 2'$. Thus, $\Omega 1$ and $\Omega 2$
are the frequencies of the energy-containing eddies. By
inspecting the vorticity field in Fig. 3, it is found that
$\Omega 2$ corresponds to the passage frequency of the
characteristic large eddy at section x=5.0, referred to as
eddy E2, while $\Omega 1$ corresponds to the passage frequency of
the characteristic eddy at section y=3.0, referred to as
E1. The frequency shift is due to the gradual growth of
each of these two eddies by entrainment. E2 is formed by
the merging of two E1 eddies, hence the size of E2 is twice
the size of E1.

U2/U1=0.333   Re=4000.   time step=0.001179   f=0.500   a=0.100

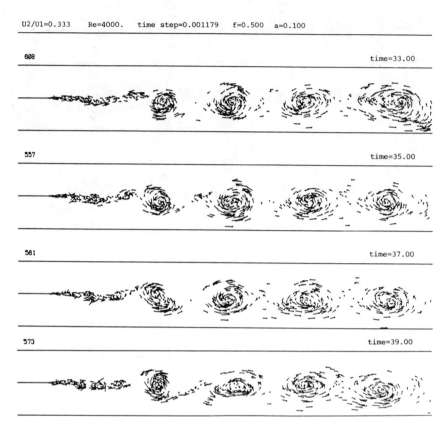

Fig. 15 Vorticity field of a shear layer, computed at the same conditions as in Fig. 3 with forcing at $\Omega 2=0.5$, depicting the lock-in mechanism and the early pairing of eddies into an eddy of the size of E2 and the frequency $\Omega 2$.

Fig. 14 depicts the vorticity field of a layer forced at frequency $\Omega 1$, causing an early roll-up of vorticity into a set of eddies having the size of E1. The response frequency – the frequency of eddy formation – is the same as $\Omega 1$, i.e. a lock-in mechanism is established downstream of the splitter plate. Pairing is inhibited for a distance downstream and the eddies grow only by entraining irrotational fluid. The effect of forcing at $\Omega 1$ dominates until the eddies grow beyond the size of E1. At this stage, the layer resumes its growth as an unforced layer. $\Omega 1$ may be characterized as the natural roll-up frequency of the energy-containing eddies, or the fundamental frequency, in the flow at the conditions of the numerical simulation.

Forcing at the frequency $\Omega 2$ causes an early onset of pairing interactions, as shown in Fig. 15. Two or three

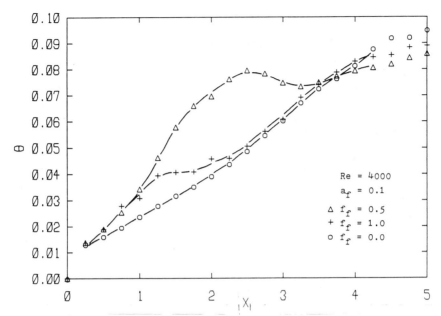

Fig. 16 Variation of the momentum thickness with downstream distance for a naturally excited layer, a layer forced at the roll-up frequency $\Omega 1$, and a layer forced at the pairing frequency $\Omega 2$.

eddies merge at a short distance downstream of the splitter plate and form eddies the size of E2. Beyond this first pairing, eddies flow without interactions, growing only by entrainment. $\Omega 2$, being the first subharmonic of the fundamental frequency, appears to be the pairing frequency of the layer. The effect of forcing at $\Omega 2$ decays approximately four channel heights downstream and the layer resumes its natural growth.

Fig. 16 shows a comparison between the growth of a naturally excited layer, a layer forced at frequency $\Omega 1$, and a layer forced at frequency $\Omega 2$, corresponding to the vorticity fields presented in Figs. 14 and 15. The significant effect of forcing at the pairing frequency $\Omega 2$ on the growth of the momentum thickness is clearly depicted by the early growth of θ, followed by a sharp slowdown until the second stage of pairing when growth is resumed at the growth rate of an unforced layer. The results of Ho and Huang (1982) show a similar behavior for a shear layer forced at the first subharmonic. In Ghoniem and Ng (1986), the response of the flow to excitations at a wider range of frequencies is reported.

Conclusions

Recent developments in the theory of vortex approximations are incorporated in a simulation of a confined shear layer. Results indicate that the accuracy of the time-integration scheme is important in controlling the diffusive error, while the spatial resolution is governed by the number of vortex elements and to a lesser extent by the core function.

The mixing of two streams initially flowing at different velocities can be attributed to the formation of a highly organized vorticity pattern. The vortex sheet originating at the splitter plate rolls up into a set of small eddies that merge into larger eddies by successive pairings. Numerical results indicate that the pairing interaction among eddies is area preserving; hence, it occurs without irrotational fluid being entrapped between the two original eddies. However, individual vortices grow by entraining surrounding irrotational fluid during their unencumbered flow, providing the necessary environment for mixing.

The two-dimensional simulation provides an accurate prediction for the average streamwise velocity and fluctuations, but it overestimates the cross-stream fluctuations. The total sum of the two autocorrelations computed numerically in two dimensions agrees with the total sum of the three autocorrelations measured experimentally. Thus, the disagreement is due to the lack of the third dimension.

Preliminary results indicate that through the superposition of small-amplitude oscillations at the roll-up frequency of the smallest energy-containing eddies of the layer, these eddies are formed earlier and their pairing is delayed. If the perturbation frequency is halved, eddy pairing is enforced closer to the origin and a substantial increase in the growth rate of the layer is achieved at the early stages of development.

Acknowledgment

This work was supported by the National Science Foundation under Grant CPE-840481 and the U.S. Air Force Office of Scientific Research under Grant AFOSR84-0356.

References

Aref, H. and Siggia, E.D. (1980) Vortex dynamics of the two-dimensional turbulent shear layer. J. Fluid Mech. 100, 705-737.

Ashurst, W.T. (1979) Numerical simulation of turbulent mixing layers via vortex dynamics. Turbulent Shear Flows (edited by F. Durst et al.), Vol. I, pp. 402-413, Springer-Verlag, Berlin.

Beale, J.T. and Majda, A. (1982) Vortex methods II: higher order accuracy in two and three dimensions. Math. Comput., 39 (159), 28-52.

Browand, F.K. and Ho, C.-M. (1983) The mixing layer: an example of quasi two-dimensional turbulence, J. Mec. (in press).

Browand, F.K. and Weidman, P.D. (1976) Large scales in the developing mixing layer. J. Fluid Mech. 76, 127-144.

Brown, G.L. and Roshko, A. (1974) On density effects and large structures in turbulent mixing layers. J. Fluid Mech. 64, 775-816.

Cantwell, B.J. (1981) Organized motion in turbulent flow. Annu. Rev. Fluid Mech. 13, 457-515.

Chorin, A.J. (1973) Numerical studies of slightly viscous flow. J. Fluid Mech. 57, 785-796.

Christiansen, J.P. (1973) Numerical simulation of hydrodynamics by the method of point vortices. J. Comput. Phys. 13, 363-379.

Clements, R.R. (1973) An inviscid model of two-dimensional vortex shedding. J. Fluid Mech. 57, (2), 321-336.

Corcos, G.M. and Sherman, F.S. (1984) The mixing layer: deterministic models of a turbulent flow, Part I: Introduction and the two-dimensional flow. J. Fluid Mech. 139, 29-65.

Fiedler, H.E. and Mensing, P. (1985) The plane turbulent shear layer with periodic excitation. J. Fluid Mech. 150, 281-309.

Ghoniem, A.F., Chen, D.Y., and Oppenheim, A.K. (1985) Formation and inflammation of a turbulent jet. AIAA Journal, 24, 224-229.

Ghoniem, A.F., Chorin, A.J., and Oppenheim, A.K. (1982) Numerical modelling of turbulent flow in a combustion channel. Phil. Trans. R. Soc. London A304, 303-325.

Ghoniem, A.F. and Gagnon, Y. (1986) Numerical investigation of recirculating flow at moderate Reynolds numbers. AIAA Paper 86-0370.

Ghoniem, A.F., and Ng, K. K. (1986) Effect of harmonic modulation on entrainment and mixing in a confined shear layer. AIAA Paper 86-0056.

Ghoniem, A.F. and Sethian, A.J. (1985) Dynamics of turbulent structures in a recirculating flow: A computational study. AIAA Paper 85-0146.

Ghoniem, A.F. and Sherman, F.S. (1985) Grid-free simulation of diffusion using random walk methods. J. Comput. Phys. 61, 1-37.

Hald, O.H. (1979) Convergence of vortex methods for Euler's equations II. SIAM J. Num. Anal. 16, 726-755.

Hald, O.H. (1984) Convergence of a random method with creation of vorticity. PAM-252. Center for Pure and Applied Mathematics, University of California, Berkeley.

Hald, O.H. (1985) Convergence of vortex methods for Euler's equations III. PAM-270. Center for Pure and Applied Mathematics, University of California, Berkeley.

Hernan, M.A. and Jimenez, J. (1982) Computer analysis of a high-speed film of the plane turbulent mixing layer. J. Fluid Mech. 119, 323-345.

Ho, C. M. and Huang, L.S. (1982) Subharmonics and vortex merging in mixing layers. J. Fluid Mech. 119, 443-473.

Ho, C.-M. and Huerre, P. (1984) Perturbed free shear layers. Annu. Rev. Fluid Mech. 16, 365-424.

Hsaio, C.C., Ghoniem, A. F., Choirn, A. J., and Oppenheim, A. K. (1985), Numerical simulation of a turbulent flame stabilized behind a rearward-facing step. The 20th Symposium (International) on Combustion, 494-504. The Combustion Institute, Pittsburg, PA.

Inoue, O. (1985) Vortex simulation of a turbulent mixing layer, AIAA J. 23, 367-373.

Keller, J.O. (1982) An experimental study of combustion and the effects of large heat release on a two-dimensional turbulent mixing layer, Ph.D. Thesis. University of California, Berkeley.

Leonard, A. (1980) Vortex methods for flow simulation. J. Comput. Phys. 37, 289-335.

Mansour, N.N. (1985) A hybrid vortex-in-cell finite-difference method for shear-layer computation. AIAA Paper 85-0372.

Nakamura, Y., Leonard, A., and Spalart, P. (1982) Vortex simulation of an inviscid shear layer. AIAA Paper 82-948.

Oster, D. and Wygnanski, I. (1982) The forced mixing layer between parallel streams. J. Fluid Mech. 123, 91-130.

Riley, J.J. and Metcalfe, R.W. (1980) Direct numerical
    simulation of a perturbed turbulent mixing layer. AIAA
    Paper 80-0274.

Roberts, F.A., Dimotakis, P.E., and Roshko, A. (1982)
    Kelvin-Helmholtz instability of superposed streams. Album
    of Fluid Motion (edited by M. Van Dyke), p. 85. Stanford,
    CA.

Roshko, A. (1976) Structure of turbulent shear flows: A
    new look. AIAA J. 14, (10), 1349-1357.

Roshko, A. (1981) The plane mixing layer: flow
    visualization results and three-dimensional effects.
    Lecture Notes in Physics, 36, 208-217. Springer-Verlag,
    New York.

Spencer, B.W. and Jones, B.G. (1971) Statistical
    investigation of pressure and velocity fields in the
    turbulent two-stream mixing layer. AIAA Paper 71-613.

Winant, C.D. and Browand, F.K. (1974) Vortex pairing: the
    mechanism of turbulent mixing-layer growth at moderate
    Reynolds number. J. Fluid Mech., 63, 237-255.

The Vortical Structure of Premixed Flames Stabilized by a Circular Cylinder

Tatsuya Hasegawa,* Shigeki Yamaguchi,† and Norio Ohiwa‡
Nagoya Institute of Technology, Nagoya, Japan

Abstract

The formation of vortical structures in a premixed flame stabilized by a circular cylinder was numerically modeled by the random vortex method. The simulations predicted the following phenomena in accordance with an increase of a uniform flow velocity: a flashback, a flame having two independent shear layers, an asymmetrical reciprocal vortical flame, and a blowoff. The experimentally observed behavior of the premixed flame agreed with the simulation. The transition of the flame structure was affected by the balance between the shear flow and combustion. The Strouhal numbers derived from the simulated asymmetrical flame patterns were twice as much as those of Kármán vortex street and increased slightly with the flow velocity.

Nomenclature

a   = radius of a circular cylinder
d   = diameter of a circular cylinder
f   = volume ratio of burned gas in a mesh cell
Δf = increment of volume ratio of burned gas in a mesh cell
$F(z)$ = complex velocity potential
h   = mesh size
M   = number of segments that approximate cylinder surface
n   = normal unit vector
p   = pressure
Q   = threshold value of corrective technique
Re   = Reynolds number defined by radius

Presented at the 10th ICDERS, Berkeley, California, August 4-9, 1985. Copyright © 1986 by the American Institute of Aeronautics and Astronautics, Inc. All rights reserved.
*Assistant Professor, Department of Mechanical Engineering.
†Professor, Department of Mechanical Engineering.
‡Associate Professor, Department of Mechanical Engineering.

Re_d = Reynolds number defined by diameter
$\mathbf{r}$ = position vector
S_b = velocity of burned gas behind the flame front
S_u = burning velocity
St = Strouhal number
$\mathbf{t}$ = tangential unit vector
t = time
Δt = time step
U = uniform flow velocity
$\mathbf{u}$ = velocity vector
u = x component of velocity vector
v = y component of velocity vector
x = x coordinate
y = y coordinate
z = complex coordinate
Δ = intensity of source blob
Γ = intensity of vortex blob
ε = source
η_1 = random walk in x direction
η_2 = random walk in y direction
ξ = vorticity
ρ = density
σ = cutoff radius
ϕ = equivalence ratio of premixed gas
ϕ = velocity potential
ψ = stream function

Subscripts

b = burned gas
f = flame
s = source
u = unburned gas

Superscript

* = mirror image

Introduction

It is well known that the transition of the vortical structure in the wake of a cylindrical rod from a symmetrical to an asymmetrical configuration occurs when the Reynolds number exceeds a critical value. The transition of the rod-stabilized flame structure from a symmetrical one to an asymmetrical one is, however, affected by combustion that produces a volumetric expansion behind the flame front and increases the kinematic viscosity of the burned gas. The conditions in which the transition occurs, and the

structure of symmetrical and asymmetrical flames, are not
yet clearly understood.

A new approach for modeling nonsteady turbulent premixed flames was developed by Ghoniem et al. (1982). Based
on the random vortex method (Chorin, 1973) and the simple
line interface calculation method (SLIC) (Chorin, 1980),
the analysis succeeded in explaining the vortical structure
of a premixed flame stabilized behind a rearward-facing
step. This procedure was improved by Sethian (1984) who
included volumetric sources on the flame front and introduced a more complicated SLIC method. Another vortex method
for simulating premixed flames was also developed by
Ashurst (1982) who used the finite difference method for
calculating flame propagation.

All of these procedures incorporate nonsteady behavior
of the separated shear layers, flame propagation, and volumetric expansion caused by heat release. The fluid dynamic
interaction between the premixed flame and the shear flow
can thereby be modeled, although the increase of kinematic
viscosity due to temperature increase is not included. This
is a remarkable theoretical advancement when it is compared
with closure modeling that can simulate only time-averaged
properties.

In this paper, some experimental observations were
conducted to clarify the conditions in which the symmetrical and asymmetrical vortical flame structure appeared. The
vortex method developed by Ghoniem et al. (1982) was then
applied to simulate the vortical structure of a premixed
flame stabilized by a circular cylinder. The structures of
premixed flames at flows of various speeds were simulated
and compared with each other to elucidate the role of
combustion.

Experimental Observations

A schematic diagram of a combustion tunnel used in
this study is shown in Fig. 1. Commercial grade gaseous
propane and air are introduced into a 500 mm long mixing
duct consisting of one perforated plate and three No.100
wire mesh screens. Through a 5:1 contraction nozzle, mixed
gas is introduced into a test section having a $22 \times 60 \text{ mm}^2$
rectangular cross section at a height of 500 mm. The side-
walls of the test section consist of quartz windows for
visual and optical observations. A cylindrical rod with a
diameter of 4 mm is set 110 mm downstream from the entrance
of the test section. The blockage ratio of the test section
is 0.07 and the velocity at the cylinder is almost the
same as the velocity at the entrance.

The regions in which flame blowoff, flashback, and the symmetrical and asymmetrical vortical structure occurred were visually observed as the equivalence ratio of the propane-air mixture and the flow velocity were changed. The obtained results are shown in Fig. 2. The asymmetrical structure appears near the blowoff limit, where the equivalence ratio deviates from unity and the flow velocity is relatively large. The symmetrical structure appears between the asymmetrical flame regions. Typical shadowgraph pictures of symmetrical and asymmetrical flames having the

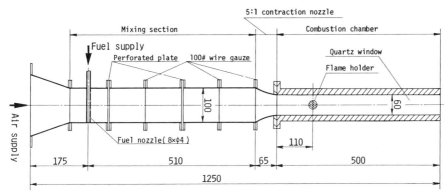

Fig. 1   Schematic diagram of combustion tunnel.

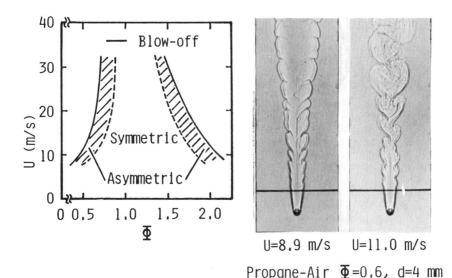

U=8.9 m/s   U=11.0 m/s

Propane-Air   Φ=0.6, d=4 mm

Fig. 2   Experimentally observed premixed flame patterns stabilized by a circular cylinder for various flow speeds and various equivalence ratios (d=4 mm, propane-air).

same equivalence ratio of 0.6 and different flow velocities
are also shown in Fig. 2. The transition of the flame
structure is obvious. A symmetrical structure, however, can
still be observed in the asymmetrical flame near the cylinder where the recirculation zone exists. The transition of
the flame structure also occurs when heat release due to
combustion decreases, i.e., the equivalence ratio deviates
from unity and the uniform flow retains a constant velocity. The transversal lines in the photographs are shadows
of joints in the quartz windows and do not affect the
flowfield.

Simulation Method

In this work, a premixed flame stabilized by a circular cylinder with a radius of "a" in a uniform flow with a
velocity of U as shown in Fig. 3 is modeled subject to the
following assumptions:
1) The flowfield is two-dimensional.
2) The uniform flow velocity U and the burning velocity S_u are much smaller than the sound velocity. Thus, the
fluid can be regarded as incompressible.
3) The turbulent scale is much larger than the flame
thickness and the flame can be regarded as a wrinkled
laminar flame. Furthermore, the flame sheet is regarded as
a discontinuity of incompressible burned and unburned gases
having different densities.
4) The burning velocity S_u is constant throughout.
5) Volumetric expansion caused by heat release is replaced by volumetric sources located on the flame sheet.

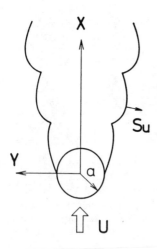

Fig. 3 Flowfield and coordinate system.

6) Variations in the burning velocity due to chemical reactions, variations of the kinematic viscosity due to temperature increase, buoyancy force, heat conduction, molecular diffusion, and three-dimensional effects are negligible.

Although some of the above assumptions represent gross simplifications of actual combustion systems, they are considered to be reasonable as a first-order approximation for the modeling of the interaction between large eddies and combustion. More complicated simulations including heat transfer, molecular diffusion, and chemical reactions were recently developed by Ghoniem and Oppenheim (1984).

With the foregoing assumptions, the governing equations, nondimensionalized by the uniform flow velocity U and the radius "a," are written as

$$\nabla \cdot \mathbf{u} = \varepsilon \tag{1}$$

$$D\mathbf{u}/Dt = -\nabla p + (1/Re)\Delta \mathbf{u} \tag{2}$$

$$D\mathbf{r}_f/Dt = S_u \mathbf{u}_f \text{ (flame propagation)} \tag{3}$$

where $D/Dt \equiv \partial/\partial t + (\mathbf{u} \cdot \nabla)$. The boundary conditions on the cylinder surface are

$$\mathbf{u} \cdot \mathbf{n} = 0 \tag{4}$$

$$\mathbf{u} \cdot \mathbf{t} = 0 \tag{5}$$

The velocity $\mathbf{u}$ has both rotational and irrotational components. Thus, when the velocity potential ϕ for irrotational velocity component and stream function ψ for rotational velocity component are introduced, the following Poisson's equations are derived from Eq. (1) and the definition of vorticity:

$$\Delta\phi = \varepsilon \tag{6}$$

$$\Delta\psi = -\xi \tag{7}$$

A flowfield at a given moment can be described by solution of Eqs. (6) and (7).

The vortex method developed by Ghoniem et al. (1982) may be used to express the complex velocity potential for the flowfield of a premixed flame stabilized by a circular cylinder immersed in a uniform flow as

$$F(z) = z + 1/z$$

$$+ i\Sigma_1(\Gamma_j/2\pi)\ln(z-z_j) + i\Sigma_2(\Gamma_j/2\pi\sigma)(z-z_j)$$

$$- i\Sigma_1(\Gamma_j/2\pi)\ln(z-z_j*) - i\Sigma_2(\Gamma_j/2\pi\sigma)(z-z_j*)$$

$$+ \Sigma_1(\Delta_j/2\pi)\ln(z-z_j) + \Sigma_2(\Delta_j/2\pi\sigma_s)(z-z_j)$$

$$+ \Sigma_1(\Delta_j/2\pi)\ln(z-z_j*) + \Sigma_2(\Delta_j/2\pi\sigma_s)(z-z_j*)$$

$$- \Sigma_1(\Delta_j/2\pi)\ln(z) \tag{8}$$

where $z*=1/\bar{z}$ is the mirror image of z ($\bar{z}$ is the complex conjugate of z), subscript j the jth vortex or source blob, σ and σ_s the cutoff radii to avoid the singularity of induced velocity, Σ_1 a summation when $|z-z_j|\geq\sigma$ and σ_s, and Σ_2 a summation when $|z-z_j|<\sigma$ and σ_s. The normal boundary condition on the cylinder surface is satisfied by this equation.

When random walks are introduced to simulate the diffusion of vorticity caused by viscosity (Chorin, 1973), the motion of vortex blob i is described as

$$d\bar{z}_i/dt = [dF(z)/dz]_{z=z_i} = u_i - iv_i \tag{9}$$

$$x_i(t+\Delta t) = x_i(t) + u_i\Delta t + \eta_1 \tag{10}$$

$$y_i(t+\Delta t) = y_i(t) + v_i\Delta t + \eta_2 \tag{11}$$

where random walks η_1 and η_2 have a variance of $(2\Delta t/Re)$ and a mean of zero.

The tangential boundary condition on the cylinder surface is satisfied by the production of vortex blobs (Chorin, 1973) with intensities of

$$\Gamma = u_k \cdot t_k 2\pi/M \tag{12}$$

and with a cutoff radius determined as

$$\sigma = 2/M \tag{13}$$

where M is the number of segments that approximate the cylinder surface, u_k the velocity at the center of kth segment, and t_k the tangential unit vector at the kth segment.

Propagation of the premixed flame can be calculated by the SLIC (simple line interface calculation) method com-

bined with Huygens' principle (Chorin, 1980). Expansion of
the burned gas is accounted for by increasing the volume
ratio of the burned gas with $(\rho_u/\rho_b-1)\Delta f$ behind the flame
front as well as by locating the source blobs on the flame
front. In this manner, the region behind the flame front is
filled with burned gas.

The intensity of the source blob is determined from a
consideration of one-dimensional flame propagation (Ghoniem
et al., 1982). Thus, the source intensity and its cutoff
radius are written as

$$\Delta = (\rho_u/\rho_b-1)h^2\Delta f/\Delta t \qquad (14)$$

$$\sigma_s = \Delta/[\pi(\rho_u/\rho_b-1)S_u] \qquad (15)$$

where h is the mesh size to be used in calculating the
flame propagation, and f the volume ratio of burned gas in
a mesh cell. When the behavior of two flame sheets propagating in opposite directions or a flame sheet propagating
away from a solid boundary is accurately simulated, the
source intensity is evaluated to be twice that used by
Ghoniem et al. (1982). This evaluation of source intensity
results in less accuracy when two flame sheets approach
each other. However, this is not a serious problem when
two flame fronts propagate in opposite directions, as would
be the case for a large-scale structure of a V-shape premixed flame stabilized by a cylinder.

A corrective technique was used to remove the pieces
of burned gas that were of too small a value to be resolved
on the SLIC grid. This technique determined the value of f
at every time step such that f=0 when f<Q and f=1 when f>
1-Q, where Q is a threshold value. The threshold value was
selected to avoid neglecting ordinary flame propagation
from a mesh cell filled with burned gas, i.e., Q was less
than $S_u\Delta t/h$.

The simulation algorithm is shown in Fig. 4. The separation points of the boundary layers are automatically
determined by the interaction of the vortex blobs and the
boundary layers represented by newly created vortex blobs
on the cylinder surface.

Simulations were performed with a radius of 2 mm, a
burning velocity of 0.4 m/s, ρ_u/ρ_b=7.5, and a kinetic
viscosity of 1.5×10^{-5} m^2/s, i.e., a propane-air mixture
with an equivalence ratio of unity. The mesh size h was
determined to be half of the radius to save calculation
time and obtain a sufficient resolution for a qualitative
study of large-scale flame structures. Four mesh cells just
behind the cylinder were ignited at a time of t=0. The
number of segments that approximated the cylinder surface
was selected to be M=24 according to our preliminary simu-

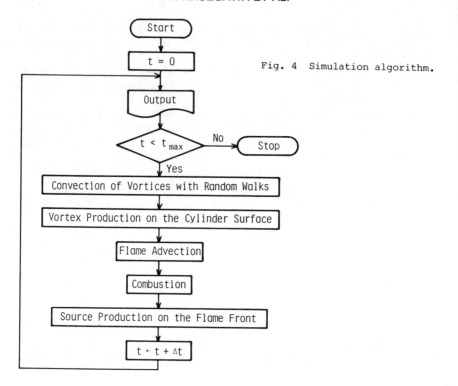

Fig. 4   Simulation algorithm.

lations of Kármán vortex street (Hasegawa et al., 1984).
The time step for convection of vortex blobs and flame
propagation by combustion was 0.2; otherwise, the time step
for flame advection was 0.01 to satisfy the stability
condition of SLIC method, $(\text{Max}|\mathbf{u}|)\Delta t/h < 1/2$ (Chorin, 1980).
The velocity of the uniform flow which was impulsively
started at t=0 was changed from 1.875 to 60 m/s.

Simulation Results and Discussions

     To determine the accuracy of the flame propagation
algorithm, an outwardly propagating flame from a square
cell filled with burned gas was first simulated. Temporal
changes of the equivalent radius derived from the burned
gas volume are shown in Fig. 5. The propagation velocity
was found to be the same as S_u when the density ratio of
the unburned and burned gases was unity. The propagation
velocity became about five times as large as S_u when the
density ratio became 7.5, although the predicted velocity
was 63% of the theoretical one, $(\rho_u/\rho_b)S_u$. These results
showed that the simulated flame behaved as a flame with
decreased heat release, i.e., a flame with an equivalence
ratio deviating from unity.

The inexact prediction of the flame's propagation velocity possibly occurred because a coarse SLIC grid was used together with the original SLIC method that reproduced a rough flame interface. The positions of the volumetric sources were therefore determined imprecisely and the velocity induced by the sources became less than that of the actual smooth flame. The burned gas, however, was almost quiescent within a few percent of the outward moving unburned gas. The corrective technique led to more accurate simulation of the propagation velocity.

A better result could be obtained if the grid size was decreased and if a more precise method was used to determine the position of the interface. (See Barr and Ashurst, 1984.) However, qualitative flame behavior can be simulated by using our method.

To test the capability of the vortex method for modeling flow phenomena, a Kármán vortex street pattern was simulated with Reynolds numbers of 1000 and 10,000. The approximate Strouhal number of 0.2 and the drag coefficient of about 1.0, as predicted by the simulations, agreed with the experimental values (Hasegawa et al., 1984). A typical example of a simulated Kármán vortex street pattern at Re=1000 is illustrated in Fig. 6.

In the simulation of premixed flames stabilized by a circular cylinder, the uniform flow velocities were determined to be 1.875, 7.5, 15, 30, and 60 m/s and the threshold values of the corrective technique were determined to be 0.08, 0.02, 0.01, 0.005, and 0.002, respectively. These threshold values were determined by using the nondimension-

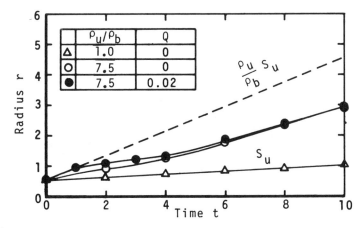

Fig. 5   Temporal change of the equivalent radius of an outwardly propagating flame.

alized burning velocity and the criterion described above. The other parameters were not changed.

When the velocity of a uniform flow is 1.875 m/s, a flame propagates upstream like a flashback, as shown in Fig. 7. The temporal behavior of the burned gas region (black region) is illustrated in this figure. Volumetric expansion, which pushes the uniform flow upstream, and flame propagation with a velocity of S_u contribute to the flashback behavior in this simulation, since the flame propagating in the boundary layer is not obvious in Fig. 7. Other effects such as pressure oscillation and wall heating, which are likely to contribute to flashback, are not in-

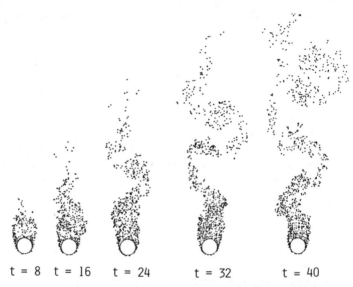

t = 8   t = 16   t = 24   t = 32   t = 40

Fig. 6   Kármán vortex street pattern at Re=1000.

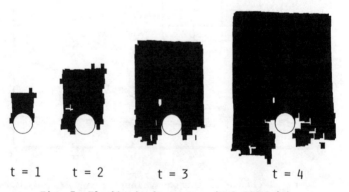

t = 1   t = 2   t = 3   t = 4

Fig. 7   Flashback phenomenon (U=1.875 m/s).

cluded in this simulation. The experimentally observed
uniform flow velocity resulting in flashback was below 2
m/s under the same conditions as in the simulation and
supported the simulated results.

When the flow velocity is 7.5 m/s, the boundary layer
separates from the cylinder surface and forms two independent shear layers on both sides of the cylinder, as shown in
Fig. 8. The burned gas region and the distribution of the
vortex and source blobs at t=16 are illustrated in this
figure. The two shear layers do not interact with each
other and are decoupled by the volumetric sources. The
flame has a wavy shape in accordance with the velocity
field produced by the shear layers.

As shown in Fig. 8, the simulation predicted two independent shear layers on both sides of the cylinder. The
shear layers, however, are not synchronized to produce a
symmetrical vortical flame as shown in Fig. 2. This possibly resulted because the increase of kinematic viscosity
due to heat release was not included in the simulation.
When heat is added to the wake region of the cylinder in a
high-speed flow, symmetrical vortices having two independent shear layers on both sides of the cylinder have been
observed experimentally (Mori et al., 1983). The transition
of the flame structure from asymmetrical to symmetrical
when combustion heat release was increased was also observed in a flame stabilized by a cylinder, as can be seen
in Fig. 2. These results suggest that increases in the
kinematic viscosity because of heat addition is important
for the formation of a symmetrical flame.

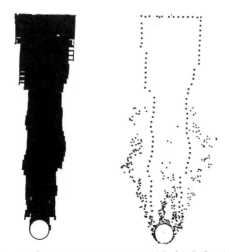

Fig. 8 Flame structure at t=16 (U=7.5 m/s).

T. HASEGAWA ET AL.

When the flow velocity is 15 m/s, the effects of combustion appear to be less important than the effects of the shear layers. The shear layers on both sides of the cylinder thus interact with each other and produce an asymmetrical flame as shown in Fig. 9. The distribution of vortex blobs shown in Fig. 9 indicates that the asymmetrical structure of the flame is the result of the reciprocal configuration of counterrotating vortices. If it is noted that the simulated flame behaves as a flame with an equivalence ratio deviating from unity (as shown in Fig. 5), the appearance of the asymmetrical flame when the uniform flow velocity is increased qualitatively agrees with the experimental observations as shown in Fig. 2.

When the flow velocity is 30 m/s, the flame's structure is also asymmetric, but the much smaller effect of

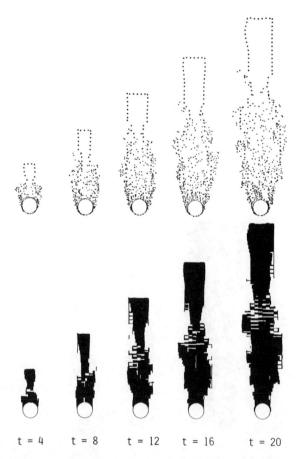

t = 4   t = 8   t = 12   t = 16   t = 20

Fig. 9   Temporal behavior of an asymmetrical flame (U=15 m/s).

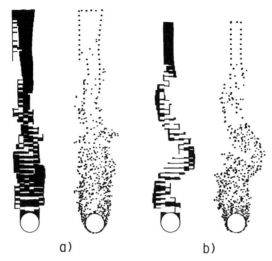

a)   b)

Fig. 10 Flame structures at t=20: a) U=30 m/s, b) U=60 m/s.

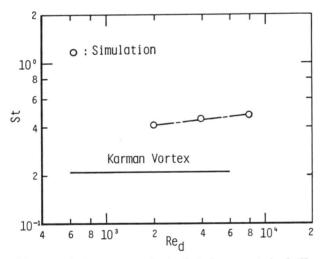

Fig. 11 Strouhal numbers of simulated asymmetrical flames.

combustion narrows the flame's width, as shown in Fig. 10a.
The unburned gas in the burned gas region is presumably
results from a coarse SLIC grid, which cannot simulate
highly wrinkled interfaces within the meshes and an increased total burning rate.
    When the flow velocity is 60 m/s, a sinusoidal flame
pattern, almost the same as Kármán vortex street shown in
Fig. 6, appears as shown in Fig. 10b. The flame is, therefore, considered to be blown off.

The variation of the Strouhal numbers derived from the simulated asymmetrical vortical flames are shown in Fig. 11. The Reynolds number in abscissa, which is derived from the uniform flow velocity and kinematic viscosity of unburned gas, merely indicates the variation of the flow velocity. The dashed line connects the simulated values and does not mean an experimental value. The Strouhal number is twice that of Kármán vortex street and slightly increases with the flow velocity. Thus, it can be concluded that the vortical structure of the asymmetrical flames stabilized by a circular cylinder is not similar to that of Kármán vortex street, but is affected by combustion.

Conclusions

The formation of vortical structures in premixed flames stabilized by a circular cylinder was modeled with the random vortex method. The obtained results are as follows:

1) The simulations predicted the following phenomena in accordance with the increase of the uniform flow velocity: when the flow velocity was small, the flame propagated upstream like a flashback; when the flow velocity was moderate, two independent shear layers appeared on both sides of the cylinder; when the flow velocity became larger and the effects of combustion became relatively smaller, the shear layers interacted with each other and an asymmetrical reciprocal vortical structure appeared; and when the uniform flow velocity became much larger, a flow pattern similar to the Kármán vortex street was predicted and the flame was considered to be blown off. The experimentally observed behavior of the premixed flames when increasing uniform flow velocity agreed with the simulation results.

2) The Strouhal numbers derived from the simulated asymmetrical flame patterns were twice as much as those of Kármán vortex street and increased slightly with the flow velocity. Thus the asymmetrical vortical structure was affected by combustion and different from the Kármán vortex street structure.

References

Ashurst, W. T. (1982) Turbulent flow structure effect on diffusion and premixed flame propagation. Proceedings of the Fourth GAMM Conference on Numerical Methods in Fluid Mechanics. Friedr. Vieweg & Sohn, Braunschweig, FRG.

Barr, P. K. and Ashurst, W. T. (1984) An interface scheme for turbulent flame propagation. SAND-82-8773, Sandia National Laboratories, Livermore, CA.

Chorin, A. J. (1973) Numerical study of slightly viscous flow.
    J. Fluid Mech. 57, 785-796.

Chorin, A. J. (1980) Flame advection and propagation algorithms.
    J. Comp. Phys. 35, 1-11.

Ghoniem, A. F., Chorin, A. J., and Oppenheim, A. K. (1982)
    Numerical modeling of turbulent flow in a combustion tunnel.
    Phil. Trans. R. Soc. London, Ser. A 304, 303-325.

Ghoniem, A. F. and Oppenheim, A. K. (1984) Numerical solution
    for the problem of flame propagation by the Random Element
    Method. AIAA J. 22, 1429-1435.

Hasegawa, T., Yamaguchi, S., and Ohiwa, N. (1984) Discrete vortex
    simulation of Karman vortex street. Bull. Nagoya Inst. Tech.
    36, 199-204.

Mori, Y., Hijikata, K., and Nobuhara, T. (1983) A fundamental
    study of symmetrical vortices generation behind a cylinder by
    heating wake or providing splitter plate or mesh. Trans. JSME
    B 49-444, 1782-1790 (in Japanese).

Sethian, J. (1984) Turbulent combustion in open and closed
    vessels. J. Comp. Phys. 54, 425-456.

Numerical Simulation of Unsteady Mixing Layers

E.S. Oran,* J.P. Boris,† and K. Kailasanath‡
U.S. Naval Research Laboratory, Washington, D.C.
and
F.F. Grinstein‡
Berkeley Research Associates, Springfield, Virginia

Abstract

The time-dependent compressible conservation equations
were solved numerically in order to examine the evolution of large-scale structures resulting from the Kelvin-
Helmholtz instability. Three two-dimensional configurations were considered: a planar splitter plate, an axisymmetric round jet, and an axisymmetric central dump
combustor. For the splitter-plate and round-jet configurations, the simulations modeled two subsonic coflowing
streams in which different initial velocity ratios, temperatures, types of gases, and ways of initiating the instability were considered. The effects of energy release
were considered in the round jet for a hydrogen-oxygen
mixture. The simulations of the dump combustor modeled
flow from a long diffuser into a wider combustor. As the
gas leaves the diffuser, it flows over a step. At the
exit nozzle, the flow becomes sonic. The splitter-plate
simulations showed that the composition of the resulting
mixing layers are asymmetric in the first few roll-ups,
with more of the faster fluid entrained by the roll-up.
The existence of this asymmetry caused the formation of
pockets of fuel-rich mixtures in the reacting round-jet
calculation. The dump-combustor calculations showed the
effects of acoustic waves and imposed pressure oscillations on the frequency of vortex shedding and merging.

Presented at the 10th ICDERS, Berkeley, California, August 4-9,
1985. This paper is a work of the U.S. Government and therefore is
in the public domain.

*Senior Scientist, Laboratory for Computational Physics.
†Chief Scientist, Laboratory for Computational Physics.
‡Research Scientist, Laboratory for Computational Physics.

Introduction

Many combustion applications involve shear flows in
which chemical reactions occur after the fluids on each
side of the layer mix. Experimental investigations of
nonreactive shear flows in idealized geometries have shown
that large coherent structures dominate the entrainment
and mixing processes[1-4]. These structures are certainly
evident at the transition to turbulence and persist downstream. This has important consequences for combustion.
Fluid from the two streams is entrained in the structures
so that when chemical reactions are possible, they occur
preferentially in regions associated with these structures. For combustion systems, it would be extremely helpful to understand the mechanisms by which mixing can be
enhanced or retarded. When the flows are confined, and
possibly even when they are free, the effects of acoustic-
vortex and acoustic-vortex-chemical interactions may be
important in determining the flow patterns and causing
flow instabilities.

Here we report the results of time-dependent, two-
dimensional numerical simulations of shear flows. The
purpose of the studies is to isolate and investigate some
of the mixing and feedback mechanisms involved in the development of reacting and nonreacting mixing layers. We
summarize the work done on the development of computational models, but emphasize and show the relations among
the mechanistic studies performed using these models. The
model development process is systematic. Complexity in
boundary conditions, flow geometry, and additional phenomenologies are added step by step. In this paper, we
describe calculations with newly developed inflow and outflow boundary conditions, confined and unconfined flows,
and acoustic forcing. We have also begun to look at the
effects of energy release on the unstable flow patterns.

This work follows the numerical studies of inflow and
outflow boundary conditions by Boris et al.[5,6] and the detailed analysis of splitter-plate simulations by Grinstein
et al.[7]. We first summarize the two-dimensional planar
simulations. Second, we present numerical studies of the
evolution of an axisymmetric gas jet emerging into a quiescent background. Finally, we look at the effects of
the natural acoustic modes of a chamber on a confined jet
flowing over a step. The combustor calculations are for
cold, nonreacting flows. These show the effects of acous-

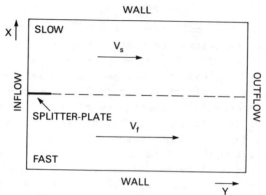

Fig. 1 Schematic diagram showing the initial and boundary conditions for the splitter-plate calculations. V_s and V_f are the fast and slow streams, respectively.

tic waves on vortex shedding from the step and subsequent vortex merging.

Numerical Model

The numerical model used to perform the simulations solves the time-dependent conservation equation for mass, momentum, and energy in two dimensions using the flux-corrected transport (FCT) algorithm[8,9]. This is an explicit, conservative, monotonic finite-difference algorithm incorporating variably spaced grids. No artificial viscosity is required for stabilization. Inflow and outflow boundary conditions[5] have been developed and tested extensively for multidimensional FCT calculations.

We have developed and included a phenomenological induction parameter model to introduce chemical ignition delays and subsequent energy release[10]. A quantity called the induction parameter is convected with each fluid element to keep track of the time history of the temperature, pressure, and stoichiometry of the fluid. When the mixture is within the flammability limits and has been heated for a long enough time, the correct amount of energy is released. This phenomenology has been well calibrated for hydrogen-oxygen mixtures. However, the ignition delay, the energy release time, and the amount of energy released can be varied through input parameters to model a wide variety of fuel-oxidizer systems.

The simulations were performed on nonuniform grids varying in size from 20×50 to 300×600. Computational zones around the edge of the splitter plate, the exit noz-

zle of the jet, and the step of the combustor were closely
spaced. Thus there were closely spaced zones in the directions across the shear layer where the large structures form. The spacing becomes farther apart as the distance from the shear layer increases. The zone separa-
tions in the streamwise direction also increase in size
moving downstream from the nozzle or splitter plate. This
takes advantage of the physical observation that the structures merge and grow downstream, and so fewer cells are
necessary to keep the same relative resolution to that
obtained near the nozzle. This assumption was tested by
varying the computational cell spacing and uniformity.
Resolution tests have been discussed by Boris et al.[5],
Grinstein et al.[7], and Kailasanath et al.[11,12].

The calculations used inflow and outflow boundary conditions as discussed previously[5,7,11,12]. The open outflow
boundary condition imposes a slow relaxation of the pressure towards the known ambient value. The inflow boundary
conditions allow feedback between the downstream vortices
and the inflowing material. In summary, the splitter-
plate calculations imposed free slip, reflecting boundary
conditions on the top and the bottom of the computational
domain at the walls of the tube (Fig. 1). Part of the inflow plenum was included in the simulation. In the round
jet, outflow boundary conditions were imposed on the sides
in order to simulate flow into a large volume (Fig. 4).
The exit of the nozzle was where the inflow boundary condition was imposed. In the dump combustor calculations,
free slip, reflecting boundary conditions were used at
all of the solid walls (Fig. 7). Inflow conditions were
used at the beginning of the long, narrow diffuser. The
flow exited through a narrow annular region where it became sonic.

In summary, we are presenting two-dimensional, compressible, inviscid calculations of the evolution of the
Kelvin-Helmholtz instability in three shear flow geometries. The processes included are convection and chemical energy release. The numerical algorithm is high order so that there is minimal numerical diffusion. The
calculations are expected to be valid for describing the
large-scale structures in a high Reynolds number flow. No
subgrid turbulence model is used. In the last section
of this paper, we present a discussion of the important
processes not yet included, such as molecular diffusion,
buoyancy, small-scale turbulence and the third dimension.

Simulations of Splitter Plate Experiments

Recent splitter-plate experiments by Koochesfahani et al.[13] indicate that there is more high-speed than low-speed fluid entrained in the coherent structures. The amount of mixing was inferred based on measurement of the amount of product of a reaction between dilute reactants in the two streams. The authors found a significant mixing asymmetry which is not predicted by models based on gradient transport and eddy diffusivity. This effect has also not been noted in spectral simulations of unsteady shear flows with periodic boundary conditions[14], or in previous simulations of the splitter-plate geometry using either vortex[15] or finite-difference[16] methods.

The system modeled is shown schematically in Fig. 1. Two subsonic coflowing streams of air are initially separated by a thin plate before entering a long channel. The simulations were initialized by assuming that both gas streams have the same initial pressure (1 atm) and temperature (298 K). The velocities of the gas streams ranged from 3×10^3 to 2×10^4 cm/s, while the velocity ratios in the calculations were in the range 3.0-10.0.

The perturbation that initiated the transition from laminar to turbulent flow in this case is a very small pressure gradient and vorticity at the shear layer, near the edge of the splitter plate. This disturbance moves downstream as the integration proceeds, generating the transverse perturbations for the exponentially growing Kelvin-Helmholtz instability. Previously, we used sinusoidal perturbations along the shear interface to start the instability[5]. The more recent approach is closely analogous to a delta function perturbation in a periodic, temporal simulation of two equal and opposite streams.

Figure 2 shows the development of the flow for the case in which the initial velocity ratio for the streams is 5:1. The velocity of the faster fluid is 10^4 cm/s, about a third of the speed of sound in the material. The material flows in on the left and flows out on the right of the computational domain. The edge of the splitter plate is indicated by an 'x' along the centerline. The faster fluid flows in the lower half of the plane. To emphasize the mixing processes in the visualization, we have contoured values of the ratio of the number density of fluid from the faster stream to the total number density

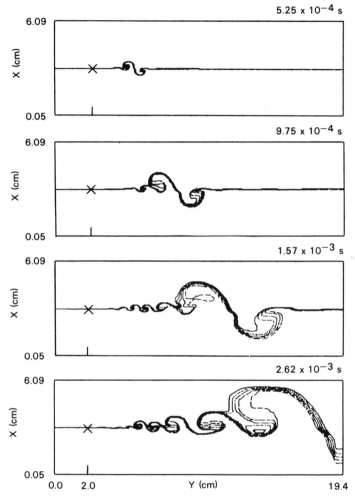

Fig. 2 Contours of the faster fluid number density ratio in the range 0.3--0.7 for the splitter-plate simulation of two streams of air, for which $V_f = 1.0 \times 10^4$ and $V_s = 2.0 \times 10^3$ cm/s.

in the range 0.3-0.7. The transition from a uniform shear flow first appears as a pair of vortices just ahead of the tip of the splitter plate, here shown at 5.25×10^{-4} s. As the original vortex pair grows and moves downstream, small vortices are still being shed at the same approximate location. These vortices grow and merge. By the last panel at 2.62×10^{-3} s, a spectrum of modes is seen suggesting a snapshot from an experimental flow visualization. Figure 3 shows isovorticity contours for the later stages in the development of the flow.

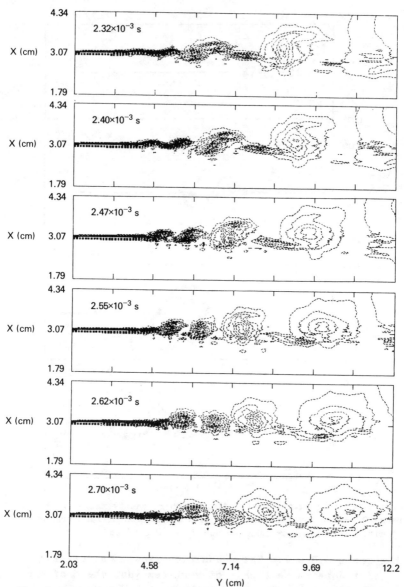

Fig. 3 Isovorticity contours for the later stages in the development of the splitter-plate flow.

Grinstein et al.[7] have shown that the mixing layer in this flow is asymmetric in composition in the first few roll-ups. This means that more of the high-speed than low-speed material is entrained in the vortices. The asymmetry increases as the velocity ratio increases and is

virtually independent of the vertical size of the chamber. These studies have also shown numerical convergence
of the calculation of the asymmetry by testing the dependence of the asymmetry on the grid spacing and parameters
in the outflow boundary condition.

Simulations of Reactive Gas Jets

We have also considered a high-speed (10^4 cm/s) mixture of molecular hydrogen and nitrogen emerging from an
axisymmetric jet into a background of molecular oxygen and
nitrogen (air), as shown schematically in Fig. 4. These
flows are also Kelvin-Helmholtz unstable and large structures form in the transition process. Both unreactive
and reactive mixtures were considered[17,5]. In the reactive mixture, the high-speed nitrogen and oxygen mixture
was 1200 K, the background oxygen and nitrogen mixture was
298 K, and the pressure was 1 atm.

Figure 5 shows vorticity contours from a calculations
of a a jet flowing through a hole in a wall into a quiescent background. The contours look similar to those
shown in Fig. 3 for the splitter plate. In both cases,
the unstable flow results in coherent structures that develop at a fixed distance from the nozzle edge. Newly
formed structures move along the interface as they merge
and grow.

The reactive calculations were initiated in the same
way as the nonreactive calculations. After a period of
nonreactive mixing, however, the material is allowed to
react and release energy. Figure 6 shows results from
a calculation that has the same geometry as the nonreactive case described in Fig. 5. Reactions were started
at 490 μs to assess their effects on the structures. The
contour plots on the left-hand side are the number density

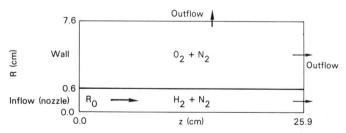

Fig. 4 Schematic diagram showing the initial and boundary conditions for the axisymmetric gas jet. The fast jet enters from the
bottom left along the r axis.

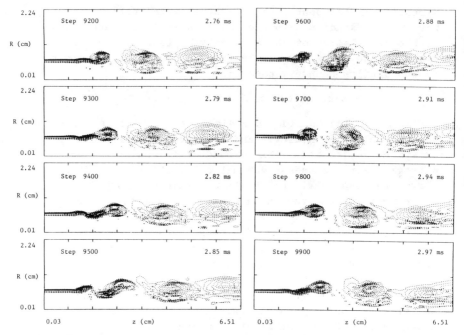

Fig. 5 Isovorticity contours for the unreacted gas jet calculation.

ratio R, where R = [H₂]/([H₂] + [O₂]) and [H₂] and [O₂] are the number densities of molecular hydrogen and oxygen, respectively. The graphs on the right-hand side show the location of the reaction zone, as determined by the induction parameter describing where the energy is being released. We see that the reaction zones move progressively outward. Reaction ceases in the centers when the oxidant is completely consumed. The material remaining there is hot and consists of hydrogen, from the high-speed stream, and product, water.

The reaction zones convecting downstream resemble expanding shells. Eventually, many of the structures that were apparent from the contours of R have been eaten away by the reactions, and the active reaction zones appear to be limited to a continuous mixing layer rather than centered in eddies. Preliminary study of selected vorticity contours indicates that vorticity decreases in reacted regions and the highest vorticity appears in regions where reaction is about to occur. The reaction centers, as shown in Fig. 6, are contained in the coherent structures. They do not, however, look in any way like these

structures and, imaged alone, give limited information
about the properties of the flow. For example, they are
not a good visualization of the coherent structures in
which they occur.

The energy release model used here is primitive. It
is strictly valid only for fast reacting flows in which
convective mixing dominates. This means that the convective time scales must be faster than any diffusive transport time scale. The flows calculated are in fact close
to the sound speed, and the energy release times are fast,
so that the flow meets the criterion. Thus we say we are
calculating the effects of convective mixing. Any mate-

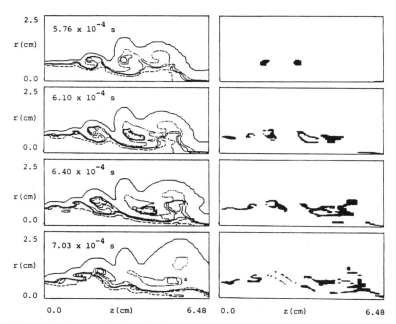

Fig. 6 Contours of the mixing ratio R (left column) and the reaction zones (right column) for the reactive axisymmetric jet calculation. The range of R is 0.0--1.0.

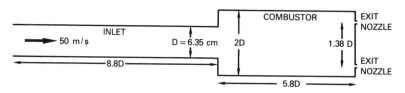

Fig. 7 Configuration of the axisymmetric diffuser and dump combustor.

rial that gets into a computational cell is assumed to be
premixed in that cell. This is, of course, an approximation that can be checked by reducing the cell size and
repeating the calculation. Our conclusions are that it is
a qualitative picture of the effects of energy release on
an axisymmetric jet flow, but it is not quantitative.

Central Dump Combustor

The problems of instability in combustors have been
attributed to complex, nonlinear interactions among acoustic waves, vortex shedding and merging, and chemical energy release. Here we describe results of numerical simulations performed to isolate and study the interaction
between acoustic waves and large-scale coherent vortex
structures in an idealized ramjet combustor. The objective is to determine the extent to which acoustic waves
influence vortex shedding and merging and to test the numerical model thoroughly before an energy release model is
incorporated to study the full instability problem.

A schematic of the central dump ramjet combustor used
in the simulations is shown in Fig. 7. The size of the
chamber and the mean flow characteristics model the experiments of Schadow et al.[18]. A cylindrical jet with a
mean velocity of 50 m/s flows into a cylindrical dump with
twice the jet diameter. The length of the dump, which
acts as an acoustic cavity, may be varied to change the
first longitudinal mode frequency. An exit nozzle at the
end of the chamber is modeled to produce choked flow. Figure 8 shows density contours for a calculation using 80×200
computational cells. An analysis of the computed data
shows that the calculated average flow properties in the
combustion chamber agree with those measured in the experiments. This is essentially a check on the boundary
conditions used in the calculations.

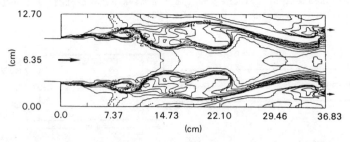

Fig. 8 Density contours for the dump combustor calculation.

40 X 100   80 X 200

DENSITY   STREAMLINES   DENSITY   STREAMLINES

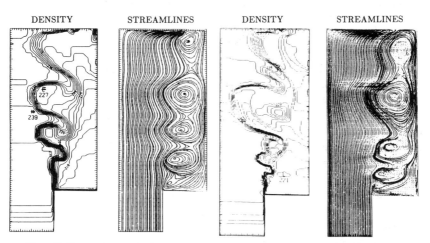

Fig. 9 Comparison of dump combustor calculations for two resolutions.

The effect of resolution was checked by comparing calculations with 20×50, 40×100, and 80×200 cells. Figure 9 shows density contours and streamlines from typical calculations. (Note that Fig. 9 is plotted as if the cells were evenly spaced. Thus the figure is not to scale.) The large-scale structures are essentially the same with the two resolutions. The shedding frequency from the step as well as the first vortex merging frequency downstream (450 and 230 Hz, respectively) both agree well with the measured values (411 and 215 Hz). The 450 Hz is consistent with the first longitudinal acoustic mode of the combustor. The dynamic structure that evolves goes to a steady state when the calculation is not resolved well enough. This is equivalent to performing a considerably more viscous calculation at a lower Reynolds number. With the finer grid, the large structures are more resolved and we also see some finer-scale structures.

Acoustic driving was simulated by introducing pressure waves of fixed amplitude and frequency at the wall near the exit nozzle. Preliminary calculations with acoustic forcing show strong coupling between the large-scale vortex structures and the acoustic waves when the first longitudinal acoustic mode was excited with various amplitude perturbations. A resonance occurs when the driven frequency is close to one of the natural frequencies of the

combustor. Our calculations show that driving the system
close to the first longitudinal mode increases the amplitude of the fluctuations.

Summary and Conclusions

This paper has presented numerical simulations of the
evolution of the Kelvin-Helmholtz instability for two-
dimensional splitter-plate, and axisymmetric round-jet
and central-dump-combustor geometries. These calculations
were performed using the Eulerian, explicit flux-corrected
transport algorithm to solve the convective transport problem on stationary, nonuniform grids. An energy release
model was coupled to the convective transport to model
chemical energy release and conversion to products occurring in combustion in the round jet.

The splitter-plate calculations showed the evolution
of the coherent structures in spatially evolving mixing
layers and emphasized the importance of mixing asymmetries in the flows. Computed asymmetries in the nonreacting flows are consistent with the experimental results
of Koochesfahani et al.[13], which show a trend in which
more fast fluid than slow fluid is entrained in the larger
structures. The asymmetry in the calculations is more apparent as the the velocity ratio of the two streams increases.

The mixing asymmetries studied in the splitter-plate
flows also appear in the first few roll-ups in the axisymmetric jet. The importance of this effect becomes evident when we consider the mixing and burning process. Because the mixing is asymmetric, hot pockets of pure fuel
and product, or oxygen and product, develop. These can
last a significant time before they are further mixed and
all of the fuel or oxidizer consumed. It is not clear how
important this process is in the dump combustor, where it
is assumed that the inflowing material is premixed. Here
the combustible gas is heated by mixing with hot products,
which are nominally fully reacted.

From the reactive round-jet simulations, we have seen
that when energy release occurs in the material entrained
in the coherent structures, the structures are smeared
out, often elongated. The reaction zones convecting downstream resemble expanding shells that, when viewed alone,
do not look like the coherent structures. Preliminary
study of selected vorticity contour plots indicates that

vorticity decreases in reacted regions and the highest
vorticity appears in regions where reaction is about to
occur.

Two coflowing fluids are unstable at their interface
when the Reynolds number is high enough and there is some
small perturbation in the system. This perturbation can
be an imposed frequency at the beginning of the calculation, or even just noise or roundoff error. In the splitter-
plate, round-jet, and some of the dump-combustor calculations, the instability was not constantly externally
driven, but a perturbation was imposed only at the beginning of the calculation. However, we note that it was not
necessary to drive the instability for it to persist and
reinitiate itself. In the splitter-plate and round-jet
calculations, the mechanism which reinitiates the instability near the splitter-plate tip or jet nozzle exit is
the feedback between downstream fluctuations and upstream
flow[19]. The source of the pressure fluctuations are the
downstream development of the coherent structures. The
dump combustor was a confined geometry, so that in addition to downstream perturbations driving the instability
as they propagate upstream, the natural acoustic modes of
the chamber also played an important part in the reinitia-
tion. They clearly controlled the shedding frequency from
the step.

Thus we postulate that vortex shedding close to the
inflow boundary is reinitiated by transverse flows caused
by pressure pulses generated by vortex roll-up and vortex merging downstream. Though these pulses are transmitted acoustically, they exist even in the incompressible
limit. This concept was tested in the calculations by
damping all acoustic pulses reaching the inflow boundary,
i.e., by setting the pressure at the boundary equal to the
ambient pressure. Then the initial instability dies out
near the nozzle or tip of the splitter plate. When the
inflow boundary condition is unrealistic and does not allow for the interaction of the pressure pulses generated
downstream with the inflowing gas, or the calculation is
too numerically diffusive, it is possible to damp the instability totally and thus have to drive it. A feedback
mechanism, through which the upstream flow is influenced
by the downstream, was proposed by Dimotakis and Brown[3] to
explain the experimentally observed existence in a mixing
layer of a long correlation time that could not be scaled
with any local flow property.

In practical combustion flows, the asymmetric mixing
can clearly be important. However, in such flows there
are many sources of perturbations that occur naturally
which can trigger the instability. The most prevalent is
noise in the inflow, or noise in the surrounding medium,
which can be generated by the combustion process itself.
The natural confinement of the dump combustor generates
its own noise, which influences the vortex shedding and
merging, as well as the stability properties of the combustor.

The dump combustor is confined, premixed, and there is
flow over a step. The confinement enhances the importance
of the acoustic modes of the system. The step produces a
recirculation zone behind it. When combustion occurs, the
function of the step is to cause hot reacted material to
mix with cold unreacted material and thus raise the temperature to ignition temperatures. Mixing then occurs
in the vortices that are shed from the step, and the energy release occurs in these vortices. An important next
step in this problem is to consider the effects of energy
release and how this couples to the vortex shedding and
merging and acoustic processes.

There are many potentially important physical processes that have not been included in these calculations.
These include boundary layers, more realistic chemical
models, the third dimension, buoyancy, and subgrid turbulence. Of these, the effects of buoyancy and the third
dimension can be calculated at the same level of detail
as the calculations presented above. Buoyancy effects
are not in the basic model, and three-dimensional breakdown can be studied given adequate computer time. The
other effects listed require introducing phenomenologies
into the model. We are currently involved in the systematic development of such phenomenologies based on more exact calculations for inclusion in these models. For example, the induction parameter model is a first step at
the chemistry. It works very well for detonation calculations and is probably fairly accurate for fast, energetic flows. However, a better model is required for very
subsonic flows with slow energy release, where diffusion
effects are important. Subgrid models for premixed turbulent flames are now being developed. These additions to
the basic model are now within our grasp and will be included in future studies.

Acknowledgments

This work was sponsored by the Naval Research Laboratory, the Office of Naval Research, and the Naval Air Systems Command. The authors would like to acknowledge the work of K. Laskey in producing and interpreting Fig. 6.

References

[1] Winant, C.D. and Browand, F.K. (1974) Vortex pairing: The mechanism of turbulent mixing layer growth at moderate reynolds number. J. Fluid Mech. 63, 237--255.

[2] Brown, G. and Roshko, A. (1974) On density effects and large structure in turbulent mixing layers. J. Fluid Mech. 64, 775--816.

[3] Dimotakis, P.E. and Brown, G.L. (1976) The mixing layer at high reynolds number: large structure dynamics and entrainment. J. Fluid Mech. 78, 535--560.

[4] Breidenthal, R. (1981) Structure in turbulent mixing layers and wakes using a chemical reaction. J. Fluid Mech. 109, 1-24.

[5] Boris, J.P., Oran, E.S., Fritts, M.J., and Oswald, C.E. (1983) Time dependent, compressible simulations of shear flows: Tests of outflow boundary conditions. NRL Memorandum Report 5249, Naval Research Laboratory, Washington, D.C.

[6] Boris, J.P., Oran, E.S., and Gardner, J.H. (1985) Direct simulations of spatially evolving compressible turbulence. Proceedings of the ninth International Symposium on Numerical Methods in Fluid Dynamics, pp. 98-102. Springer-Verlag, New York.

[7] Grinstein, F.F., Oran, E.S., and Boris, J.P. (1986) Numerical simulations of asymmetric mixing in planar shear flows. J. Fluid Mech. (to be published).

[8] Boris, J.P. and Book, D.L. (1976) Solution of continuity equations by the method of flux-corrected transport. Methods in Computational Physics, Vol. 16, Chap. 11. Academic Press, New York.

[9] Boris, J.P. (1976) Flux-corrected transport modules for solving generalized continuity equations. NRL Memorandum Report 3237, Naval Research Laboratory, Washington, D.C.

[10] Oran, E.S., Boris, J.P., Young, T.R., et al. (1980) Simulation of gas phase detonations: Introduction of an induction parameter model. Memorandum Report 4255, Naval Research Laboratory, Washington, D.C.

[11] Kailasanath, K., Gardner J., Boris J., and Oran E. (1985) Acoustic-vortex interactions in an idealized ramjet combustor. Paper presented at the 22nd JANNAF Combustion Meeting.

[12]Kailasanath, K., Gardner J., Boris J., and Oran E. (1986)
Numerical simulations of acoustic-vortex interactions in an
idealized ramjet combustor. AIAA paper, to be presented at
the AIAA 24th Aerospace Sciences Meeting, Reno, Nev.

[13]Koochesfahani, M.M., Dimotakis, P.E., and Broadwell, J.E.
(1983) Chemically reacting turbulent shear layers. AIAA Paper 83-0475. AIAA, New York.

[14]Riley, J.J. and Metcalfe, R.W. (1980) Direct numerical simulation of a perturbed, turbulent mixing layer. AIAA Paper
80-0274. AIAA, New York.

[15]Ashurst, W.T. (1979) Numerical simulation of turbulent mixing layers via vortex dynamics. Turbulent Shear Flows I edited
by F. Durst, B.E. Launder, F.W. Schmit, and J.H. Whitelaw,
pp. 402-413. Springer-Verlag, New York.

[16]Davis, R.W. and Moore, E.F. (1985) A numerical study of vortex merging in mixing layers. Phys. Fluids 28, 1626-1635.

[17]Laskey, K.J., Grinstein, F.F., Oran, E.S., Kailasanath, K.,
and Boris, J.P. (1984) Numerical simulation of reacting,
exothermic mixing layers. Paper presented at the 1984 Technical Meeting of the Eastern Section of the Combustion Institute.

[18]Schadow, K.C., Wilson, K.J., Crump, J.E., Foster, J.B., and
Gutmark, E. (1984) Interaction between acoustics and subsonic
ducted flow with dump. AIAA Paper No. 84-0530. AIAA, New
York.

[19]Grinstein, F.F., Oran, E.S., Boris, J.P. (1986) Numerical
simulation of unforced spatially-developing mixing layers.
J. Fluid Mech. (to be published).

Velocity and Density Measurements in a Planar Two-Stream Turbulent Mixing Layer

I.G. Shepherd,* J.L. Ellzey,† and J.W. Daily‡
University of California, Berkeley, California

ABSTRACT

Stationary and spectra statistics of the gas density and velocity have been measured in a two-stream plane mixing layer using Rayleigh scattering and laser Doppler anemometry. The conditions in the lower stream were kept constant at 5 m/s and 1770 K. In the upper stream of premixed propane/air at ambient temperature, the velocity was varied between 5 and 20 m/s and the equivalence ratio between 0.0 and 0.6. An argon ion cw laser operated at 488 nm and with 1.5 W power was used as the Rayleigh light source. The probe volume was a cylinder of dimensions 1 mm in length by 0.3 mm in diameter. The laser Doppler anemometer system was of the conventional two-color forward-scatter type with a fringe volume of 0.14 mm in the mean-flow direction by 2 mm. Mean density and velocity profiles were obtained from traverses through the mixing layer at various positions downstream of the splitter plate. The layer thickness, derived from these results, was found to grow linearly with increasing downstream location. Spectra of density and the transverse component of velocity were also measured. The scalar spectra have a much higher degree of organization than the velocity spectra. Characteristic peaks appear in all the spectra, indicating that the mixing layer is being driven by acoustic resonances in the combustion chamber. Close to the splitter plate mixing of the two streams takes place at small scales that decay rapidly as the layer grows, probably due to pairing of the initial vortices. Farther downstream the velocity and density spectra decouple as the growing layer starts to interact with

Presented at the 10th ICDERS, Berkeley, California, August 4-9, 1985. Copyright © 1986 by the American Institute of Aeronautics and Astronautics, Inc. All rights reserved.
*Research Engineer, Department of Mechanical Engineering.
†Assistant Professor, Department of Mechanical Engineering, University of Wisconsin, Madison, Wisconsin.
‡Associate Professor, Department of Mechanical Engineering.

the wall boundary layer and the velocity fluctuations characteristic of the vortical motion are damped. At higher shearing rates the velocity spectrum shows the layer to be much more disorganized, but the scalar field still has dominant peaks at the characteristic frequencies.

INTRODUCTION

The two-dimensional mixing layer has been the object of much research, stemming from the pioneer work of Brown and Roshko (1974) and continued by other workers, Dimotakis and Brown (1976) and Keller and Daily (1985), for example. They have demonstrated the importance of large-scale structures in the interpretation of many such flows. In this paper, aspects of the mean statistics and spectra of density and velocity measurements in such a system will be investigated and compared.

Laser based techniques are especially useful in combustion systems. Two such methods will be used here: laser Doppler anemometry (LDV) for measuring velocity and Rayleigh scattering for density. LDV is now a standard research tool and its advantages and limitations are well established. Rayleigh scattering is an attractive method for investigating scalar variables due to its essentially nonperturbing character, Namazian et al.(1982). If the scattering cross sections of the constituent molecules are known, the local total number density can be obtained and, by careful matching of the cross sections across the flame, variations of Rayleigh scattering intensity may be interpreted as variations in temperature, Namer and Schefer (1985). In the present investigation, which considers propane/air mixtures, there are significant differences in cross section between burnt and unburnt states and these changes must be accommodated in the interpretation of the data, Shepherd and Daily (1984). Two factors arising from the elastic nature of the scattering, i.e. with no shift in wavelength, that are important potential contaminants of the signal are the presence of stray laser light and particulate scattering. These problems make it necessary to perform separate LDV and Rayleigh experiments and their impact on the present density measurements is considered below.

The Rayleigh scattering intensity I is given by:

$$I = CI_L N \sum_i X_i \sigma_{ri} \tag{1}$$

where C is an optical calibration constant, I_L the incident light intensity, N the total number density of the gas, X_i

the mole fraction of species i and σ_{ri} the Rayleigh cross section of species i. This cross section is given by:

$$\sigma_{ri} = \frac{4\pi^2}{\lambda 4}((\mu_i-1)/N_o)^2 \sin^2\theta \qquad (2)$$

where λ is the laser wavelength, μ_i the index of refraction of species i, N_o the Loschmidt number, and θ the collection angle of the scattered light measured relative to the plane of polarization of the laser light. Equation (1) can be nondimensionalized by conditions in the unburnt gas (designated by subscript o):

$$\frac{I}{I_o} = \frac{\rho}{\rho_o} \cdot \frac{\sigma}{\sigma_o} \cdot \frac{M_o}{M} \qquad (3)$$

where M is the mean molecular weight and ρ the density. When the signal has been corrected in this way,

$$\rho \, \alpha \, I_r = I_p - I_b \qquad (4)$$

where I_r, I_p and I_b are the mean Rayleigh signal, the measured intensity and the background light, respectively. Hence

$$\rho/\rho_o = I_r/I_{ro} \qquad (5)$$

where I_{ro} is the intensity at the reference conditions where there are no density fluctuations.

EXPERIMENTAL CONDITIONS

The combustion chamber in which the present measurements were made is illustrated in Fig. 1. It is of rectangular cross section (122.2 mm wide by 56.9 mm high) and consists of two streams of gas separated by a splitter plate (4 deg. wedge angle). Immediately upstream of the contraction (2.7:1) are fine stainless-steel, wire mesh screens upon which a flat preheat flame can be stabilized. The four walls of the working section are made of quartz to facilitate optical access.

LDV Measurements

The two-component LDV measurements were made with a conventional forward scatter, two-color (argon ion laser), four-beam system with Bragg cell shifting (2 MHz) of the transverse velocity component to resolve flow direction ambiguities. The fringe volume was 140 μm in diameter by 2

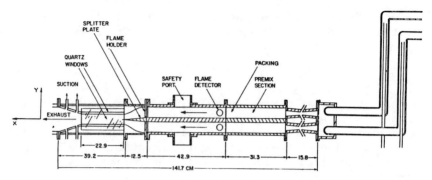

Fig. 1   Combustion chamber.

mm in length and the signal, derived from nominally 0.05 μm
diameter Al_2O_3 particles, was processed by two Macrodyne
counters with simultaneous-validation circuitry. Mean
statistics were calculated from 2048 point probability density functions (pdfs) stored on a PDP 11/34 computer. Velocity spectra were obtained from no less than 128 samples of
1024 points processed on a Genrad 2512 spectrum analyser.

Rayleigh Measurements

        The light source for the Rayleigh measurements was a
Spectra Physics argon ion cw laser operated at 488 nm and
1.5 W. The scattered light was collected perpendicularly
to the plane of polarization of the laser beam; the beam
dimensions were 1x0.3 mm (shorter dimension in the mean
flow direction). The signal from the photomultiplier tube
was passed to an inverting operational amplifier incorporating a low-pass filter set at 15 kHz; the output of the
amplifier was sampled at 5 kHz by an A/D converter.   There
was little spectral activity above 2 KHz and evidence from
high-speed schlieren photography, Keller and Daily (1985),
indicates that the passing frequencies for vortical structures were mostly less than 1 KHz. Pdfs of signal intensity
of 32,000 determinations were generated and stored in the
computer. In this way, profiles (y direction) through the
mixing layer of pdfs of signal intensity at various downstream positions (x direction) were obtained. Spectra at
the position of maximum signal variance were taken as
outlined above.
        To determine the contribution of background laser
light and flame luminosity to the signal at each measurement position, the collection optics were moved to focus at
a point a small distance away from the laser beam and the
mean intensity was measured. Due to the presence of many

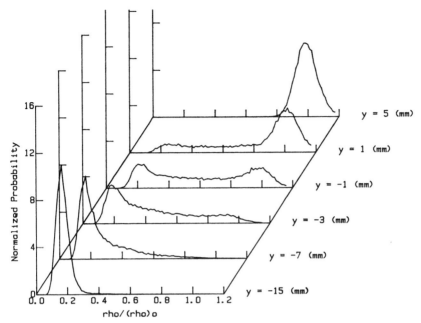

Fig. 2 Density pdfs. Traverse through mixing layer.
U_1 = 15 m/s, ϕ_1 = 0.6, x = 100 mm.

scattering surfaces in the experimental apparatus much
effort was devoted to the elimination of stray light. It
was found to be imperative to keep the windows clean as any
imperfection in their surfaces greatly increased the background light. This contamination of the signal rendered
measurements close to the splitter plate or near to walls
impossible. However, in the measurements reported here,
background light was reduced to at most 10 percent of the
Rayleigh signal.

By filtration and drying of the inlet gases it was
possible to clear the system of solid particles. It proved
much more difficult to eliminate the low concentrations of
sub-micron oil droplets which existed in the 'house' air
supply. The oil did not, however, survive the preheat flame
and the burnt gas was free of contamination. In the
unburnt mixture almost all signal due to Mie scattering
appeared in the final bin of the light intensity pdf and
this was excluded for statistical analysis.

In these experiments conditions in the lower stream,
designated by subscript 2, were kept constant at 1770 K and
a mean flow velocity, U_2, of 5m/s by stabilizing a premixed
propane/air flame at an equivalence ratio of ϕ_2 = 0.64 on
the flameholder. In the upper stream, designated by sub-

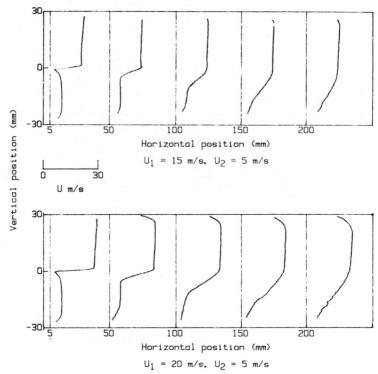

Fig. 3   Transverse velocity profiles. ϕ_1 = 0.0.

script 1, a propane/air mixture at ambient temperature was
varied in velocity and equivalence ratio. The confined
nature of the flow results in a streamwise pressure gradient which is increased by combustion. Keller and Daily
(1985) have reported the magnitude and effects of this
pressure gradient in the present experimental facility.

RESULTS AND DISCUSSIONS

Fig. 2 shows a typical set of Rayleigh scattered light
intensity pdfs uncorrected for noise, for a traverse
through the mixing layer with U_1 = 15 m/s and ϕ_1 = 0.6 at a
position 100 mm downstream of the splitter plate. The Gaussian pdfs at y = -15 and 5 mm are fully burnt and unburnt
gas, respectively. The pdfs between these extremes at high
signal variance (e.g. y = -1), although broadened by noise,
have a very pronounced bimodality arising from the presence
of a thin flame interface between regions of constant density, the burnt and unburnt gases. When ϕ_1 = 0.0, although
a flame no longer exists between the hot and cold gases,

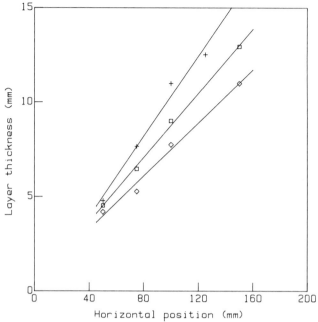

Fig. 4 Layer thickness from mean density profiles. $\diamondsuit U_1 = 5$ m/s, $\phi_1 = 0.0$; $\square U_1 = 15$ m/s, $\phi_1 = 0.0$; $+ U_1 = 15$ m/s, $\phi_1 = 0.6$.

the density pdfs at this location are still strongly bimodal, indicating the importance of large scale mixing. Profiles of mean density were obtained from these traverses.

Velocity profiles were also measured by LDV. A range
of results at differing shear rates is shown in Fig. 3.   By
fitting an error function through these profiles it was
possible to obtain a measure of the vorticity thickness of
the layer. Similar results have been reported by Keller and
Daily (1983). Using the maximum gradient method, the layer
thickness also could be calculated from the density profiles (Fig. 4).Comparison of these two sets of results shows
that the scalar thickness was larger than the vorticity
thickness as previously indicated by Batt (1977), and for a
backward-facing step by Pitz (1981).   Fig. 4 also shows
that for "wake" flow, i.e. $U_1 = U_2$, and a typical shearing
rate case with and without combustion, the layer grows
linearly with increasing downstream location. The lines in
Fig. 4 are least mean square fits. Increases in both
shearing rate and heat release augment the spreading rate
of the layer.   The detailed effects of velocity ratio
differences on the present mixing layer have been reviewed
elsewhere by Ellzey (1985).

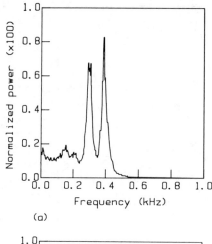

(a)

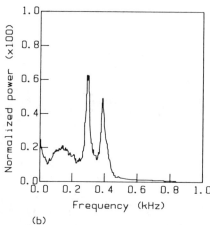

(b)

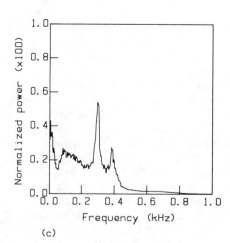

(c)

Fig. 5 Density spectra.
U_1 = 5 m/s, φ_1 = 0.0;
(a) x = 50 mm,(b) x = 100 mm,
(c) x = 150 mm.

At each downstream location, a spectrum of the density
fluctuations at maximum signal variance was recorded and
the spectra were normalized so that the area under the
curve was unity. It should be noted that no attempt has
been made to measure the 'turbulence spectrum' down to
scales characteristic of the flame front thickness (1 mm)
and that the investigation is directed towards the resolution of the structures observable by schlieren. For "wake"
flow the spectra are highly organized around peaks at 300
and 385 Hz, and there is no spectral activity above 500 Hz,
(Fig. 5). These peaks persist as the most dominant feature
with some shift to lower frequencies at downstream locations as the layer grows. Fig. 6 and 7 illustrate shearing
cases with and without combustion, respectively. These
spectra have the same spectral features close to the
splitter plate with peaks at 1.2 kHz, 625 Hz and 400 Hz.
The pdfs at this location are also very similar. The
existence of clearly defined burnt and unburnt gas is much
less obvious, and intermediate states are in evidence.
High velocity gradients exist here and, under these intense
shearing conditions, mixing may be taking place without
combustion. Moving downstream, the 1.2 kHz peak rapidly
disappears probably due to vortex pairing and is not found
at x = 100 mm in any of the spectra. In the nonflame case,
at the farthest downstream position (Fig. 7c), the spectrum
has broadened around peaks at 250 and 400 Hz while with
combustion (Fig. 6c), it remains highly organized with peaks
at 275, 295 and 350 Hz.

Significant differences are apparent when these spectra are compared with those obtained from the transverse
component of velocity (see Fig. 8 and Keller and Daily,
1985). Although very similar close to the splitter plate,
the scalar and velocity spectra become rapidly decoupled
farther downstream. Distinct spectral features quickly
disappear and the energy in the velocity spectra moves to
lower frequencies. Schlieren evidence from Keller and
Daily (1985) indicates that the velocity fluctuations are
associated with rotating vortical structures, and that when
the growing layer interacts with the wall boundary layer,
(see Fig. 3), the frequencies that characterize this motion
decay rapidly (Fig. 9c). The scalar structures, however, are
little affected by this interaction and are convected downstream by the mean flow.

It will be observed that the same characteristic peaks
appear in all the spectra independent of the convection
speed or equivalence ratio, and are also found in the power

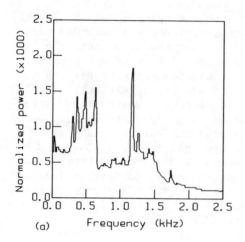

(a)

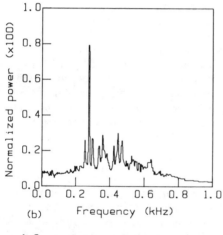

(b)

Fig. 6   Density spectra.
U_1 = 15 m/s, ϕ_1 = 0.6;
(a) x = 50 mm, (b) x = 100 mm,
(c) x = 150 mm.

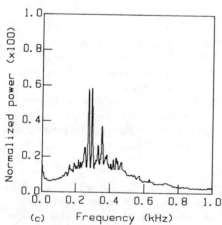

(c)

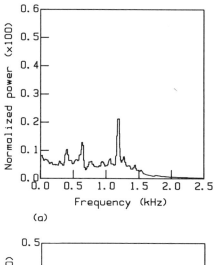

(a)

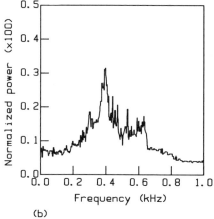

Fig. 7 Density spectra.
U_1 = 15 m/s, ϕ_1 = 0.0;
(a) x = 50 mm,(b) x = 100 mm,
(c) x = 150 mm.

(b)

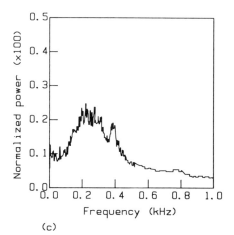

(c)

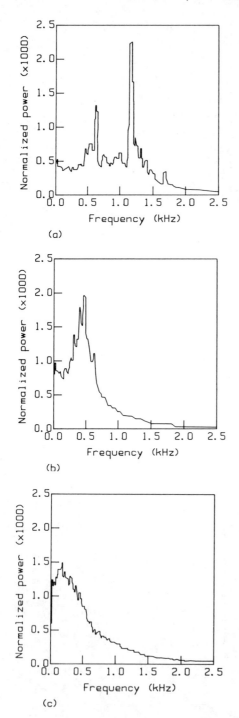

Fig. 8   Velocity spectra.
U_1 = 15 m/s, ϕ_1 = 0.0;
(a) x = 50 mm, (b) x = 100 mm,
(c) x = 150 mm.

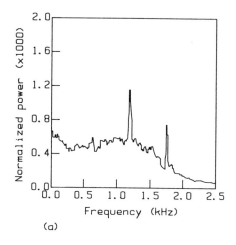

(a)

Fig. 9   Density spectra.
$U_1 = 20$ m/s, $\phi_1 = 0.0$;
(a) x = 50 mm, (b) x = 150 mm.

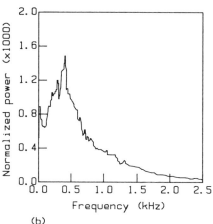

(b)

spectra of the pressure field obtained from a wall-mounted
pressure transducer that was isolated from mechanical
vibrations of the chamber, Ellzey (1985). This, and the
sharpness of the peaks, indicate that the layer is being
driven by acoustic resonances within the combustion
chamber. Comparing the density and velocity spectra at
higher shearing rates (Fig. 9a and 9b, and Fig. 10a and 10b,
respectively), similar differences are observable. In this
case, the velocity spectrum has no characteristic peaks close
to the splitter plate (Fig. 10a), indicating that the layer
is in a highly randomized state at this position. The scalar
spectrum shows some structure, however, with peaks at 1.2
and 1.7 kHz (Fig. 9a), which decay to a broad peak at 400
Hz (Fig. 9b).

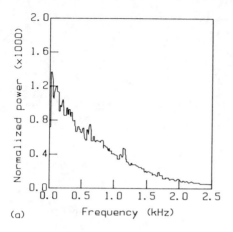

Fig. 10 Velocity spectra.
$U_1 = 20$ m/s, $\phi_1 = 0.0$:
(a) x = 50 mm, (b) x = 150mm.

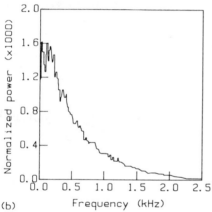

CONCLUSIONS

Spectra and mean statistics of the gas density have been measured in a two-stream, planar, turbulent mixing layer using Rayleigh scattering and compared with aspects of similar measurements of velocity by laser Doppler anemometry. Mean profiles of density and velocity were obtained from traverses through the mixing layer at various positions downstream of the splitter plate. The mixing layer thickness derived from these results was found to grow linearly with increasing downstream location.

The bimodal character of many of the density pdfs, even where there is no combustion, indicates the importance of large-scale entrainment and mixing in these flows. When a flame is present the bimodality becomes very prominent and both states, burnt and unburnt, are found throughout the layer.

Spectra of density and velocity fluctuations were also obtained. The scalar spectra had a much higher degree of organization than the velocity spectra, and characteristic peaks appear in all the spectra. This is indicative of the layer being driven by acoustic resonances in the combustor.

Close to the splitter plate, mixing of the two streams takes place at small scales which decay rapidly as the layer grows, probably due to pairing of the initial vortices. Farther downstream, the velocity and density spectra decouple as the growing layer starts to interact with the wall boundary layer and the velocity fluctuations characteristic of the vortical motion are damped. At higher shearing rates, the velocity spectrum shows the layer to be highly random but the acoustic interactions still organize the scalar field at the characteristic frequencies.

ACKNOWLEDGMENTS

We gratefully acknowledge support from the Army Research Office, contract ARO-5-82, through Alfred Buckingham and Wigbert Siekhaus at Lawrence Livermore National Laboratory and from the National Science Foundation, grant ENG 77-02019. We also wish to thank G. Hubbard for stimulating conversations and help in the laboratory.

REFERENCES

Batt, R.G. (1977) Turbulent mixing of passive and chemically reacting species in a low-speed shear layer. Journal of Fluid Mechanics, 82, 53-95.

Brown, G. and Roshko, A. (1974) On density effects and large structures in turbulent mixing layers. Journal of Fluid Mechanics, 64, 775.

Dimotakis, P.E. and Brown, G. (1976) The mixing layer at high Reynolds number. Journal of Fluid Mechanics, 78, 553-560.

Ellzey, J.L. (1985) Experimental and numerical study of a two-stream, planar, turbulent mixing layer. Ph.D. Thesis, University of California, Berkeley, California.

Keller, J.O. and Daily, J.W. (1985) The effects of highly exothermic chemical reaction on a two-dimensional mixing layer. AIAA J. 23 (12), 1937-1945.

Namazian, M., Talbot, L., Robben, F. and Cheng, R.K. (1982) Two-point Rayleigh scattering measurements in a v-shaped turbulent flame. Nineteenth Symposium (Int'l) on Combustion, The Combustion Institute, Pittsburgh, Pennsylvania, pp. 487-493.

Namer, I. and Schefer, R.W. (1985) Error estimates for Rayleigh scattering density and temperature measurements in premixed flames. Experiments in Fluids, 3, 1-9.

Pitz, R.W. (1981) An experimental study of combustion: The turbulent structure of a reacting shear layer formed at a rearward-facing step. Ph.D. Thesis, University of California, Berkeley, California.

Shepherd, I.G. and Daily, J.W. (1984) Rayleigh scattering measurements in a two-stream mixing layer. Paper 84-45. Western States Meeting of the Combustion Institute, Boulder, Colorado.

Two-Dimensional Imaging of C_2 in Turbulent Nonpremixed Jet Flames

R.W. Dibble,* M.B. Long,† and A. Masri‡

Sandia National Laboratories, Livermore, California

Abstract

We report the first two-dimensional imaging of diatomic carbon C_2. This new C_2 fluorescence technique promises to be a superb marker of the flame front position, since the C_2 molecule is a short-lived transient that is generated and consumed in a very thin zone near the fuel-rich side of the flame front. The images are obtained from a C_2 fluorescence that is excited by a planar sheet of light from a tunable dye laser. The intensity is maximized by tuning the dye laser to $\lambda = 512$ nm, near the (0,0) and (1,1) bandheads of the C_2 Swan system, $d^3\pi_g - a^3\pi_u$. The resulting emission is comparable in intensity to Rayleigh scattering from nitrogen in room air. Images of this fluorescence are readily obtained with an intensified vidicon imaging system which includes an optical interference filter passing the wavelength ($\lambda = 565$ nm) of the most intense Stokes transition of the Swan system (1,0), $d^3\pi_g - a^3\pi_u$.

1. Introduction

Most fossil-fuel energy is released in combustion devices in which the air and the fuel are introduced into the combustion chamber separately. This mode of combustion is called nonpremixed. While the combustion chamber can be quite large, the actual reactions between oxygen and fuel occur in thin, highly contorted sheets that fluctuate in space and time. At any instant, the volume

---

Presented at the 10th ICDERS, Berkeley, California, August 4-9, 1985. This paper is declared a work of the U.S. Government and is not subject to copyright protection in the United States.

*Member of Technical Staff, Sandia National Laboratories.

†Professor, Department of Mechanical Engineering, Yale University.

‡Graduate Student, Department of Mechanical Engineering, University of Sidney, Sidney, Australia.

of these zones represents only a small fraction of the
total volume of the combustion chamber. Accordingly,
efficient numerical modeling of nonpremixed combustion
attempts to separate the detailed chemistry within the
zone from the spatial and temporal fluctuations of the
sheet within the combustion chamber. The statistics and
topology of this zone are of great interest in current
discussions concerning the fundamental physics and subsequent modeling of nonpremixed combustion (Peters 1984).

Progress toward this understanding has improved with
use of nonintrusive laser diagnostics. Laser scattering
from a point has provided the statistics of the flame
zone. However, improved understanding of the topological
nature of the flame zones demands multipoint measurements.
Progress toward this understanding is being made through
the use of instantaneous images from the scattered light
from a laser pulse that passes through the reaction chamber. Examples include one- and two-dimensional images of
Raman scattering, Rayleigh scattering, and OH fluorescence. However, when these images are from nonpremixed
flames, the precise location of the flame zone is elusive.
In the case of Raman scattering (which with present laser
energies is limited to images of major species), a logical
species for imaging is the reaction product H_2O, since its
mole fraction peaks in the flame zone at the stoichiometric contour. However the concentration of H_2O, and
hence the Raman scattering intensity, is nearly constant
over a wide range of fuel-air mixtures. This is because
an increase in H_2O mole fraction, as stoichiometric conditions are approached, is compensated by a decrease in the
total number density; consequently, images of H_2O concentration are an imprecise indication of the flame zone
position. Imaging of laser induced fluorescence from OH
radicals (Kychakoff et al. 1984) offers an improvement in
locating the flame front. However, the equilibrium concentration of OH in the combustion products is only a
factor of 2 less than the peak concentration of OH at the
flame front, and hence it is difficult to distinguish
between flame front and products mixing with cold reactants.

More positive identification of flame front location
may be provided by images of short-lived transient species
that exist in the flame zone but have very low concentrations in both the reactants and the products. Images of O
atoms and H atoms marginally meet this criteria, since the
ratios of the peak concentrations to equilibrium product
concentration are approximately 10 and 20, respectively.
Hydrocarbon radicals, such as CH_3, CH_2, CH, C_2, CH_3CH_2,
etc., easily meet the criterion since the ratio of their

peak concentrations to the equilibrium concentrations is
many orders of magnitude. For example, these ratios for
CH and for C_2 are

$$(10^{-6}/10^{-18}) = 10^{12} \text{ and } (10^{-9}/10^{-24}) = 10^{15},$$

respectively.

In spite of low peak concentrations of C_2 in flames,
one-dimensional images of single-pulse laser-excited fluorescence of C_2, as well as OH, have been made by Alden et
al. (1982). In this paper, we present the first single-
laser-pulse two-dimensional images of C_2 fluorescence as
well as the simultaneous image of Rayleigh scattering from
a turbulent nonpremixed flame.

2. Experimental

A schematic of the experimental apparatus is presented in Fig. 1. Two aspects of this apparatus are
unique: 1) the Combustion Research Facility flashlamp-
pumped dye laser which delivers over a joule of tunable
laser light in a 2 μs pulse with a linewidth of $\Delta\lambda=0.4$ nm,
and 2) the multipass cell which effectively increases
laser energy by a factor of 30. This cell has cylindrical
mirrors that have a radius of curvature that is equal to
half the distance between the two mirrors. The two mirrors are slightly canted so that a laser beam entering at
the top of the cell reflects numerous times and ultimately
emerges out the top of the cell. These multiple passes

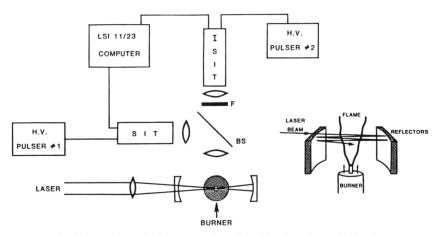

Fig. 1   Schematic of the apparatus for the imaging of the laser
excited fluorescence from C_2 in the flame.

overlap and thereby produce a quasicontinuous plane of
light between the two mirrors.

Scattering from this sheet of light is collected by a
lens and imaged through an interference filter ($\lambda = 565$ nm)
onto an intensified vidicon tube. A beam splitter in
front of the interference filter directs 8% of the light
to a second intensified vidicon which detects Rayleigh
scattering.

The flame, which is intersected by the plane of laser
light, is stabilized on a vertical tube surrounded by
quiescent room air. The inside diameter of the tube is
2 mm, and the fuel is a mixture consisting of 30% hydrogen
in methane. The addition of hydrogen was necessary for
flame stabilization as well as suppression of soot
formation.

3. Results and Discussion

Green radiant emission from fuel-rich premixed flames
results from the well-known Swan bands of C_2. The emission is due to a transition to the electronic ground state
from the first excited electronic state. The transition
occurs at $\lambda = 512$ nm, near the (0,0) and (1,1) bandheads
of the C_2 Swan system, $d^3\pi_g - a^3\pi_u$. By tuning the laser to
this transition, fluorescence is observed at $\lambda = 565$ nm
which is the (1,0) $d^3\pi_g - a^3\pi_u$ Stokes transition of the Swan
system. Images of this fluorescence are readily obtained
with an intensified vidicon imaging system which includes
an optical interference filter passing the wavelength
($\lambda = 565$ nm). The intensity of this fluorescence is comparable to that of Rayleigh scattering from room air.

The simultaneous Rayleigh scattering and C_2 fluorescence from a laminar nonpremixed flame are shown, respectively, in Figures 2a and 2b. The high-intensity zone in
the Rayleigh image is due to scattering from relatively
cold methane which is in the center of the jet. On either
side of this methane column, the Rayleigh scattering
intensity decreases as the flame zone is approached.
Beyond the flame zone, the Rayleigh scattering intensity
increases as the ambient air is approached; however, this
is outside the field of view. Corresponding to this image
of the Rayleigh scattering is the simultaneous image of
C_2 fluorescence shown in Fig. 2b. Comparison of Figs. 3a
and 3b shows that the maximum of the C_2 fluorescence
occurs in a narrow zone which is slightly to the fuel-rich
side of the flame front (maximum temperature contour).
These observations regarding the relative location of
C_2 and the flame front are supported by the recent point
measurements of Smyth et al. (1985).

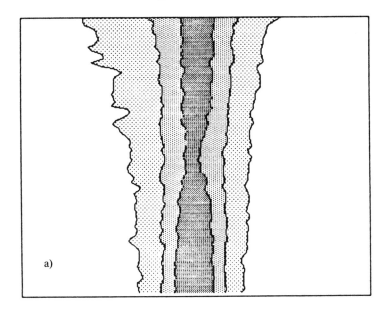

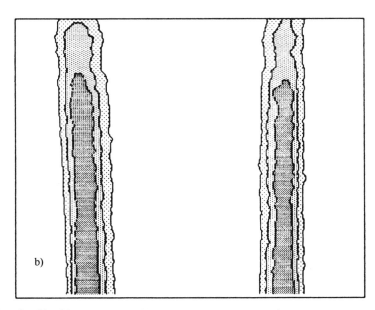

Fig. 2 Simultaneous two-dimensional images of a) Rayleigh
scattering and b) C$_2$ fluorescence centered 15 diameters downstream
from the nozzle exit in a laminar flame from a fuel consisting of
30% hydrogen in methane. The image dimension is 5x5 diameters.
For this flame, the jet Reynolds number is 5000. In both images,
the signal intensity increases in steps of 20% of maximum
intensity.

Increasing the fuel velocity of the laminar jet shown in Fig. 2 results in a slightly turbulent jet, images of which are shown in Fig. 3. This flame is particularly interesting since it shows a turbulent methane core (Fig. 3a) which is surrounded by a laminar flame (Fig. 3b). The increase in kinematic viscosity, associated with the heat release at the flame front, is sufficient to damp the turbulence. As the fuel velocity is further increased, the images show greater convolutions of the fuel jet (Figs. 4a, 5a, 6a), and the images of the fluorescence from C_2 (Figs. 4b, 5b, 6b) are no longer continuous. If one assumes that breaks in the images of the C_2 fluorescence are holes in the flame front, then Figs. 5 and 6 (Re=25,000) are clear evidence of local flamelet extinction. As the fuel velocity is further increased, the occurence of flamelet extinction increases until complete flame blowout occurs at Re=30,000.

As these images indicate, turbulent mixing has a profound effect on flame chemistry. Departures from chemical equilibrium occur when mixing rates are comparable to chemical kinetic rates. Insufficient time for complete combustion results in a depressed flame temperature, which, in turn, reduces the kinetic reaction rates; hence, products of incomplete combustion begin to appear in the exhaust. As the mixing rate is further increased, the temperature is further depressed until ultimately the kinetic rates become so slow (frozen) that the flame blows out.

Most models of nonpremixed combustion do not have a mechanism for predicting phenomena such as flame blowoff, and they incorrectly predict equilibrium values for exhaust pollutants such as oxides of nitrogen, oxides of carbon, and particulate carbon. These pollutants generally occur because of chemical reactions that are slower than the primary fuel consumption channels. The missing mechanism in these models is one which allows for the rate of mixing of fuel to compete with the rate at which fuel and air react, i.e., which allows for departures from chemical equilibrium.

Models with such a mechanism have emerged (Janicka and Kollmann 1978; Bilger and Starner 1983). While these models allow the chemical kinetics to fall behind the fast chemistry limit, a finite rate is always maintained. Thus, these models do not allow for flame extinction, even locally within the flame. Recent models (Liew et al. 1984, Peters 1985) are addressing the possibility that from time to time, the mixing rate exceeds the chemical kinetic rate, and local flame extinction is possible. It is

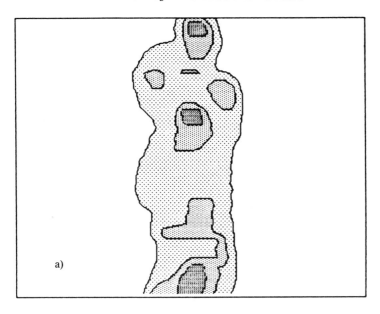

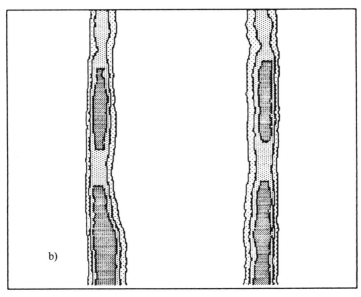

Fig. 3 Simultaneous two-dimensional images of a) Rayleigh
scattering and b) C$_2$ fluorescence collected 30 diameters
downstream from the nozzle exit in a nonpremixed turbulent flame
of a fuel consisting of 30% hydrogen in methane. The image
dimension is 5x5 diameters. For this flame, the jet Reynolds
number is 8000. In both images, the signal intensity increases in
steps of 20% of maximum intensity.

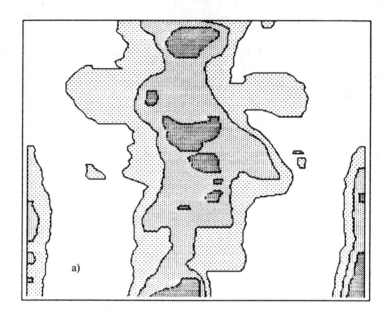

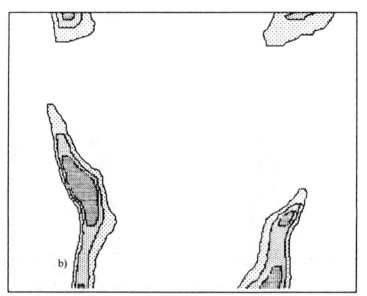

Fig. 4 Same as for Fig. 3 except this flame has a jet Reynolds number of 15,000. Note that the image of C_2 is not connected.

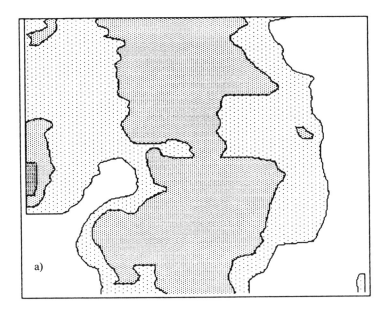

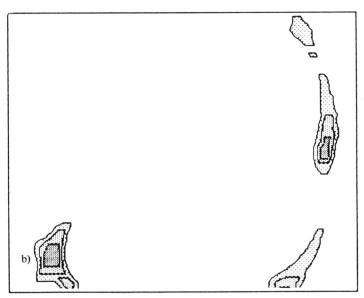

Fig. 5   Same as for Fig. 3 except this flame has a jet Reynolds
number of 20,000.   Note that the image of C$_2$ is not connected.

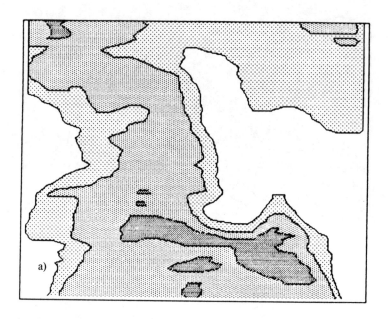

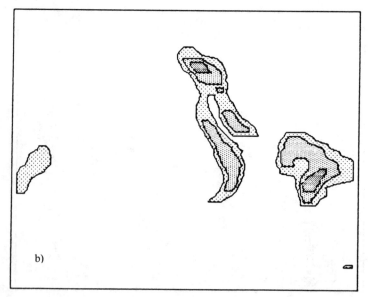

Fig. 6 Same as for Fig. 3 except this flame has a jet Reynolds number of 25,000. Note that the image of C_2 is not connected.

expected that models that contain a mechanism for local flame extinction through inclusion of finite rate chemical kinetics may have improved predictions of other flame properties such as pollutants and flame radiation. Measurements, such as described in this paper, will be useful in refining these new models.

4. Summary

The paper reports the first two-dimensional imaging of diatomic carbon C$_2$ using a single laser pulse. This new C$_2$ fluorescence technique promises to be a superb marker of the flame front position, since the C$_2$ molecule is a short-lived transient that is generated and consumed in a very thin zone near the fuel-rich side of the flame front.

5. Acknowledgement

This work was supported by the U.S. Department of Energy, Office of Basic Energy Sciences.

References

Alden, M., Edner, H., and Svanberg, S. (1982) Simultaneous spatially resolved monitoring of C$_2$ and OH in a C$_2$H$_2$/O$_2$ flame using a diode array detector. Appl. Phys. B B29 (2), 93-97.

Bilger, R.W. and Starner, S. (1983) A simple model for carbon monoxide in laminar and turbulent hydrocarbon diffusion flames. Combust. Flame 51 (2), 155-176.

Janicka, J. and Kollmann, W. (1978) Method for predicting reacting turbulent shear flows with chemical non-equilibriums. W. Waerme Stoffuebertrag Thermo Fluid Dyn. 11 (3), 157-174.

Kychakoff, G., Howe, R.D., Hanson, R.K., et al. (1984) Visualization of turbulent flame fronts with planar laser-induced fluorescence. Science 224 (4647), 382-384.

Liew, S.K., Bray, K.N.C., and Moss, J.B. (1984) A stretched laminar flamelet model of nonpremixed combustion. Combust. Flame 56 (2), 199-213.

Peters, N. (1984) Laminar diffusion flamelet models in non-premixed turbulent combustion. Prog. Energy Combust. Sci. 10 (3), 319-339.

Smyth, K.C., Miller, J.H., Dorfman, et al. (1985) Soot inception in a methane/air diffusion flame as characterized by detailed species profiles. Combust. Flame 62 (2), 157-181.

Simulation of Turbulence Development at Propagating Flame Fronts

Takashi Tsuruda* and Toshisuke Hirano†
University of Tokyo, Tokyo, Japan

Abstract

The development of turbulence at the front of propagating premixed flame has been modeled by simulating flame front movement based on the calculated flowfield and laminar burning velocity. When the flow is induced by the combustion reaction, the flowfield can be predicted by an inviscid theory in which the thermal expansion of gas due to combustion is replaced by a surface source at the flame front. The amplitude of the flame front turbulence is indicated to continue increasing, and the rate of turbulence development is shown to increase as the scale of turbulence decreases or the amplitude or burning velocity increases. Sections of the flame front become convex toward the unburned mixture with cusp-shaped lines connecting these convex parts as a consequence of the flowfield induced by the curved flame front. The flowfield change caused by an overall pressure gradient is also evaluated. Acceleration of the gas in the direction of flame propagation increases the amplitude of the flame front turbulence, while deceleration decreases it.

Introduction

The turbulence at the front of a propagating premixed flame induces its acceleration. Since the intensity of a gas explosion depends mainly on the rate of pressure rise, which is caused by flame propagation, the prediction of turbulence development is indispensable for the evaluation of gas explosion effects or the assessment of gas explosion hazards. Also, as the performance of a spark ignition

---

Presented at the 10th ICDERS, Berkeley, California, August 4-9, 1985. Copyright © 1986 by American Institute of Aeronautics and Astronautics, Inc. All rights reserved.
*Graduate Student, Department of Reaction Chemistry.
†Professor, Department of Reaction Chemistry.

engine depends largely on the flame behavior, the prediction of turbulence development at the flame front of a premixed flame is extremely important. Thus, many studies have been performed on the growth of the turbulence at premixed flame fronts(Landau and Lifshitz 1959; Markstein 1964; Williams 1982;Sivashinsky 1976; Joulin and Clavin 1979). Most of these studies are based on the concepts of the flame induced turbulence and/or flame-turbulent flow interaction.

In the studies of the instability appearing at premixed flame fronts(Landau and Lifshitz 1959; Markstein 1964; Williams 1982) and the measurements of propagating turbulent premixed flames(Markstein 1964; Solberg 1980; Wagner 1982; Tsuruda et al. 1986), a number of facts have been revealed concerning the characteristics of the instability and the growth of flame front turbulence in various cases. However, very few data on the flame front behavior during the turbulence growth are available, although such data are essentially required for practical purposes(Yao 1974; Zalosh 1979; Solberg 1980; Hirano 1982; Jones and Whitelaw 1984).

Hirano (1982) and Tsuruda et al. (1986) have identified the various mechanisms by which turbulence develops in the front of a propagating flame. An important process involved in some of these mechanisms is the growth of flame front turbulence due to the flow induced by the combustion reaction at the flame front. Further, the growth of flame front turbulence has been found to be enhanced in an accelerating, or suppressed in a decelerating, flowfield (Tsuruda et al. 1986; Hirano 1985). In the present study, the growth of flame front turbulence, when the effects of other causes than combustion reaction are negligible or when an overall pressure gradient accelerates or decelerates gas, has been explored by simulating flame front movements.

For practical purposes, the prediction of the behavior of developed turbulent flame fronts is also important. However, knowledge of the growth of flame front turbulence must be accumulated before an attempt to solve such a complex problem is made. Thus, the simulation performed in the present study is confined mainly to the initial and succeeding stages of the turbulence growth.

Model and Basic Equations

Since the model established in this study is applied to the initial and succeeding stages of the development of turbulence at the front of a propagating premixed flame, the amplitude of the turbulence at the front is assumed to be small compared to the scale of the turbulence. In such a situation, the wrinkled laminar flame assumption must be

valid, and the effect of flame curvature on the local
structure across the flame front must be negligible.
Further, the flame front can be assumed to be a surface
without thickness (flame sheet).

The turbulence in a shear flow is caused by the
viscosity of gas, and a flame propagating through a shear
flow region becomes turbulent. In previous studies, the
scale of flame instability is closely related to the
viscosity of gas (Markstein 1964; Oran et al. 1982). In the
flowfield at the initial and succeeding stages of the
turbulence growth at the flame front, the velocity gradient
would be slight except for the region very close to the
flame front. Therefore, to simplify the problem, an
inviscid flowfield is assumed except for the region very
close to the flame front.

The behavior of a local flame front of a wrinkled
laminar flame depends on the burning velocity and flowfield
(Fig. 1). The local flame front velocity is equal to the
sum of the burning velocity and unburned gas velocity just
ahead of the flame front:

$$\mathbf{v}_f \cdot \mathbf{n} = \mathbf{v}_u \cdot \mathbf{n} + S_L \qquad (1)$$

where $\mathbf{v}_f$ and $\mathbf{v}_u$ are the velocity vectors representing the
movements of the flame front and gas just ahead of it,
respectively, $\mathbf{n}$ is the unit vector normal to the flame front
and pointing to the unburned mixture region, and S_L is the
normal burning velocity. The variation of the flame front
configuration is predictable if the flowfield of unburned
gas is known.

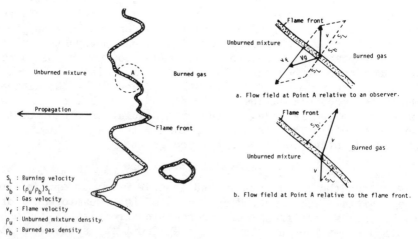

S_L : Burning velocity
S_b : $(\rho_u/\rho_b)S_L$
v : Gas velocity
v_f : Flame velocity
ρ_u : Unburned mixture density
ρ_b : Burned gas density

a. Flow field at Point A relative to an observer.

b. Flow field at Point A relative to the flame front.

Fig. 1   Instantaneous aspects of a propagating turbulent premixed
flame.

In general, the flame velocity during the growth of flame front turbulence is much less than the speed of sound, so that the gas can be assumed to be incompressible. To evaluate the flowfield, the density change across the flame front should be taken into account. If the temperature and composition of the unburned mixture are uniform, those of burned gas must be uniform, too. As the flowfield before and after the flame front can be treated as isothermal flowfields, the energy equation is not necessary for the solution.

Under these assumptions, the flowfields, except for the discontinuities of the gas velocity and density at the flame front, are described by the following equations for inviscid and irrotational flow:

$$\nabla(\mathbf{v}) = 0 \tag{2}$$

$$\rho\frac{D\mathbf{v}}{Dt} = -\nabla p \tag{3}$$

$$p = \frac{\rho RT}{M} \tag{4}$$

where t is the time, ρ the density, $\mathbf{v}$ the velocity vector representing the gas movement, p the pressure, R the gas constant, T the temperature whose value is $T_{-\infty}$ for x < 0 and $T_{+\infty}$ for x > 0, and M the mean molecular weight of gas.

The scale and amplitude of the turbulence at a turbulent flame front are in general distributed, and the flame front configuration is three dimensional. To simulate the movements of flame front of an arbitrary shape is difficult and can be accomplished only with the loss of the generality of the problem. Therefore, the simulation in this study has been performed for flames of simple two-dimensional shapes. An example of initial flame front configuration, the variation of which will be examined in the following analysis, is shown in Fig. 2. The frame of reference is attached to the flame front, so that the unburned mixture approaches with a velocity equal to the flame speed. x and y are the distances from the center point of the flame front along the directions opposed and normal to that of flame propagation, respectively.

The boundary and initial conditions, which are needed to evaluate the two dimensional flowfield for the situation shown in Fig. 2 by solving Eqs. (2) and (3), are as follows: As x → - ∞, y → ± ∞;

$$\mathbf{v} = -\mathbf{v}_f \quad \rho = \rho_u \quad p = p_u \tag{5a}$$

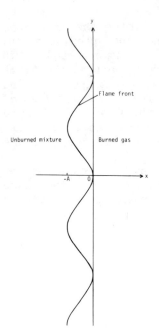

Fig. 2   Model representing the initial flame front configuration for numerical simulation of flame front movements.

For t = 0,

$$\mathbf{v} = \mathbf{v}_i(x,y) \quad p = p_i(x,y) \tag{5b}$$

where $\mathbf{v}_f$ is the flame velocity. $\mathbf{v}_i$ and p_i are the initial velocity and pressure, respectively, and must be given in the unburned and burned regions to satisfy Eq. (1) at the flame front.

Analysis

For the cases in which the gas velocity changes only because of the combustion reaction at the flame front, the entire flowfield can easily be expressed by a complex stream function, in which the thermal expansion of gas is included. The thermal expansion of gas is replaced by an imaginary line source of equivalent strength at the flame front (Uberoi 1963; Hirano et al. 1981). For a given flame front configuration, the corresponding line source can be assumed. The strength of the source is determined from the equation

$$m = (\rho_u/\rho_b - 1) \, S_L \tag{6}$$

where m is the strength per unit length of the source, and ρ_u and ρ_b are the densities of the unburned mixture and burned gas, respectively.

To start the calculation, the negative value of the flame velocity, i.e., the unburned mixture velocity $v(-\infty,y)$ at $x = -\infty$ approaching to the flame front is assumed as a value given by the following equation:

$$v(-\infty,y) = -v_f = (S_L \int_{y_1}^{y_2} \frac{\partial s}{\partial y} \, dy)/(y_2 - y_1) \qquad (7)$$

where s is the distance along the flame front; $y = y_1$, $y = y_2$ are arbitrary points; and the length $y_2 - y_1$ is much larger than the scale of the flame front turbulence.

Thus, for a given flame front configuration and given values of ρ_u, ρ_b, and S_L, the entire flowfield as well as the pressure distribution can be calculated, i.e., the replacement of the thermal expansion of gas by a source is equivalent to the determination of $v_i(x,y)$ and $p_i(x,y)$.

Once the gas velocity just ahead of the flame front is evaluated, the movement of each fragment of it can be estimated from Eq. (1), so that the flame front configuration at the next moment can be determined. From the calculation of the flame front configuration at succeeding small time intervals, the deformation of the front with time is obtained. When the gradient ∇p_* of the pressure deviation induced by other causes than combustion reaction is constant and in the direction of the flame propagation, the flame behavior is examined as an example of the effect of a finite value of ∇p_*.

The component of the pressure gradient ∇p_s in the direction parallel to the flame front:

$$\nabla p_s = n \times (n \times \nabla p_*) \qquad (8)$$

accelerates or decelerates the fluid elements in the direction along the flame front (Fig. 3). As the densities of the unburned mixture and burned gas are different, a velocity difference is induced across the flame front. This velocity difference leads to a vortex sheet along the flame front. As the flowfield except at the flame front is assumed to be irrotational, the flowfield can be described by a complex stream function as long as the vortex sheet is located at the flame front. Thus, the flowfield results from the superposition of the irrotational flowfield and that induced by the vortex sheet (Lamb 1932; Goldstein 1965).

The rate $d\omega/dt$ of vortex generation at the flame front is:

$$\frac{d\omega}{dt} = 1/\delta \, (1/\rho_b - 1/\rho_u) \, n \times (n \times \nabla p_*) \qquad (9)$$

where δ is the thickness of the flame front across which the gas temperature changes from that of unburned mixture to the flame temperature. For a given flame front configuration and ∇p_*, $d\omega/dt$ can be calculated, and $\mathbf{v}$ is obtained as the combination of the complex stream functions of the line source and the vortex sheet at the flame front. Based on the obtained value of $\mathbf{v}$, the flame front movement can be estimated with Eq. (1), and the flame front configuration at the next small time interval can be determined. From successive calculation, the behavior of the flame front is examined.

Numerical Examples and Discussion

Typical examples of calculated flame front configurations are shown in Figs. 4-6. The center points of the flame fronts are fixed at a definite point (0,0) in each figure. In these calculations, the initial configuration of the flame front is sinusoidal (Fig. 2). In these calculations, the following nondimensional quantities are adopted:

$$\xi = x/L \quad \eta = y/L \quad \lambda = \Lambda/L \tag{10a}$$

$$A = 2\alpha/L \quad B = S_L/S_{LO} \tag{10b}$$

$$T = S_L(\rho_u/\rho_b - 1)t/nL \quad K = \omega\delta^2/\mathbf{nj} \tag{10c}$$

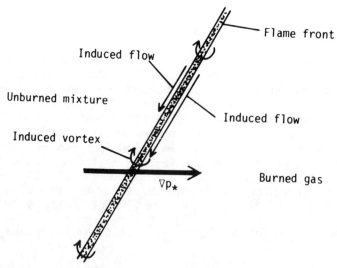

Fig. 3 Physical description of the flowfield near the flame front under a constant, overall pressure gradient.

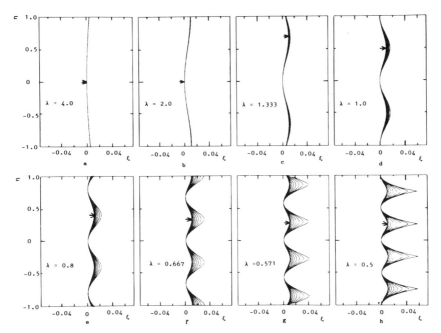

Fig. 4 Numerically calculated flame front configuration. The scale in x direction is enlarged to be 10 times as large as that in y direction. The arrow in each figure points the initial flame front configuration. A = 0.01, B = 1.0, and $\Delta T = 0.001$.

where L is the scale of reference; the length of the frame of each figure is 2L; Λ is the scale of initial turbulence; α is the amplitude of the initial turbulence; S_L and S_{L0} are the burning velocity of the unburned mixture to be calculated and that of reference, respectively; n is the total imaginary source strength at the flame front in the range of $0 < \eta < 1$ on the $\xi - \eta$ plane; t is the time; and **j** is the unit vector in the direction y.

Figure 4 shows the changes of flame front configurations for different turbulence scales for a given initial amplitude of turbulence. The rate of turbulence growth is found to increase as the scale of turbulence decreases, or as the value of B increases. Figure 5 shows the change of flame front configurations when various scales of turbulence appear closely. In the region of small-scale turbulence, the rate of turbulence development is larger, and consequently the flame velocity becomes larger.

Although the initial configuration of the flame front is assumed to be sinusoidal, the departure from a sinusoidal one increases as the turbulence develops. Sections of the front become convex toward the unburned mixture with cusp-

shaped lines connecting these sections (Figs. 4 and 5d).
While this final configuration is nearly identical to that
observed in several previous studies (Karlovitz 1953; Fox and
Weinberg 1962; Williams et al. 1969) and described in the
text book of Gaydon and Wolfhard (1979), the postulated
process of flame front deformation is different.   The
results of the present study indicate that the flowfield
induced by the curved flame front causes this phenomenon.

    A typical result of the calculation of flame front
configurations for the case of a finite, constant value of
∇p_* is shown in Fig. 6.   When ∇p_* is an appreciable negative
value, the flame decelerates because of the overall gas
deceleration (or acceleration) induced in the same direction
as (or the opposed direction to) the flame propagation. The
amplitude of the flame front turbulence decreases and
approaches a definite value (Fig. 6a) and the flame front
turbulence is suppressed in this case.   At later stages, the
deformation of the front configuration due to the gas flow
induced by the expansion of gas at the curved flame front
cancels that due to ∇p_*.   If the initial amplitude of flame
front turbulence is smaller than this asymptotic value, the
amplitude increases and approaches to the limiting value.
This situation would occur appear for a small negative value
of ∇p_*.

    For a positive value of ∇p_*, i.e., for a situation of
overall flame acceleration, the amplitude of the flame front
turbulence increases (Fig. 6c).   In the accelerative flame
propagation induced by a pressure gradient in the same
direction as that of flame propagation, the flame front
turbulence continues to intensify and finally the front
"bursts".

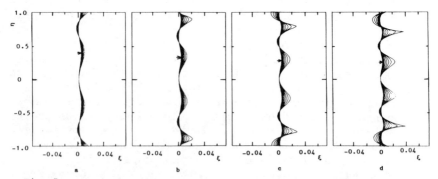

Fig. 5 Numerically calculated flame front configurations when
various scales of turbulence appear close. The arrow in each figure
points to the initial flame front configuration. A=0.004, B=1.0,
ΔT=0.001.

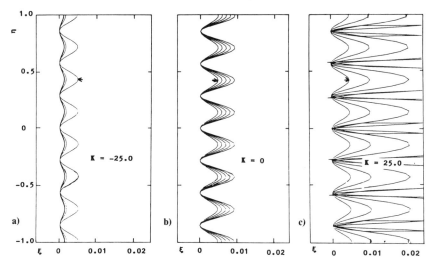

Fig. 6 Numerically calculated flame front configurations when a
finite, constant value of the overall pressure gradient induces
acceleration or deceleration of the gas flow. K is the nondimen-
tional pressure gradient ($= \omega\sigma^2/nj$). The arrow in each figure
points to the initial flame front configuration. $\lambda=0.286$, B=1.0,
A=0.005, $\Delta T=0.0005$.

These results are in accord with the experimental
results. Turbulence suppression in a decelerative flame
propagation and enhancement in an accelerative flame
propagation were pointed out previously (Tsuruda et al.
1986). Also, in a recent experimental study, the flame
front "burst" was observed when a flame propagated through
an acceleratively flowing premixed gas (Hirano 1985).
The instability of a premixed flame or turbulence
growth at a premixed flame front has been studied in a
number of previous studies (Landau and Lifshitz 1959;
Markstein 1964; Williams 1982; Sivashinsky 1976; Joulin and
Clavin 1979). Through these studies, most experimentally
observed phenomena could be consistently explained.
However, these results provide very poor images of flame
behavior. The results of numerical simulations like that
adopted in the present study usually present detailed images
of flame behavior as shown in Figs. 4-6.
Comparison of the results of the present study with
those of previous theoretical studies (Landau and Lifshitz
1959; Markstein 1964; Williams 1982; Sivashinsky 1976;
Joulin and Clavin 1979) for linear cases indicates that the
initial stage of the turbulence growth predicted in the
present study agrees well with that predicted in the previous studies. From Eq. (10) and Fig. 4, it is seen that at

the initial stage the growth rate is proportional to S_L as well as (p_u/p_b-1). Other aspects of the turbulence growth at the initial stage also are in accord with those predicted in the previous theoretical studies for linear cases.

Because of the adopted assumptions and computational precision, these results can be used to explain some aspects of the phenomena, and many refinements remain to be developed.

The initial scale of turbulence used at the start of calculation is based on the results of the discussion on the instability of a premixed flame (Landau and Lifshitz 1959; Markstein 1964; Williams 1982; Sivashinsky 1976; Joulin and Clavin 1979). The scales of turbulence concerned in the present study are much larger than the flame front thickness composed of reaction and preheat zones, because the heat conduction and molecular diffusion in the direction normal to that of flame propagation are neglected.

Various scales of flame front turbulence are observed in the flame zone of a turbulent flame except for that under limiting or special conditions. The method of the present study can be applied for these cases as shown in Fig. 5. The interaction between sections of the flame front may be examined, and the overall flame velocity for a given distribution of the scale of turbulence may be obtainable at higher compution speeds.

As shown in Figs. 4-6, this method can be applied for the stages of nonlinear behavior of turbulent flame if the assumptions adopted in the present study are valid. However, for the case when the unburned mixture flow is turbulent, this method cannot be applied. To develop a method of numerical simulation applicable to the case of a turbulent unburned mixture flow is a future task.

Concluding Remarks

A model is proposed for the prediction of turbulence development at the front of propagating premixed flames. The model is applied to a two-dimensional flame front configuration. When the flow is induced by only the combustion reaction, the flowfield at each moment is predicted by an inviscid theory in which the thermal expansion of gas due to combustion is replaced by a line source at the flame front.

The rate of turbulence development is found to increase as the scale of turbulence decreases, or as the amplitude or burning velocity increases. It is shown that as the flame front turbulence develops, sections of the flame front become convex toward the unburned mixture with cusp-shaped lines connecting these convex parts. This phenomenon is caused by the flowfield induced by the curved flame front.

The flame behavior under a constant, overall pressure gradient, which induces a constant acceleration or deceleration of gas, is examined. In an accelerating or decelerating flowfield, the turbulence development is promoted or suppressed, respectively.

Acknowledgment

This work was supported by a grant-in-aid for special project reaserch and for scientific research from the Ministry of Education, Science, and Culture of Japan.

References

Fox, M.D. and Weinberg, F.J. (1962) An experimental study of burner-stabilized turbulent flames in premixed reactants. Proc. R. Soc. London Ser. A 268 (1333), 222-239.

Gaydon, A.G. and Wolfhard, H.G. (1979) Flames, Their Structure, Radiation and Temperature, 4th ed. Chapman and Holl, London.

Goldstein, S. (1965) Modern Developments in Fluid Dynamics. Dover Publications, New York.

Hirano, T., Suzuki, T., and Mashiko, I. (1981) Flame behavior near steps bounding layered flammable mixtures. 18th Symposium (International) on Combustion, pp.647-655. The Combustion Institute, Pittsburgh, Pa.

Hirano, T. (1982) Some problems in the prediction of gas explosions. Fuel-Air Explosions (edited by J.S.H. Lee and C.M. Guirao), pp. 823-839. University of Waterloo Press, Waterloo, Canada.

Hirano, T. (1985) Turbulence growth at the flame fronts propagating through an accelerating (decelerating) premixed gas. Report for grant-in-aid for scientific research from Ministry of Education, Science, and Culture of Japan, University of Tokyo, Japan.

Jones, W.P. and Whitelaw, J.H. (1984) Modeling and measurements in turbulent combustion. Paper 26 presented at the 20th Symposium (International) on Combustion, Ann Arbor, Mich.

Joulin, G. and Clavin, P. (1979) Linearized stability analysis of nonadiabatic flames: Diffusion-thermal model. Combust. Flame 35 (2), 139-153.

Karlovitz, B. (1953) Open turbulent flames. Fourth Symposium (International) on Combustion, pp. 60-67. Williams and Wilkins, Baltimore, Md.

Lamb, H. (1932) Hydrodynamics. Cambridge University Press, London.

Landau, L.D. and Lifshitz, E.M. (1959) Fluid Mechanics (translated by J.B. Sykes and W.H. Reid). Pergamon Press, Oxford, England.

Markstein, G.H. (1964) Nonsteady Flame Propagation. Pergamon Press, Oxford, England.

Oran, E.S., Boris, J.P., Young, T.R., et al. (1982) Numerical simulation of fuel-air explosions: Current method and capabilities. Fuel-Air Explosions (edited by J.S.H Lee and G.M. Guirao), pp. 447-474. University of Waterloo Press, Waterloo, Canada.

Sivashinsky, G.I. (1976) On a distorted flame as a hydrodynamic discontinuity. Acta Astronaut. 3 (11-12), 889-916.

Solberg, D.M. (1980) Experimental investigations of flame acceleration and pressure rise phenomena in large scale vented gas explosions. Paper 80-004 presented at The Third International Symposium on Loss Prevention and Safety Promotion in the Process Industries, Basle, Switzerland.

Tsuruda, T., Harayama, M., and Hirano, T. (1986) Growth of flame front turbulence. Transactions of ASME, J. Heat Transfer(in press).

Uberoi, M.S. (1963) Flow fields of flame propagating in channels based on the source sheet approximation. Phys. Fluids 6, 1104-1109.

Wagner, H.G. (1982) Some experiments about flame acceleration. Fuel-Air Explosions (edited by J.S.H. Lee and C.M. Guirao), pp.77-99. University of Waterloo Press, Waterloo, Canada.

Williams, F.A. (1982) Laminar flame instability and turbulent flame propagation. Fuel-Air Explosions (edited by J.S.H Lee and C.M. Guirao), pp.69-76. University of Waterloo Press, Waterloo, Canada.

Williams, G.C., Hottel, H.C., and Gurnitz, R.N. (1969) A study of premixed turbulent flames by scatterd light. 12th Symposium (International) on Combustion, pp. 1081-1092. The Combustion Institute, Pittsburgh, Pa.

Yao, C. (1974) Explosion venting of low-strength equipment and structures. Loss Prev., 8, 1.

Zalosh, R.G. (1979) Gas explosion tests in room size vented enclosures. Loss Prev., 13, 98.

The Effect of Large-Scale Fluctuations on Flame Radiation

W.L. Grosshandler*
Washington State University, Pullman, Washington
and
P. Joulain†
Université de Poitiers, Poitiers, France

Abstract

The relation between the nature of hydrocarbon/air diffusion flames and the radiative heat transfer from such systems is examined in this paper. A computer model has been developed that couples the combustion to a nonisothermal radiation model. Account is taken of the major infrared active combustion gases and soot, all of which are assumed to vary in composition with time and distance along a given line of sight through the flame. The significance of the large-scale fluctuations is determined by comparing the predicted radiant intensity based upon the time mean temperature and composition profiles along the line of sight, to that based upon a temporal average of the intensity based upon the instantaneous temperature and composition profiles. The ratio of these two intensities is most sensitive to the amplitude of the periodic fluctuations and to the expression used to model the soot formation process; and the ratio is least sensitive to the functional form of the fluctuations, the randomness of the fluctuations, and the scale of the flame. The actual intensity is found to exceed that based upon time-mean temperatures by as much as a factor of two; but in some strongly sooting situations, the calculations suggest that the actual intensity can, in fact, be less than that implied from the time-mean profiles.

Presented at the 10th ICDERS, Berkeley, California, August 4-9, 1985. Copyright © 1986 by the American Institute of Aeronautics and Astronautics, Inc. All rights reserved.
*Associate Professor, Department of Mechanical Engineering.
+Maitre de Recherche, Chimie Physique de la Combustion, Centre National de la Recherche Scientifque.

Nomenclature

a	= absorption coefficient
b	= stochiometric fuel/oxidizer mass ratio
c_p	= specific heat, kJ/kg·K
c_1, c_2, c_3	= constants in steady-state soot expression
f	= amplitude fluctuation parameter
F	= random fraction
$G(\theta)$	= probability distribution function
$g(t)$	= periodic function for intermittency
$I(t)$	= instantaneous radiant intensity, kW/m^2·sr
I_1	= implied radiant intensity, kW/m^2·sr
I_2	= actual time mean intensity, kW/m^2·sr
L	= characteristic length, m
$\dot{m}$	= mass flow rate, kg/s
P	= pressure, kPa
t	= nondimensional time
T	= temperature, K
X	= Shvab-Zel'dovich parameter
Y	= mass fraction
β	= proportionality constant in Eq. (25)
γ	= constant in Eq. (5)
ε	= emittance
λ	= wavelength, m
σ	= Stefan-Boltzmann constant
θ	= nondimensional temperature
ψ	= intensity ratio

Subscripts

A	= adiabatic
c	= correlated
f	= pertaining to fuel
g	= pertaining to gas
i	= pertaining to inert
M	= maximum
o	= pertaining to oxidizer
p	= pertaining to products
r	= random
rms	= root mean square deviation
s	= pertaining to soot
λ	= monochromatic quantity

Superscripts

()*	= inlet value
$\overline{()}$	= time mean value
()'	= fluctuating

Introduction

That radiation makes a significant contribution to the overall heat transfer in flames of moderate scale has been well established by experiments in a variety of configurations (de Ris, 1979; Orloff et al., 1975). Even so, modelers of combustion systems often choose to treat radiation superficially because of the complexity of the process and the resultant burden put upon numerical computations. The complexity arises from two distinct sources. The first is the field nature of the radiant flux vector, leading to an integration over solid angle in an otherwise differential equation. This incompatibility makes assigning a numerical grid cumbersome, since the divergence of the radiant flux vector depends upon conditions all the way to the boundary as well as in adjacent cells. The second source of complexity is the strong spectral dependence of the absorption coefficient and the blackbody function, which makes an analytical integration over the wavelength virtually impossible.

The penalty for treating radiation superficially varies from almost nothing to quite severe, depending upon the distribution of temperature and radiating species present, the turbulence level, and the scale of the flame. The application of the optically thin radiation limit (Kinoshita and Pagni, 1981; Sibulkin et al., 1981), the uniform gray gas assumption (Orloff et al., 1977; Modak, 1977), and semi-empirical methods (Tamanini, 1978) to account for flame radiation in various laminar and turbulent combustion geometries is extremely useful since these techniques allow the modeler to concentrate on understanding the turbulent transport processes within the flame.

The total transmittance nonhomogeneous (TTNH) model is a method for calculating the radiant intensity in nongray, nonisothermal combustion gas mixtures. The so-called TTNH model (Grosshandler, 1979) is based upon the original thought of Hottel that the radiant heat flux can be written as an integral of the gradient of total emittance along the nonhomogeneous path (Hottel, 1954). The technique was applied by Leckner (1973) to nonisothermal CO_2 and H_2O gas clouds and it has since been refined and extended to include mixtures of N_2, O_2, CO_2, H_2O, CO, CH_4, and soot within black enclosures of arbitrary temperature (Grosshandler and Nguyen, 1985; Grosshandler and Modak, 1981).

Whichever of these models is chosen, it is customary to make the radiation calculations based upon time mean values of temperature and composition, even in turbulent

flames. Because the physical parameters controlling the
radiative heat transfer interact in a highly nonlinear
fashion, such calculations can be subject to serious
errors.

Foster (1969) investigated this effect on the time
mean transmittance. He concluded that the absorption
coefficient inferred by assuming the flame to be static
could be as much as five times less than the true time mean
absorption coefficient. His conclusion was experimentally
verified by measurements of transmittance through and
radiation from the 4.3 µm band of CO_2 in a tunnel burner
(Tan and Foster, 1978) and measurements using laser attenuation of soot in a turbulent propane flame (Amin and
Foster, 1973) . In the work of Amin and Foster (1973), a
Maxwellian distribution of optical depth was found to
closely represent the probability distribution of the
measured transmittance and the absorption coefficient due
to the soot was found to be underestimated significantly
when based upon time mean transmittance measurements. This
effect decreased when the average transmission was high,
and increased with the extent of intermittency [defined by
Foster (1969) to be the fraction of time at which the flame
transmittance is 1.0].

The interaction between concentration fluctuations and
variations in radiative flux and temperature within a jet
diffusion flame was studied by Becker (1975). His radiation model, based upon CO_2 and H_2O emission, indicated that
the local rms fluctuation of radiative flux along the jet
centerline was between 20 and 500% of the mean radiative
flux at the same location. He made no attempt to determine
the effect of the localized fluctuations on the overall
intensity leaving the boundary of the flame.

This paper studies in more detail the physical parameters that are likely to control the time-varying radiation
leaving the flame boundary. Specifically, a hydrocarbon/
air diffusion flame is analyzed in which the temperature
and composition are varying in space as well as time. A
combustion/radiation model is developed using the Shvab-
Zel'dovich formulation and the TTNH radiation calculation
technique. It is shown that it can be applied to predict
the essential features of fluctuating diffusion flames.

The Combustion Model

Radiation calculations such as those mentioned above
require knowledge of the temperature and composition
throughout the combustion field. It is desirable to reduce
the amount of detailed information necessary to describe
the flame, while still yielding an accurate prediction of

the radiative heat transfer. A model of the combustion process has been developed with this as its primary purpose. With this model, the experimental measurement of intensity in turbulent diffusion flames can be analyzed or synthetic flames can be generated that include the relevant characteristics of real, fluctuating combustion processes.

The combustion model chosen is that which has been applied traditionally to laminar diffusion flames: the classical thin flame model (Williams, 1965). If appropriate assumptions are made, the model can yield estimates of the local temperature and product concentration. The fuel and the oxygen are assumed to react instantaneously to form a predictable set of products at the adiabatic flame temperature.

The conservation equations for the fuel and oxidizer can be combined to eliminate the source term that arises from the infinitely fast chemical reaction. The natural variable which results is the Shvab–Zel'dovich parameter X, which is defined in terms of the mass fractions of fuel Y_f and oxygen Y_o as

$$X = Y_f - (b \, Y_o/Y_o^*) \qquad (1)$$

where b is the stoichiometric mass ratio of fuel to oxidizer mixture and the starred quantity represents the mass fraction of oxygen in the oxidizer mixture as it enters the flame. (Any diluents are assumed to be inert, with a molecular weight similar to the oxidizer.) The mass fraction of the combustion products Y_p is easily determined from X once Y_o and Y_f are known. The temperature, when properly normalized, behaves with X exactly as Y_p.

The gaseous species important to thermal radiation are CO_2, H_2O, CO, and the hydrocarbon fuel. Equilibrium thermodynamics can be used to estimate the relative fraction of these species once Y_p is known. Such a calculation can also yield the adiabatic temperature T_A, which will generally exceed the peak temperature measured in the flame due to radiative losses to the boundaries. A more appropriate maximum temperature T_M has been suggested by Grosshandler and Vantelon (1985) to be

$$T_M = \frac{T_A}{1 + 4 \, L^2 [1 - \exp(-aL)] \sigma T_M^3 / (c_p \dot{m}_p)} \qquad (2)$$

where L is the characteristic size of the flame, a the effective gray absorption coefficient of the flame, c_p the specific heat of the product mixture, and $\dot{m}_p$ the total mass flow rate of the combustion products.

A comprehensive combustion/radiation model must be able to predict, at least qualitatively, the level of soot within the flame. Recently, an hypothesis was put forth based upon the assumption that the soot concentration is controlled by steady-state kinetics (Grosshandler and Vantelon, 1985). That is, the reactions forming the soot occur as rapidly as those responsible for soot destruction through oxidation. The soot spectral absorption coefficient can then be shown to be approximated by the following global expression:

$$a_{\lambda,s} = C_1 \, (P/T)^{C_2+1} \, Y_p \, Y_f^{C_2} \, \exp \, (C_3/T) \tag{3}$$

The empirical constants C_1, C_2, and C_3 are found from measurements of the temperature, composition, and absorption coefficient in laminar diffusion flames. For the purpose of the present analysis, it will be assumed that Eq. (3) is valid in turbulent diffusion flames as well, if they are composed of the same fuel and oxidizer combination as when the empirical constants were determined.

The Model for Flame Fluctuations

"Intermittent" is a term to describe the large-scale fluctuations that lead to gross changes in the flame structure as a function of time. In such a flame, the temperature and composition profiles change drastically from instant to instant; however, it will be assumed that when averaged over a long period of time relative to the frequency of the large-scale fluctuations, a meaningful time mean structure can be determined. Orloff (1981) has measured the radiant intensity leaving a pool fire and found that it can be represented by a composite flame shape that is invariant over relatively long times. The large-scale fluctuations are distinct from the turbulence controlling the small-scale transport within the flame. The local turbulence is usually associated with temperature fluctuations of much higher frequency but of lesser amplitude. Although progress has been made in the modeling of real turbulent diffusion flames (e.g., Drake, et al., 1982), a more naive approach to turbulence will be used here so as to concentrate on the radiative transport.

The simple combustion model is used to describe the instantaneous flame conditions, after assuming that a statistical distribution of the Shvab-Zel'dovich parameter about its time mean, $\bar{X}$, is given at every point within the combustion volume. As an example, consider the case where the stoichiometric mass ratio b is equal to one. If the fluctuations in X are periodic and symmetric about the mean, one can write

$$X(t) = \bar{X} + f(1 - |\bar{X}|) \, g(t) \qquad (4)$$

where $g(t)$ is a periodic function chosen to model the nature of the intermittency and f a fluctuation parameter between zero (no fluctuation amplitude) and one (maximum amplitude of fluctuations).

The Shvab-Zel'dovich parameter is physically constrained to the closed interval $(-b,1)$, so that the periodic function $g(t)$ must vary from -1 to $+1$ for the case of $b = 1$. The root mean square deviation of X, ΔX_{rms}, can then be shown to be related to the fluctuation parameter by

$$\Delta X_{rms} = \gamma \, f(1 - |\bar{X}|) \qquad (5)$$

where γ is a constant depending upon the functional form of $g(t)$.

Different functional forms for $g(t)$ can be chosen based upon the observed flame behavior. A close to sawtooth form for the fuel mass fraction has been observed in experiments (Yee, 1982) that yields a constant probability distributed about the mean value of X up to $\pm \sqrt{3} \, \Delta X_{rms}$. The probability of an excursion beyond this point is zero. A square-wave yields a nonzero probability only $\pm \Delta X_{rms}$ about $\bar{X}$ and a sinusoidal wave form is in between these two extremes. A Gaussian plus intermittent spike probability distribution function (pdf) can also be modeled with the appropriate form of $g(t)$. Such a pdf approximates the experimental measurements of Drake et al. (1982).

The pdf's for the fuel, oxygen, and temperature will be identical to those of X as long as $X(t)$ does not change its sign over the oscillation period. This will be the case when $|\bar{X}| > f(1 - |\bar{X}|)$. As $\bar{X}$ approaches zero, a portion of the period will be spent in which $X(t)$ is of the opposite sign. In this region, the pdf's for the physical parameters will either be truncated (fuel and oxygen) or enhanced (products and temperature). For example, Fig. 1 shows the distribution function $G(\theta)$ for the nondimensional temperature, where θ is defined as

$$\theta \equiv \frac{T_M - T(t)}{T_M - T^*} \qquad (6)$$

and T* is the inlet temperature. In Fig. 1, $\bar{X}$ is varied over its full range assuming sawtooth oscillations with a

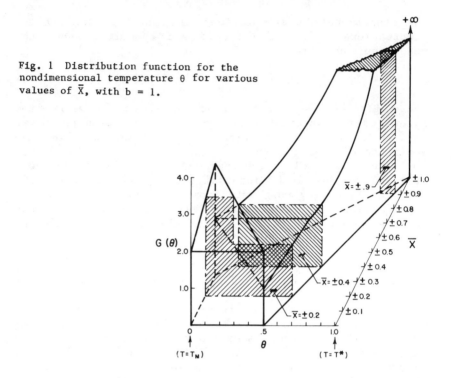

Fig. 1 Distribution function for the nondimensional temperature θ for various values of $\bar{X}$, with b = 1.

value of f = 0.5. When $|\bar{X}| > 0.5(1 - |\bar{X}|)$, the pdf is the same as for X. As $|\bar{X}|$ approaches 1.0, $G(\theta)$ approaches a Dirac delta function centered at θ = 1.0 (i.e., T = T*). The wave cannot penetrate far enough into the fuel or oxidizer region; thus, the temperature is fixed at its initial value. When $|\bar{X}| < 0.5 (1 - |\bar{X}|)$, X(t) is multi-valued in time, which accounts for the step increase in $G(\theta)$ at the maximum temperature. The peak broadens as $\bar{X}$ decreases, until at $\bar{X}$ = 0, a uniform distribution is again obtained.

The periodic oscillations associated with different locations in the turbulent flame may or may not be correlated. Two extremes can be visualized to exist along any particular line of sight: either the oscillations are completely in phase or the oscillations are completely random. Intermediate degrees of correlation can be introduced with a randomly generated phase lag t_r. That is, let g(t) be evaluated at the nondimensional time

$$t = (1 - F_r) t_c + F_r t_r \qquad (7)$$

where F_r is the fraction of randomness and t_c the fraction of the oscillation period elapsed assuming that all of the eddies fluctuate in phase. F_r equal to zero implies 100% correlations, while F_r equal to one implies complete randomness to the fluctuations.

The TTNH radiation model (Grosshandler, 1979) is coupled to the combustion model in a new computer code called radiation from intermittent flames (RIF) (Grosshandler, 1984). The radiant intensity is computed twice. First, the time averaged values of temperature and composition are found based upon the specified time variation of the Shvab-Zeldovich parameter. (These values represent the type measured with large thermocouples and gas probe sampling systems, which are unresponsive to the local fluctuations.) Using this profile, TTNH computes what is called the implied radiant intensity $I_1(\bar{X})$. The second computation is for the instantaneous radiant intensity $I(t)$, based upon the value of X everywhere along the line of sight at any given time. The integral of $I(t)$ over the period of oscillation is the actual intensity registed by a relatively slow response radiation flux meter. It is this term, defined as $\overline{I_2(X)}$, that is of interest to heat flux calculations. Because the radiation depends upon temperature and composition in a highly nonlinear fashion, $I_1(\bar{X})$ will, in general, deviate from $\overline{I_2(X)}$. The magnitude of the difference is represented by the intensity ratio Ψ, which is defined as

$$\Psi \equiv I_1(\bar{X})/\overline{I_2(X)} \qquad (8)$$

The parameters describing the combustion system are read into RIF from a data file. These include the initial fuel and mass fractions, temperatures, and mass fluxes; the total pressure; the equilibrium adiabatic temperature and composition of CO_2, H_2O, and CO; the empirical constants for the soot concentration [Eq. (3)]; the fluctuation amplitude fraction f and the random fraction F_r; and the Shvab-Zel'dovich parameter as a function of pathlength along a given line of sight. If the program is being used to interpret available experimental data, then the mean and rms fluctuating temperature can also be inserted.

When the mean temperature and rms fluctuation are specified, the corresponding values for the mean and rms fluctuation of the local Shvab-Zel'dovich parameter can be calculated. This is straight forward if the fluctuating temperature is small relative to the local mean temperature, in which case $\bar{X}$ is equal to the nondimensional mean temperature $\bar{\theta}$ and ΔX_{rms} is equal to the nondimensional rms fluctuating temperature θ'. That is,

$$\bar{X} = \bar{\theta} \equiv \frac{T_M - \bar{T}}{T_M - T^*} \tag{9}$$

and

$$\Delta X_{rms} = \theta' \equiv \frac{\Delta T_{rms}}{T_M - T^*} \tag{10}$$

During the oscillation period, if ΔT_{rms} is sufficiently large, the instantaneous value of X changes sign.
However, $\bar{\theta}$ and θ', always remain nonnegative. In the
region of large fluctuations, which is found when $\bar{\theta}/\theta'$ is
less than $\sqrt{3}$, the following expressions must be substituted
for Eqs. (9) and (10):

$$\Delta X_{rms} = \theta'\left\{ \frac{\sqrt{3}}{2} \left(\frac{\theta}{\theta'}\right) \pm \left[\frac{1}{4} \left(\frac{\theta}{\theta'}\right)^2 - \frac{1}{2}\right]^{1/2}\right\} \left(\frac{\theta}{\theta'} < 3\right) \tag{11}$$

$$\bar{X} = \pm \left[\bar{\theta}^2 + \theta' \left(1 - \frac{\Delta X_{rms}}{\theta'}\right)\right]^{1/2} \tag{12}$$

The proper sign is chosen for $\bar{X}$ based upon whether or not
the local conditions are fuel-rich or fuel-lean.
   With $\bar{X}$ and ΔX_{rms} established for a given position
along the path, the program then proceeds to evaluate the
mean value of the fuel mass fraction. This, too, varies
depending on the magnitude of the sawtooth fluctuation.

$$\bar{Y}_f = \bar{X}, \qquad\qquad\qquad \bar{X}/\Delta X_{rms} \geq \sqrt{3} \tag{13}$$

$$\bar{Y}_f = \frac{(\bar{X} + \sqrt{3}\Delta X_{rms})}{4\sqrt{3}\Delta X_{rms}}, \qquad \sqrt{3} > \frac{\bar{X}}{\Delta X_{rms}} > -\sqrt{3} \tag{14}$$

$$\bar{Y}_f = 0, \qquad\qquad\qquad -\sqrt{3} \geq \bar{X}/\Delta X_{rms} \tag{15}$$

The remaining mass fractions follow directly,

$$\bar{Y}_o = (\bar{Y}_f - \bar{X})\, Y_o^* \tag{16}$$

$$\bar{Y}_i = 1 - (Y_o^* - 1)\bar{X}/2 - (Y_o^* + 1)/2 \tag{17}$$

$$\bar{Y}_p = 1 - (\bar{Y}_f + \bar{Y}_o + \bar{Y}_i) \qquad (18)$$

where $\bar{Y}_i$ accounts for inert species such as N_2 or Ar.

The local mean temperature, if not inserted as experimental data, is then calculated. The parameter t_o is the fraction of the oscillation period that X is positive, in which case the temperature is nondimensionalized relative to T_f^*. When t_o is less than unity, the oxidizer temperature T^* is included as the reference temperature. The result is that

$$\bar{T} = t_o T_f^* + (1-t_o)T_o^* + (T_M-T_f^*)(t_o-\bar{Y}_f) + (T_M-T_o^*)(1-t_o-\bar{Y}_o/Y_o^*)$$

$$(19)$$

The local soot absorption coefficient is given in terms of the instantaneous values of Y_f, Y_p, and T. The expression must be integrated numerically to find the time mean since a closed form expression does not exist even for the simple sawtooth oscillation. In terms of $\bar{X}$ and g(t), the mean soot absorption coefficient is given by

$$a_{\lambda,s} = C_1(1+Y_o^*) \int_0^{t_o} [\bar{X}+\sqrt{3}\Delta X_{rms} g(t)]^{C_2} [1-\bar{X}-\sqrt{3}\Delta X_{rms} g(t)]$$

$$\times \exp(\frac{C_3}{\bar{T}})dt \qquad (20)$$

Finally, the time mean partial pressures of CO_2, H_2O, CO, and fuel are found from the product and fuel mass fractions, the molecular weight of the mixture, and the total pressure. The fuel is treated as if it has the radiative properties of methane on a mass basis, so that for higher molecular weight fuels, the effective partial pressure is greatly increased. This technique is necessitated by the lack of radiative models for the hydrocarbon species. The partial pressures, along with the mean temperature and soot absorption coefficient, are determined for all positions along the nonhomogeneous line of sight.

The instantaneous mass fractions and temperatures are found by using Eq. (4) and the sawtooth expression for g(t), if the value of X is specified.

$$g(t) = 1 - 4 (t_c + F_r t_r), \quad 0 < (t_c + F_r t_r) < .5 \qquad (21)$$

Only one half of the period is required since the oscillations are symmetric in time. The random component of time is found from a simple random number generator.

The partial pressures are established as before so that TTNH can then be called for each instant in time. The

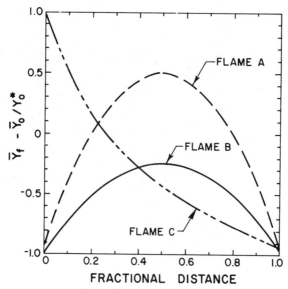

Fig. 2 Profiles of $(\bar{Y}_f - \bar{Y}_o/Y_o^*)$ plotted vs. nondimensional distance through the flame.

instantaneous value of radiance is saved and the values are summed over the half-oscillation period to determine I_2, the true time mean intensity. The rms fluctuating component of $I(t)$ is also found and the ratio of $I_1:I_2$ is taken to establish the fraction of the true intensity represented by a calculation based on the time mean profiles. A complete program listing and additional details are given by Grosshandler (1984).

Numerical Results

The program RIF was used to generate three different synthetic flames. The profiles of $(\bar{Y}_f - \bar{Y}_o/Y_o^*)$ chosen can be seen in Fig. 2. For flame A, the profile is symmetric with a fuel-rich center. Flame B is also symmetric, but nowhere does $(\bar{Y}_f - \bar{Y}_o/Y_o^*)$ exceed zero. In flame C, an asymmetric structure was chosen, with a cool fuel-rich boundary on one side. The fluctuation amplitude was assumed to be 0.5 for the flames and the number of elements along the line of sight was chosen as 20. Note that since, in general, $b \neq 1$, the mean Shvab-Zel'dovich parameter is found by subtracting $(1-b)\bar{Y}_o/Y_o^*$ from the parameter plotted in Fig. 2.

Figures 3-5 show the mean temperature, mole fraction of products, mole fraction of fuel, and soot volume frac-

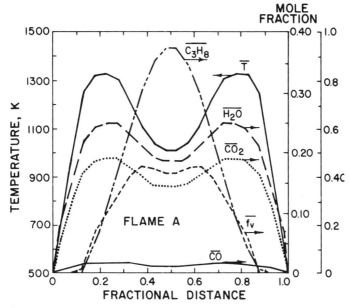

Fig. 3   Mean temperature and composition distribution corresponding to flame A.

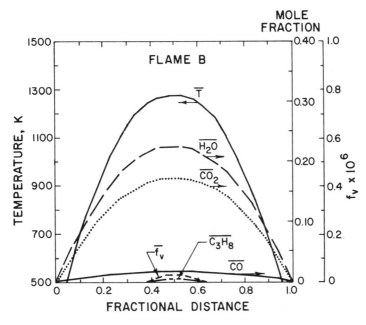

Fig. 4   Mean temperature and composition distribution corresponding to flame B.

tion vs. the fractional distance along the optical path as
calculated by RIF. In generating the profiles, the fuel
was treated as propane entering at 400 K, the oxidizer as
55% O_2 in N_2 at 350 K, and a system pressure of 0.95 atm.
The volume fraction of soot was found from the soot absorption coefficient with C_1 = 8.0 m^{-1}, C_2 = 1.0, and C_3 =
-1800 K.

The peak average temperature in flame A (Fig. 3) is
1330 K, well under the adiabatic equilibrium temperature.
This decrease in temperature is due to a combination of the
heat loss and the averaging effect of the intermittency.
The peak temperature occurs 20 and 80% along the optical
line of sight. At the fuel-rich center, the mean temperature drops to 1010 K and the mean fuel mole fraction
reaches its maximum of 0.38. The products of combustion
all follow the shape of the temperature profile, reaching
their lowest values at the near and far boundaries. The
soot volume fraction, which depends strongly on temperature
but also varies directly with the fuel and product mass
fractions, reaches an average maximum of 0.44×10^{-6} at a
location inside of the peak temperature but offset from the
centerline.

The average composition and temperature profiles of
flame B (Fig. 4) are singly peaked at the centerline and
drop symmetrically to their lowest values at the edges of

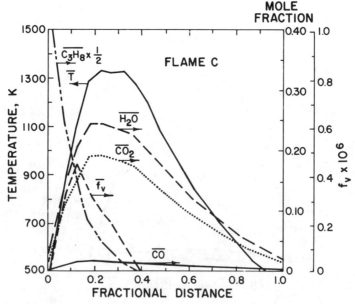

Fig. 5 Mean temperature and composition distribution corresponding to flame C.

the flame. Notice that, while the value of $(\bar{Y}_f - \bar{Y}_o/Y^*_o)$ on the centerline is considerably negative (indicating that the flame has excess oxidizer, on the average, at this point), a nonzero concentration of soot and fuel also exists. This is contrary to what can occur with a laminar diffusion flame, in which case the thin flame model requires that fuel and oxidizer cannot coexist in space. However, experimental measurements often show a finite value for oxygen on the fuel-rich side of a turbulent flame and a finite amount of fuel in some oxidizer-rich regions.

The average profiles for flame C are shown in Fig. 5. The mole fraction of propane approaches 1.0 on the left bounding surface (to which the radiant intensity is to be calculated), and is completely consumed within 40% of the total pathlength through the flame. The temperature reaches a peak of 1330 K about one-quarter of the way along the line of sight, then drops monotonically to the oxidizer temperature. When the six curves in Fig. 5 (or Figs. 3 and 4) are compared to the corresponding single curve in Fig. 2, the simplicity of the Shvab-Zel'dovich formalism and the attraction of the combustion model chosen are apparent.

Figures 6 and 7 are examples of instantaneous calculations that were made by having RIF print out intermediate

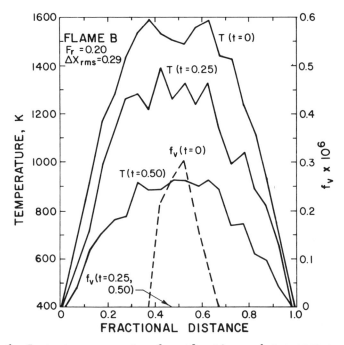

Fig. 6   Instantaneous soot volume fraction and temperature distribution corresponding to three different times in flame B.

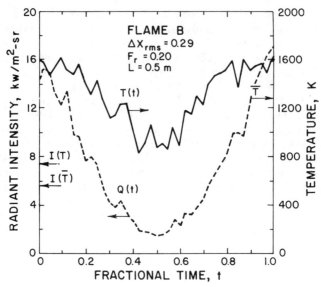

Fig. 7 Instantaneous radiant intensity and centerline temperature, as a function of time, for flame B.

values of temperature, concentration, and intensity. The profile corresponding to flame B has been used along a 0.5 m pathlength. For this calculation, the random fraction was assumed to be 0.20. In Fig. 6, the instantaneous temperature and soot volume fraction have been plotted as a function of position for three different fractions of the oscillation period: t = 0, 0.25, and 0.50. The temperature can be seen to be affected greatly by the oscillations and at all locations decreases with time. The soot volume fraction is nonzero only early in the cycle, where it reaches a maximum about 10 times that of its peak average (compare to Fig. 4).

    The solid line in Fig. 7 is the instantaneous temperature on the centerline of the flame plotted vs. the fractional time over one complete period. The intensity decreases during the first half of the period about in phase with the large scale sawtooth oscillations of $\bar{X}$ and then increases back up to a point close to the initial temperature during the second half of the period. The temperature varies from a minimum of 820 K to a maximum of 1610 K during one period, with an average centerline temperature of 1270 K. There is a significant small-scale structure on top of the large sawtooth oscillations caused by the 20% randomness. This ensures that the temperature profile will be asymmetric in time (Fig. 7), as well as in space (Fig. 6). While the temperature trace does not look unlike many thermocouple measurements in real flames, it has to be remembered that it is somewhat arbitrary since the large- and small-scale structures can be manipulated at will by adjusting the number of time intervals, the fluctuation amplitude, and the random fraction.

The instantaneous radiant intensity leaving the boundary of flame B is plotted as the dotted line in Fig. 7. The intensity also follows in phase with $\bar{X}$. Much of the small scale fluctuation exhibited by the temperature is damped in the intensity curve since it represents a spatial average of 20 elements along the optical path. The implied intensity $I_1(\bar{X})$, based upon the mean temperature and composition, is 5.6 $kW/m^2 \cdot sr$, while the actual mean intensity is 7.4 $kW/m^2 \cdot sr$, or $\Psi = 0.76$. The fluctuations are thus responsible for a 24% increase in the radiant emission.

The sensitivity of the radiation model to the following parameters has been investigated in a systematic analysis:

  1) Oscillation form (sawtooth, sinusoidal, and square wave).
  2) Random fraction ($F_r = 0 - 1.0$).
  3) Oscillation amplitude ($f = 0 - 1.0$).
  4) Total pathlength (0.25 - 2.0 m).
  5) Soot absorption (no soot, light soot, and heavy soot).

The baseline condition is defined as having sawtooth oscillations, a random fraction of 0.0, a fluctuation parameter of 0.5, a light soot loading, and a pathlength of 0.5 m. The calculations have been performed on each of the three flame structures (A-C) to ensure that any conclusions are general in nature.

The significance of the random fraction can be seen in Fig. 8, where the radiant intensity and intensity ratio are plotted vs. F_r. Flame C emits the highest level of radiation, irrespective of the random fraction, followed closely by flame A. Flame B emits less than half as much and is most sensitive to F_r. Its intensity increases by 18% as the oscillations go from perfectly correlated ($F_r = 0$) to completely random ($F_r = 1.0$). The intensity ratio Ψ drops over the same range of F_r from a high of 0.83 to a low of 0.68. The corresponding increase in intensity and decrease in Ψ for flame C is considerably less. For flame A, the value of Ψ increases only sightly with F_r. The conclusion is that, all else being equal, the intensity ratio is not very sensitive to whether or not the fluctuations are in phase or random. However, as indicated by the results from flame B, there may be circumstances where modeling the

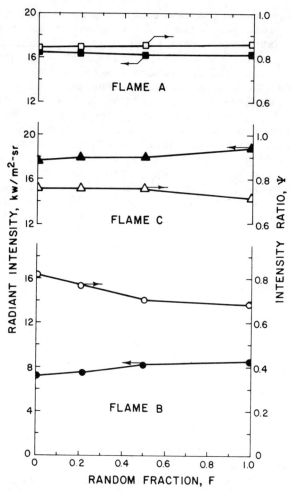

Fig. 8   Significance of the random fraction as it effects the radiant intensity and intensity ratio for each of the three flames.

correlation function more accurately is important for improved radiation predictions. (Note that the implied intensity $I_1(\bar{X})$ is not affected by F_r since the randomness must average to zero over a sufficient number of oscillation periods.)

Three functions were chosen to model the periodic oscillations. The analysis of the effect of the waveform on the intensity was made with all other conditions set to the baseline. The results are presented in Table 1. The sawtooth and sinusoidal oscillations yielded essentially the same values for actual intensity and intensity ratio in all three flames. The radiant intensities assuming square-wave oscillations were slightly less, while Ψ was slightly greater. The conclusion is that wave form is unimportant for the conditions considered here and that it can probably be ignored as a significant parameter for the radiative heat transfer in general.

The oscillation amplitude causes an increase in radiant intensity over that implied by the time-mean profiles. Figure 9 quantifies this for each of the three flame structures. The random fraction is fixed at zero and sawtooth oscillations are maintained as the fluctuation parameter is varied between 0 and 1. In all three flames, the intensity ratio drops monotonically from Ψ = 1.0 at f = 0 to a value near 0.5 at f = 1. Flame C is the most dramatic, reaching a minimum of 0.43, but the curves are strikingly similar in shape. This is not the case for the actual radiant intensity. In flames A and C, $\overline{I_2(X)}$ decreases 2 - 3 kW/m$^2 \cdot$ sr with increasing amplitude of fluctuation, but flame B increases its total emission about 50% over the same variation in f.

The fluctuation parameter is returned to its base value of 0.5 for the evaluation of pathlength shown in Fig. 10. The significant effect of pathlength on actual intensity is as expected, with the emission increasing by a factor of two to three in flames A-C as the optical path is increased from 0.25 to 2.0 m. This variation is in contrast to the nearly fixed values of Ψ computed for each flame as the pathlength is changed. The conclusion to be drawn from this result is clear, if not somewhat surprising: the optical depth alone is irrelevant in formulating the intensity ratio.

All of the flames have been modeled as nonsooting, lightly sooting (the base case), and heavily sooting for a fixed pathlength of 0.5 m, a random fraction of 0, a fluctuation amplitude of 0.5, and a sawtooth wave form. These results are summarized in Table 2. The degree of sootiness

Table 1   Effect of oscillation form

	Flame	Sawtooth	Sinusoidal	Square-wave
		\multicolumn{3}{c|}{Oscillation form}		
Radiant	A	16.49	16.40	15.79
intensity,	B	7.08	7.07	6.68
kW/m^2 sr	C	17.51	17.43	17.22
Intensity	A	0.839	0.839	0.861
ratio,	B	0.807	0.791	0.842
	C	0.775	0.776	0.779

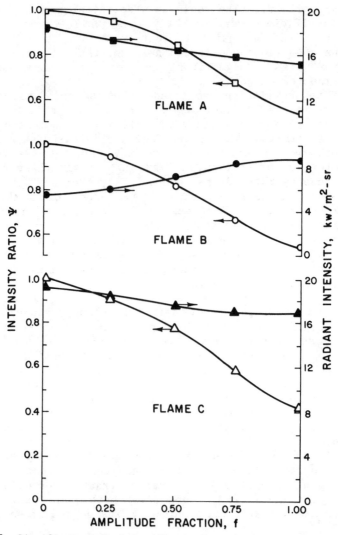

Fig. 9 Significance of the amplitude fraction as it effects the radiant intensity and intensity ratio for each of the three flames.

has been controlled by varying the constant of proportionality in Eq. (3). The heavily sooting flame A emits four times the radiation as its nonsooting counterpart, but the intensity ratio is identical for the two calculations. The insensitivity of Ψ to the soot absorption coefficient is evident in flames B and C as well, with two notable exceptions: the heaviest sooting cases of each of the two flames. For flame B, Ψ decreases substantially at the heaviest loading to .713, while the opposite trend is observed in flame C, where Ψ approaches one at the most sooting condition.

Examination of the profiles in Figs. 4 and 5 leads to a plausible explanation for the above conflicting results. For flame B, a twentyfold increase in soot absorption occurs just at the highest temperature point in the flame, which also corresponds to the the region with the absolute maximum in fluctuations. The flame is relatively thin near its edges so that the combined effects of the higher soot loads and the fluctuations are clearly visible at the boundary. On the other hand, Fig. 5 shows that the greatest effect of increasing the soot volume fraction in flame C is limited to a relatively cool, absorbing region near the point from which the radiation is emanating. As the fuel mole fraction is quite high, the absolute magnitude of the temperature fluctuation is relatively low. Thus, the effective rms deviation is less in this case.

Measurements of the soot absorption in a laminar polymethylmethacrylate flame (Grosshandler and Vantelon, 1985) suggest that the soot level may be more sensitive to the concentration of fuel and oxidation rate than the expression used in the previous analysis would indicate. Additional calculations have been performed with the exponential term in the soot absorption coefficient changed from −1800 to +1700 K and the first-order dependence on Y_f replaced with a 2.5 power. The proportionality constant C_1 has been varied between 0 and 200. The flame used for comparison, flame D, is similar in form to flame A, but the range in variation of X is a bit less, as can be noted in the description of the flame given in Table 3.

The intensity ratio, actual intensity, total emittance and corresponding radiative temperature T_R are plotted vs.

Table 2 Effect of soot absorption coefficient

| | | Soot proportionality constant, C_1 | | |
	Flame	0.0 (no soot)	8.0 (light soot)	160 (heavy soot)
Radiant intensity, kW/m²-sr	A	12.71	16.49	44.80
	B	6.71	7.08	11.68
	C	16.66	17.51	21.65
Intensity ratio,	A	0.823	0.839	0.823
	B	0.830	0.807	0.713
	C	0.750	0.775	0.974

Table 3 Description of flame D as put into RIF program

Fuel Conditions
　　Fuel: methane
　　Inlet fuel temperature: 400 K
　　Steady-state expression for soot:

$$a_s = 5 \ Y_f^{2.5} Y_p \ \exp(1700/T), \ m^{-1}$$

Environmental Conditions:
　　Inlet air temperature: 300 K
　　Operating pressure: 0.95 atm
　　Inlet oxygen mass fraction: 0.55
　　Wall temperature: 300 K

Estimated Maximum Intermittent Temperature: 1725 K

Fluctuation Parameters:
　　Sawtooth oscillations
　　Random fraction: 0.00
　　Fluctuation parameter: 0.50

Inlet Temperature Profile:

Distance, m	T,K	ΔT_{rm}	$\bar{Y}_f - \bar{Y}_o/Y_o^{\star}$
0	656	102.8	−0.750
0.05	1169	250.9	−0.390
0.10	1395	203.3	−0.110
0.15	1404	202.3	0.090
0.20	1387	246.1	0.210
0.25	1365	259.4	0.250
0.30	1387	246.1	0.210
0.35	1404	202.3	0.090
0.40	1395	203.3	−0.110
0.45	1169	250.9	−0.390
0.50	656	102.8	−0.750

C_1 in Fig. 11. The total emittance is computed in TTNH and $T_R = (\pi I_2/\varepsilon_R \sigma)^{1/4}$. While the effective radiative temperature decreases with C_1, all the other parameters increase monotonically with the soot level. The most interesting point is that Ψ actually exceeds unity, reaching a value of 1.68 at the largest value of C_1.

Conclusions regarding the effect of the soot on the intensity ratio are difficult to draw. As in the case of pathlength, it appears that it is not the optical thickness alone that exerts the most influence. However, if the absorption coefficient in one region of the flame, which is visible to the boundary, changes significantly more than another when the overall optical thickness of the flame is modified, it can appear as if it is the optical thickness that is controlling Ψ. In general, it is not possible to

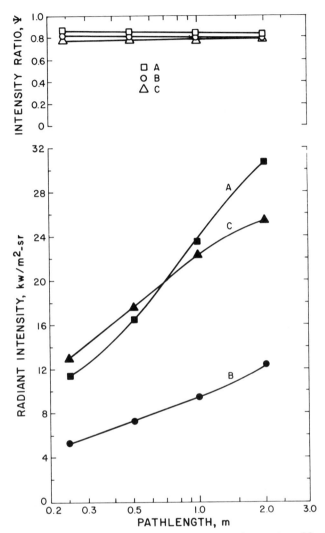

Fig. 10 Significance of the total pathlength as it effects the intensity and intensity ratio in each of the three flames.

predict whether increasing the soot level will increase, decrease, or not effect the intensity ratio, since it is a combination of the nonhomogeneity, the fluctuating nature, and the soot chemistry in the flame that must be considered.

Discussion of Results

The intensity ratios of all the cases investigated are plotted together in Fig. 12 vs. the fluctuation parameter

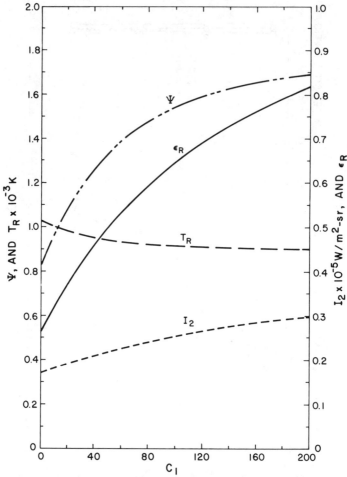

Fig. 11 Significance of the sooting tendency as it effects the radiant intensity, intensity ratio, the effective radiating temperature, and flame emittance for flame D.

f. There is a rather well-defined lower limit to the data, decreasing from 1.0 at f = 0 to less than 0.5 at f = 1.0. This trend and the fact that some of the calculations predict a value for Ψ greater than one both need explanation.

Consider a homogeneous, gray flame of emittance ε undergoinging periodic fluctuations in its Shvab-Zel'dovich parameter. The intensity ratio can be written in terms of the time mean values as

$$\Psi = I_1(\bar{X}) / \overline{I_2(X)} = \overline{\varepsilon \bar{T}^4} / \overline{\varepsilon T^4} \tag{22}$$

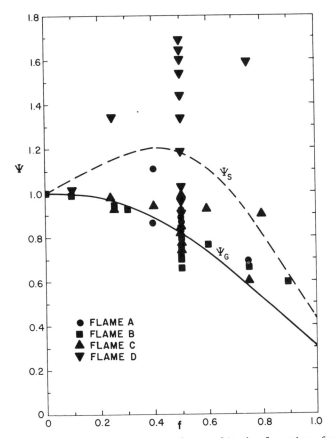

Fig. 12   Intensity ratio versus the amplitude fraction for all flame conditions considered (Ψ_s and Ψ_g represent the limiting cases for pure soot and pure gaseous radiation, respectively).

If the flame is strongly sooting, the greatest sensitivity of the emittance to the fluctuations occurs as X changes sign. Assume that the absorption coefficient is dominated by the soot rather than the gaseous products, but that the flame is not optically thick. Then the emittance can be written as being related to the mass fractions of fuel and product. To facilitate understanding, let b equal 1.0, choose the relation to be first order in both Y_f and Y_p, and neglect any exponential dependence upon temperature. With $Y^*_o = 1.0$, one may then write the emittance due to the soot as

$$\varepsilon_s(t) \propto X\,(1 - X), \quad X > 0 \qquad (23)$$

and

$$\varepsilon_s(t) = 0, \quad X \leq 0 \tag{24}$$

The expression for the instantaneous value of X (Eq. 4) yields

$$\varepsilon_s(t) = \beta\{\bar{X} + f(1 - |\bar{x}|)g - \bar{x}^2 - [f(1 - |\bar{x}|)g]^2$$

$$- 2\bar{X}f(1 - |\bar{x}|)g\} \tag{25}$$

where β is the proportionality constant. The region of most interest is where X is near zero. Confining the analysis to this region and using the sawtooth expression for g, the emittance can be approximated by

$$\varepsilon_s(t) = \beta \ f(1 - 4t) \ [1 - f(1 - 4t)] \tag{26}$$

The mean value for ε_s is found by integrating Eq. (26) over the entire interval $t = 0 - 0.5$ recognizing that $\varepsilon_s = 0$ for $t > 0.25$. The result is

$$\overline{\varepsilon_s} = \beta \ (f/4 - f^2/6) \tag{27}$$

The instantaneous temperature can be written as

$$T(t) = T_M[1 - |\bar{x}| + f(1 - |\bar{x}|) \ (1 - ft)] \tag{28}$$

if T_M is much greater than T^*_o or T^*_f. The mean temperature over the entire oscillation period is found as follows for the particular case of X = 0:

$$\bar{T} = \frac{T_M}{1/2} \ [\int_0^{1/4} (1 - f + 4 \ ft)dt + \int_{1/4}^{1/2} (1 + f - 4ft)dt] \tag{29}$$

which integrates to

$$\bar{T} = T_M \ (1 - f/2) \tag{30}$$

The denominator of Eq. (22) is found in a corresponding manner from the product of Eq. (26) and fourth power of Eq. (28).

$$\overline{\varepsilon_s T^4} = \frac{1}{1/2} \ [\int_0^{1/4} \beta f(1-4t)(1-f+4ft)T_M^4 \ (1-f+4ft)^4 dt + \int_{1/4}^{1/2} 0 \ dt] \tag{31}$$

or

$$\overline{\epsilon_s T^4} = \frac{\beta T_M^4 f}{2} (\frac{1}{4} - \frac{5f}{6} + \frac{5f^2}{4} - f^3 + \frac{5f^4}{12} - \frac{f^5}{14}) \qquad (32)$$

The intensity ratio based upon soot radiation alone Ψ_s, thus, reduces to

$$\Psi_s = \frac{(1 - f/2)^4 (3 - 2f)}{(3 - 10f + 15f^2 - 12f^3 + 5f^4 - 6f^5/7)} \qquad (33)$$

Equation (33) yields values for Ψ_s greater than or equal to unity for all values of f less than 0.68, with a maximum of 1.20 near f = 0.40. In other words, there are cases where the intensity based upon the time mean concentration and temperature profiles will exceed the actual average intensity; those are most likely to occur in flames in which the soot absorption coefficient diminishes quickly as X approaches 0. It is the fluctuations that are responsible for this effect, but the value of Ψ_s can be enhanced by the nonisothermal nature of the media.

When the gaseous radiation dominates the flame emission, the emittance is about proportional to the mass of the products of combustion along the optical path. This increases directly with Y_p and inversely with the temperature (due to density changes). These counterbalancing forces yield a constant value for the gas emittance $\epsilon_g(t)$ during the fluctuations. This means that it is necessary to evaluate only the integral of the fourth power of the instantaneous temperature, given by Eq. 28, to determine the actual intensity. For the general case of any value for $\bar{X}$, it can then be shown that the intensity ratio for a nonluminous flame Ψ_g is given by

$$\Psi_g = \frac{5[2f(1 - |\bar{x}|) - \bar{x}^2 - f^2(1 - |\bar{x}|)^2]^4}{[2f(1-|\bar{x}|)]\{2-[1-(\bar{x}+f(1-|\bar{x}|))]^5 - [1+(\bar{x}-f(1-|\bar{x}|))]^5\}} \qquad (34)$$

To compare this expression to the expression resulting from the soot radiation, $\bar{X}$ can be set to zero, so that one gets

$$\Psi_g = \frac{5f(1 - f/2)^4}{1 - (1 - f)^5} \qquad (35)$$

Equations (33) and (35) are plotted in Fig. 12. The simplified analysis assuming a homogeneous gray media with an average Shvab-Zel'dovich parameter of zero is able to explain the trends observable in the data. However, the nonhomogeneous calculations sometimes give values for Ψ

considerably greater than the maximum predicted by Eq. (33). The drop in Ψ with f is also not as severe as indicated by Eq. (35). The large spread in values for Ψ at intermediate values of f (0.66 - 1.66 at f = 0.5) indicates that a realistic model for soot concentration and a knowledge of the fluctuation amplitude fraction are required before accurate prediction of the radiant intensity will be possible.

Conclusion

Fluctuations of concentration and temperature are known to be severe within many flame structures of practical interest. The time-varying and average radiant intensities resulting from fluctuating diffusion flames have been investigated by coupling a combustion model to a nonhomogeneous radiation model in a new computer code called RIF (radiation from intermittent flames). A parametric study using synthetic flames has been performed to assess the importance of the amplitude, the randomness, and the waveform of the fluctuations in determining the actual radiant intensity. Also investigated have been the significance of the form of the expression for the soot formation, the scale of the flame, and the average flame structure. The program RIF can be used either to generate synthetic flames to test other intermittency models or to analyze experimental results when the mean values of temperature and their rms deviation have been measured.

The parametric study has shown the following:

1) The actual intensity radiating from a fluctuating flame can exceed the estimated intensity based upon the time mean temperature and concentration profiles by more than a factor of two.

2) In some strongly sooting flames, it is possible for the actual intensity to be significantly less than that which is implied from the time mean profiles.

3) The ratio of the implied intensity to the actual time mean value is sensitive to the rms deviation in temperature and the expression used for the formation of soot.

4) The ratio of the implied intensity to the actual time mean value is not very dependent upon the functional form of the fluctuations, their randomness, or the scale of the flame.

Acknowledgment

This work has been supported largely by the Groupe de Recherches de Chimie Physique de la Combustion, ERA No. 160 of the Centre National de la Recherche Scientifique. The

authors also acknowledge the financial assistance provided
by the National Science Foundation through Grant MEA-
8101953, Dr. Win Aung, Heat Transfer Program Director.

References

Amin, M. and Foster, P.J. (1973) Fluctuations in the transmittance
    of a turbulent propane jet flame. European Symposium on
    Combustion, pp. 530-535.

Becker, H.A. (1975) Effects of concentration fluctuations in
    turbulent diffusion flames. Fifteenth Symposium (Interna-
    tional) on Combustion, pp. 601-615. The Combustion Institute,
    Pittsburgh, PA.

de Ris, J. (1979) Fire radiation--A review. Seventeenth Symposium
    (International) on Combustion, p. 1004. The Combustion Insti-
    tute, Pittsburgh, PA.

Drake, M.C., Bilger, R.W., and Starner, S.H. (1982) Raman measure-
    ments and conserved scalar modeling in turbulent diffusion
    flames. Nineteenth Symposium (International) on Combustion,
    pp. 459-485. The Combustion Institute, Pittsburgh, PA.

Foster, P.J. (1969) The relation of time-mean transmission of
    turbulent flames to optical depth. J. Inst. Fuel 42, 179-182.

Grosshandler, W. (1979) Radiative heat transfer in nonhomogeneous
    fires. RC 79-BT-9, Factory Mutual Research Corp.

Grosshandler, W. (1984) Computing radiation in real flames.
    Final Project Report, Groupe de Recherches de Chimie Physique
    de la Combustion, CNRS, Universite de Poitiers, France, June
    1984.

Grosshandler, W. and Modak, A. (1981) Radiation from nonhomoge-
    neous combustion products. Eighteenth Symposium (Interna-
    tional) on Combustion, p. 601, The Combustion Institute,
    Pittsburgh, PA.

Grosshandler, W. and Nguyen, H. (1985) Application of the total
    transmittance nonhomogeneous radiation model to methane
    combustion. J. Heat Transfer 107, 445-450.

Grosshandler, W.L. and Vantelon, J.P. (1985) Predicting soot
    radiation in laminar diffusion flames. Combust. Sci. Tech-
    nol. 44, 125.

Hottel, H. (1954) Heat Transmission (edited by W.H. McAdams) 3rd
    ed. McGraw-Hill Book Co., New York.

Kinoshita, C. and Pagni, P. (1981) Stagnation-point combustion
    with radiation. Eighteenth Symposium (International) on
    Combustion, p. 1415. The Combustion Institute, Pittsburgh,
    PA.

Leckner, B. (1973) Some elements of radiative heat transfer calculations in flames and gases. Arch. Procesow Spalania 4, 387.

Modak, A. (1977) Thermal radiation from pool fires, Combust. Flame 29, 177-192.

Orloff, L. (1981) Simplified radiation modeling of pool fires. Eighteenth Symposium (International) on Combustion, p. 549. The Combustion Institute, Pittsburgh, PA.

Orloff, L., de Ris, J., and Markstein, G. (1975) Upward turbulent fire spread and burning of fuel surfaces. Fifteenth Symposium (International) on Combustion, p. 183. The Combustion Institute, Pittsburgh, PA.

Orloff, L., Modak, A., and Alpert, R. (1977) Burning of large-scale vertical surfaces. Sixteenth Symposium (International) on Combustion, p. 1345. The Combustion Institute, Pittsburgh, PA.

Sibulkin, M., Kulkarni, A., and Annamalai, K. (1981) Effects of radiation on the burning of vertical fuel surfaces. Eighteenth Symposium (International) on Combustion, p. 611. The Combustion Institute, Pittsburgh, PA.

Tamanini, F. (1978) A numerical model for the prediction of radiation controlled turbulent wall fires. RC 78-BT-20, Factory Mutual Research Corp.

Tan, E. and Foster, P.J. (1978) Radiation through a turbulent medium. Paper presented at 6th International Heat Transfer Conference, Toronto.

Williams, F. (1965) Combustion Theory. Addison-Wesley, Reading, MA.

Yee, D. (1982) An experimental study of turbulent mixing in nonburning and burning hydrogen-air coaxial jets. PhD Dissertation, Dept. of Mechanical Engineering, Washington State University, Pullman.

Chapter II. Heterogeneous Combustion

Lean Flammability Limits of Hybrid Mixtures

M. Gaug,* R. Knystautas,** and J.H.S. Lee**
McGill University, Montreal, Canada
and
L.S. Nelson,† W.B. Benedick,‡ and J.E. Shepherd§
Sandia National Laboratories, Albuquerque, New Mexico

Abstract

This paper investigates the lean flammability limits of gaseous fuel-dust-air mixtures. The gaseous fuels used were methane and hydrogen. Two different dusts were used in an attempt to determine the influence of volatile content on hybrid limits. Cornstarch was selected for its high volatile content, while iron dust was chosen for its complete lack of volatiles. The experiments were performed in two different apparatus: a 0.18-m^3 cylindrical vessel, and to investigate the effects of scaling, a 5.1-m^3 cylindrical tank. The dust was dispersed by means of an air blast into a hemispherical dust receptacle placed at the bottom of the apparatus. The criterion for self-sustained flame propagation was the evidence of an overpressure in excess of an arbitrarily set limit. For hydrogen-air-cornstarch mixtures, the lean limit of H$_2$ alone was found to be 4.25% by volume, while that of cornstarch alone was observed to be 70 g/m^3. A line joining these two limits on a graph represents the Le Chatelier limit law (formulated for gaseous homogeneous fuels), which is based on energetics. For the range of mixtures studied, the

---

Presented at 10th ICDERS, Berkeley, California August 4-9, 1985.
  * Research Engineer, Dept. of Mechanical Engineering; currently with SPAR Aerospace.
  **Professor, Dept. of Mechanical Engineering.
  † Research Scientist, Severe Accident Containment Response Division.
  ‡ Research Scientist, Shock Wave and Explosive Physics Division.
  § Research Scientist, Fluid Mechanics and Heat Transfer Division.

155

extension of Le Chatelier's law to hybrid mixtures was shown
to be invalid. The deviation of the results from the Le
Chatelier law indicates that hybrid lean limits depend more
on the mechanism of propagation than on energetics. In
fact, it can be seen that in traversing the entire hybrid
mixture range from pure dust to pure gas, a crossover region
can be identified in which the mode of propagation switches
from dust to gas.

Introduction

The study of flammability limits and combustion
behavior of dust-gas hybrid mixtures is of great practical
importance. It is well known that coal dust and methane gas
mixtures can produce violent explosions although the
individual concentrations of both fuels may be below their
respective lean flammability limits. This has been shown by
Nagy and Portman (1960), who demonstrated that the minimum
explosive concentration of coal dust decreases almost
linearly as the concentration of methane increases from 0%
to 5%. Bartknecht (1981) observed similar behavior for
mixtures of PVC dust and either methane or propane. He
demonstrated that for hybrid mixtures, the peak explosion
pressure, and the maximum burning rate do not occur at the
same mixture composition, which is generally true for
premixed gas explosions.

Generally, however, comparatively few investigations
have been carried out for hybrid lean flammability limits.
Although Le Chatelier formulated a rule for determining
flammability limits for homogeneous gas mixtures in the late
1800's, no such formulation exists today for determining
hybrid mixture flammability limits. However, Le Chatelier's
rule has been used as a point of departure for examining the
behavior of hybrid limits. Le Chatelier's law is based on
energetics and can be stated as follows: given a mixture of
flammable gases, if the sum of all individual fuel
concentrations C_i divided by their respective lean limit
concentrations L_i is greater than or equal to 1, then that
particular mixture is flammable ($\sum_i C_i/L_i > 1$). Due to its
normalizing character, this law may be readily extended to
include dust-gas hybrid mixtures. The linear behavior of
this law has been shown by Nagy and Portman (1960) as
mentioned previously. Hertzberg et al. (1981) also observed
linear trends for the flammability limits of methane gas
admixed with various coal dusts. However, both Nagy and
Hertzberg performed their investigations in very small
volume apparatus, a 0.33-ft^3 cylinder and 8-liter cylinder,
respectively. The size of these apparatus raises serious

doubts as to whether a true self-sustained flame propagation
was observed. Thus, the present study was performed in both
small-scale (.18 m³) and large-scale (5 m³) apparatus for
the purposes of determining the influence of chamber volume
and geometry on the flammability limits of the mixtures
studied. The large-scale apparatus was the principal
apparatus of interest, as its dimensions with respect to a
dust flame thickness were considered to be large enough to
eliminate the adverse effects of small scale. However, due
to the difficulties and extreme costs involved with running
tests in a 5-m³ vessel, the majority of tests were performed
in the small-scale .18-m³ chamber. An attempt was made to
investigate and explain the mechanisms inherent in
determining the flammability limits for hybrid mixtures.
Based on these results, a simple model is proposed to
explain the flammability limit behavior for hybrid dust-gas
mixtures.

Experimental Details

The present study was performed in two separate
apparatus, one large scale (5.1 m³) and one laboratory scale
(.18 m³).
The laboratory scale apparatus has a volume of 180
liters and is cylindrical in shape with rounded dome shaped
ends. The cylinder itself consists of a straight portion

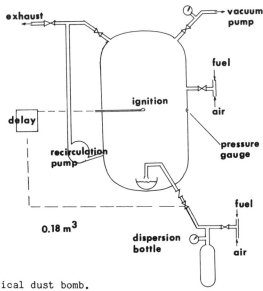

Fig. 1   0.18 m³ cylindrical dust bomb.

(48.2 cm in diameter, by 54.2 cm long) capped by two dome-shaped ends that are each 48.2 cm in diameter, with a mean equivalent cylindrical length of approximately 22 cm each. An overall view of the apparatus is shown schematically in Fig. 1. In a given experiment, the gaseous mixture in the vessel is prepared by manometric measurements of partial pressures. The contents are recirculated for 30 min with a Bellows type pump to ensure a uniform mixture. The dust is deposited in a circular cup and placed at the bottom of the vessel. The dust is then dispersed with an air blast from a 1-liter vessel pressurized to 1.5 MPa. The composition of the gas in the dispersion vessel is exactly that of the vessel itself. Ignition is achieved centrally via a glow wire wrapped with 1 g of black powder. Diagnostics involved pressure measurement in the explosion vessel via a PCB 113A24 piezoelectric transducer. A solid state delay generator was used to vary the time between dispersion and ignition. All tests were performed at atmospheric pressure.

The large scale experiments were performed in a subterranean steel pressure vessel, 5.1 m³ in volume and cylindrical in shape. The tank measures 1.2 m in diameter by 4.8 m in height and is oriented vertically. Its ends are dome shaped. The 5.1-m³ tank is equipped with an array of six aerosol dispersal cups mounted in racklike fashion along two vertical columns of three cups each about 50 cm off-axis. The cups have a vertical separation of 1.2 m with the uppermost being about 30 cm below the upper flange (Fig. 2). The dispersal is achieved by the discharge of a

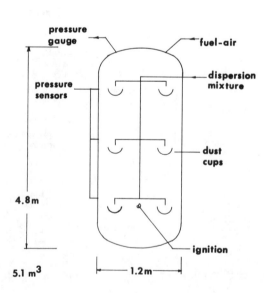

Fig. 2   5 m³ cylindrical apparatus.

100-liter pressurized vessel (2 MPa) through a tube array
into the cups. The dispersal system contains the same
mixture as that in the 5.1-m^3 tank under the test
conditions. All experiments were performed at standard
atmospheric pressure. The mixture composition was
determined by partial pressure measurements. The gas was
then mixed with a sparkless air-driven fan installed in the
upper dome of the tank, for a period of 30 min. Ignition
was achieved via an exploding length of pyrofuse (aluminum-
palladium wire) typically 100 cm in length and tightly
coiled. A delay of 1 s was used between dispersion and
ignition. Diagnostics included 18 thermocouple stations
dispersed throughout the volume of the tank and six Koolite
pressure sensors along the walls of the tank.

In the small-scale tests, both hydrogen and methane
were used as gaseous fuels. For the large-scale tests,
hydrogen alone was used. For both small-and large-scale
tests, both cornstarch and sponge iron powder were used as
the dust fuels. The cornstarch dust used had a mean
particle size of 15 μm. The mean particle size for the
iron dust used was in the range of 30-50 μm.

Results and Discussion

Constant-volume combustion properties of explosive
mixtures as well as their flammability limits are strongly
dependent on the apparatus and criteria used. Thus, it is
important to first carry out a number of tests well within
the flammability limits to determine the characteristics of
the current apparatus. For the homogeneous gas mixtures,
the particular cases of 15% H_2 in air and 9.5% methane in
air were investigated. Peak overpressures of 4 and 7 bars
respectively, were recorded as compared to the theoretical
values of 4.57 and 7.82 bars (Lee et al. 1979). Researchers
using similar apparatus have reported comparable values.
Nagy and Portman (1960) have obtained overpressures of 6.3
bars for 9.5% methane-air in a .3-ft^3 cylinder. In a
separate study, Nagy et al. (1971) reported an overpressure
of 6.8 bars for a turbulent mixture of 9.5% methane in a 1-
ft^3 cube. Benedick et al. (1984), obtained overpressures of
3.8 bars for a turbulent mixture of 15% hydrogen-air and
3.75 bars for the quiescent case in a 5-m^3 cylindrical tank
(1.2 m diam, 4.8 m long). Given the range of geometries and
volumes used, along with the degree of turbulence present
during combustion, the range of values reported is quite
reasonable.

It should be noted, that in the present experiments
mixtures were burned in a turbulent environment to simulate

the initial conditions of the hybrid mixtures studied later,
where turbulence is necessary to suspend the dust.
Following this, a pure dust mixture, also well within the
limits (i.e., 250 g/m^3 of cornstarch in air) was burned
under the same initial conditions of ignition and
turbulence. The peak overpressure recorded was 5.6 bars.
Cocks and Meyer (1979) reported a value of 3.6 bars for the
same concentration in a 20-liter sphere. Jacobson et al.
(1961) obtained overpressures of 5.6 bars in a 1.23 liter
cylinder (Hartmann bomb). Previous work performed at McGill
University (Bond et al. 1984) in a 333-liter sphere,
indicated overpressures of 5.4 bars. It is evident that for
the case of dust combustion, where the uniformity of
concentration is very difficult to guarantee, relatively
larger scatter of results is expected. Nonetheless, the
results obtained in the present apparatus, compare very
favorably with those previously mentioned.

Flammability limits of the individual fuel components
were first determined using strong pyrotechnic igniters
(100-g black powder ignited by a glow wire). The order of
magnitude of energy deposited by these igniters was about 1
kJ, determined from the overpressure generated in the vessel
by the igniter alone. The level of ignition energy was
rendered sufficiently high to ensure that true limits of
flammability were being measured (as opposed to the relative
ignitability of a dust for a given ignition energy). In
selecting the flammability criteria, it was noted that
overpressures decreased rapidly with decreasing
concentration. At some point near the lean limit, the
overpressures leveled off, essentially registering the
pressure generated by the pyrotechnic igniter. An
overpressure of 20.7 KPa was then selected as the criterion
to establish the flammability limit. This value was chosen
because it corresponded to the pressure generated by a
quiescent mixture of 4.0% hydrogen in air in the present
apparatus. 4.0% is the lean upward propagation limit for
hydrogen in air as reported by Coward and Jones (1952) and
Hertzberg (1981). Similar pressures were also obtained for
limit mixtures of 5.0% methane-air (Coward and Jones 1952;
Hertzberg 1981).

Having established the quiescent flammability limits for
the gas components, limits were then determined under the
conditions of the turbulence level generated by the
dispersion bottle discharge (henceforth referred to as
dispersion turbulence). For the case of hydrogen,
flammability limits were found to be narrower. With the
dispersion turbulence present, the hydrogen lean limit was
found to be 4.5% using the same pyrotechnic ignition source.

However, for the case of methane-air mixtures, the same lean limit of 5.0% was found for both the initially quiescent and turbulent cases. This can be explained by examining the structure of a lean limit H_2-air. At the lean limit, H_2-air flames are inherently unstable and propagate as a collection of individual flamelets. These pockets of hot gas enriched by selective diffusion are surrounded by cold unburned mixture. In a turbulent environment, the convection of large quantities of cold unburned gas into the flamelets would lower the effective concentration and hence quench the flame. This is not the case for the lean limit methane flame. At its limit, the methane-air flame was observed to still propagate as a uniform flame sheet. Hence, the increase in transport rate would tend to manifest itself as an increase in burning rate. Of course, given a high enough level of turbulence, the lean limit flame can also be quenched.

Based on the same limit criterion, the lean flammability limit of cornstarch was found to be 80 g/m^3. This is approximately double the value obtained by Jacobson et al. (1961) in a 1.23-liter Hartmann bomb. However, in the small volume of the Hartmann bomb used by Jacobson and coworkers, it is not at all certain that a self-sustained flame is present. Therefore, it is to be expected that a slightly higher lean limit dust concentration will be evident in a larger bomb. The lean limit for Fe dust in air was found to be 400 g/m^3, roughly four times that listed by Field (1982). However, Field does not report the details of vessel size or configuration, or the particle size range. It should be noted that all results are largely dependent on apparatus geometry, volume, and on ignition energy and placement. Thus, there is always a certain degree of arbitrariness associated wtih any flammability limit criterion. However, results obtained with a given criterion are meaningful within the context of its selection.

Similar steps were followed in establishing the lean limit criteria for tests performed in the large-scale 5-m^3 cylindrical tank. The criterion required to establish the lean flammability limits was much more straightforward than that for the 180-liter vessel. In all tests performed, there was a very sharp distinction noted between a "go" and a "no go." Namely, either a large overpressure was noted, or no overpressure was evidenced whatsoever. In the 5-m^3 tank, the igniter is located near the bottom of the bomb. Along with its high length to diameter ratio of about 4, the 5-m^3 tank more closely resembles a vertical flammability tube than a closed volume spherical bomb. As expected, this resulted in higher percentages of the bomb volume being

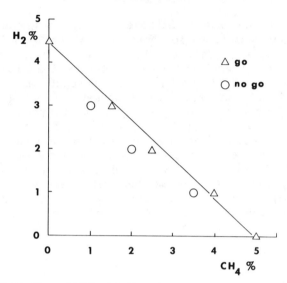

Fig. 3   Lean flammability limits of turbulent H_2-CH_4-air mixtures.

burned for a given mixture by the upward propagating flame than was the case for the .18-m^3 square cylindrical bomb.

Numerous tests performed previously in the tank showed that the lean limit for quiescent H_2-air mixtures was approximately 4.1% H_2. When the turbulence required to suspend the dust was introduced, a 4.25% H_2 mixture was found to be the minimum H_2 concentration to support combustion. The lean limits for cornstarch dust and iron dust were 70 and 275 g/m^3, respectively. Both of these values were slightly below those determined for the 180-liter bomb as expected. Evidently, more of the mixture is available for burning due to the slightly different geometry. Before determining flammability limits for hybrid mixtures, homogeneous gaseous mixtures of hydrogen and methane in air were first investigated in the present apparatus to check the validity of the Le Chatelier limit law.

Fig. 3 displays the results of flammability limit studies done in a turbulent environment for mixtures of hydrogen and methane. It can be seen that within the limits of experimental error, Le Chatelier's law is adhered to very closely. For gaseous fuels, Nagai (1929) has shown that Le Chatelier's limit law holds primarily for mixtures whose components have approximately the same lean limit flame temperature, or in terms of energetics, roughly the same limit heat of combustion per unit volume. The hydrogen-methane results shown here would seem to violate the

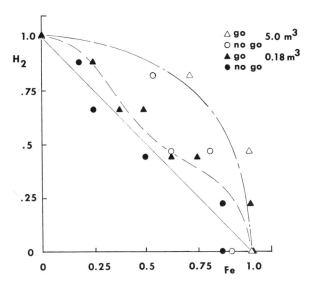

Fig. 4   Flammability limits of H_2-Fe-air hybrid mixtures for large and small scale.

observation of Nagai, since the theoretical lean hydrogen limit flame temperature is approximately 700 K, while that of methane is 1500 K. These temperatures and the energetics they represent are quite disparate. This can be explained in part by examining the mechanism by which a 4.0% H_2 flame propagates. At this concentration H_2 flames cannot propagate as uniform flame sheets; instead, they propagate as a collection of individual flamelets. Due to the very high mass diffusivity of hydrogen as compared to air, any convex curvature of the flame, however slight, will allow the hydrogen molecules to diffuse preferentially and consequently enrich the flame locally. This process of selective diffusion is believed to play a major role in the propagation of lean limit H_2 flames up to the downward propagation limit of 8.0%. This is the minimum concentration required for propagation as a flame sheet rather than as a collection of flamelets. In a hydrogen-methane air limit flame, selective diffusion will play a role, along with the kinetic characteristics of hydrogen, in allowing a flame to propagate below the energetics requirement dictated by the methane lean limit.

Figs. 4 and 5 display the results of a series of hybrid mixture flammability tests performed for both small and large scale. In both the case of H_2-Fe-air and H_2-cornstarch-air, the trends displayed in the small-scale results are amplified in the large-scale results, largely

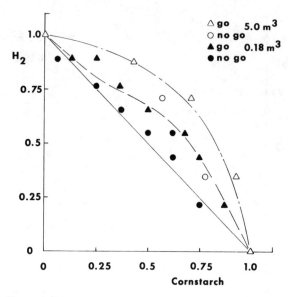

Fig. 5   Flammability limits of H_2-cornstarch-air hybrid mixtures for large and small scale.

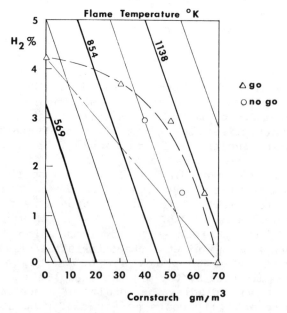

Fig. 6   5 m^3 H_2-cornstarch flammability limits with flame temperature isotherms.

due to the greater ease in determining the presence of a
self-sustained flame in the large-scale apparatus. The
greater L/D ratio and volume result in higher overpressures
for near limit flames, thus lowering the determinable dust
limit. The obvious conclusion reached upon examination of
the large-scale results is that Le Chatelier's law is simply
not valid for dust-gas hybrid mixtures. The data are
normalized with respect to the individual fuel components
lean limit concentration in a particular apparatus.
Although the dust lean limits differ for large and small
scale, normalizing the data eliminates any arbitrariness
associated with geometry or scale. Both large-and small-
scale results demonstrate conclusively that hybrid lean
flammability limits are not dependent on energetics alone.
In fact, while energetics most certainly play a role, the
mechanisms of propagation inherent in the hybrid flame
determine flammability limit behavior.

     This can be shown in Figs. 6 and 7, where the large-
scale results have been plotted over flame temperature
isotherms calculated for adiabatic isochoric combustion of
the hybrid mixture. If we examine the hybrid region near
the dust limit, we see that the results fall on isotherms
corresponding to the lean limit flame temperature of the

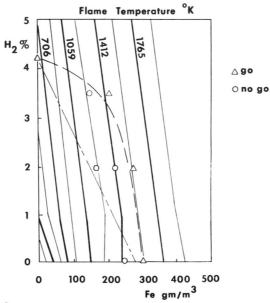

Fig. 7   5 m^3 H$_2$-Fe flammability limits with flame temperature
isotherms.

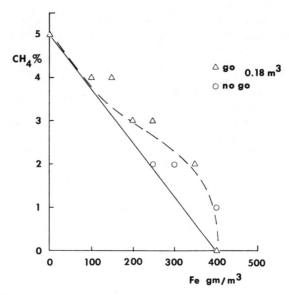

Fig. 8   Flammability limits of CH_4-Fe-air hybrid mixtures.

dust component alone, up to the point where the dust is no longer the primary fuel. Evidently, the addition of gas to a dust flame does not alter the mechanisms of propagation at all. In short, it simply ensures that the minimum flame temperature required for the propagation of the dust flame is maintained. This is analogous to what was shown by Nagai (1929) for gas flames, that all things being equal, the lean limit flame temperature is the critical parameter. Thus, for mixtures which contain fuels with two very different propagation mechanisms, as long as the dust propagation mechanism is dominant, the amount of gas fuel required to ensure the flammability of the mixture is that which will maintain the dust lean limit flame temperature. Comparing mixtures whose dust components either do or do not contain volatiles, the trend is the same. Therefore, the influence of volatiles is visible only insofar as it influences the lean limit of the dust itself. Once the dust is part of a hybrid mixture, the aforementioned mechanism is the crucial one.

For small-scale methane-iron and methane-cornstarch hybrid tests (Figs. 8 and 9), the same behavior is noted, leading to the conclusion that large-scale trends would match those of H_2-Fe and H_2-cornstarch. It becomes evident therefore, that in attempting to analyze hybrid behavior, small-scale results can be misleading. This is plausible

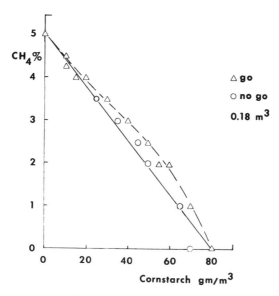

Fig. 9   Flammability limits of CH_4-cornstarch-air hybrid mixtures.

and, in fact, sensible when one considers the thickness of a
dust flame in comparison to the integral length scale of a
small-scale Hartmann-bomb-type apparatus.

Conclusion

An extension of Le Chatelier's limit law to include
hybrid mixtures has been shown to be invalid for the hybrid
mixtures studied. However, much can be learned about the
propagation mechanisms inherent in hybrid mixtures in
examining the degree of departure from the law evidenced by
the particular mixtures studied. The dust lean limit flame
temperature has been shown to be an important parameter in
assessing the lean flammability limits of hybrid mixtures
near the pure dust limit. In addition, it has been
determined that results culled from small-scale apparatus
may be very misleading when dust mixtures are studied due to
the approach of the lean limit dust flame thickness to the
integral vessel dimension. It is recommended that large-
scale apparatus be used whenever possible when studying dust
or dust-gas hybrid lean flammability limits.

Acknowledgments

The authors are grateful to A. Makris and C. Daniel for
their help in the experiments. At McGill University, the

work was sponsored by the National Sciences and Engineering Research Council (NSERC) of Canada under grants A-7091 and A-3347 and by Energy, Mines and Resources (EMR) of Canada under contract number OSV82-00039. At Sandia Laboratories, the work was supported by the Nuclear Regulatory Commission.

References

Bartknecht, W. (1981) Explosions. Springer-Verlag, Berlin.

Benedick, W.B., Cummings, J.C., and Prassinos, P.G. (1984) Combustion of Hydrogen Air Mixtures in the VGES Cylindrical Tank. SAND 83-1022, Sandia National Laboratories, Albuquerque, N.M.

Bond, J.F., Fresko, M., Knystautas, R., and Lee, J.H.S. (1984) Influence of Turbulence on Dust Explosion. Proc. First International Colloquium on Explosibility of Industrial Dusts, Warsaw, Poland.

Cocks, R.E. and Meyer, R.C. (1979) Fabrication and use of a 20-liter Spherical Dust Testing Apparatus. Process Specialties Section, Engineering Division, Proctor & Gamble, Cincinnati, Ohio.

Coward, H.F. and Jones, G.W. (1952) Limits of Flammability of Gases and Vapors. Bulletin 503, US Bureau of Mines, Pittsburgh.

Field, P. (1982) Dust Explosions. Elsevier Scientific Publishing Co., Amsterdam.

Hertzberg, M., Richmond, J.K. and Cashdollar, K. (1981) Flammability and the Extinguishment of Explosions in Gases, Dust and their Mixtures. Colloque International Berthelot-Vieille-Mallard-Le Chatelier, Universite de Bordeaux, France, pp. 202-210.

Hertzberg, M. (1981) The Flammability Limits of Gases, Vapors and Dusts: Theory and Experiment. Proc. International Conference on Fuel-Air Explosions, McGill University, Montreal, Canada. University of Waterloo Press, pp. 3-48.

Jacobson, M., Nagy, J., Cooper, A.R., and Ball, F.J. (1961) Explosibility of Agricultural Dusts. RI 5753, US Bureau of Mines, Pittsburgh.

Lee, J.H.S., Guirao, C.M., Chiu, K.W., and Bach, G.G. (1979) Blast Effects from Vapor Cloud Explosions. Proc. AIChE Loss Prevention Symposium, Houston, Tex.

Nagai, Y. (1929) Flame Propagation in Gaseous Mixtures. J. Fuel Soc., Jpn. 8 (2), 17-18.

Nagy, J. and Portman, W.M. (1960) Explosibility of Coal Dust in an Atmosphere Containing a Low Percentage of Methane. RI 5815, US Bureau of Mines, Pittsburgh.

Nagy, J., Seiler, E.C., Conn, J.W., and Verakis, H.C. (1971) Explosion Development in Closed Vessels. RI 7507, US Bureau of Mines, Pittsburgh.

Flame Structure in Dusted-Air and Hybrid-Air Mixtures New Lean Flammability Limits

R. Klemens* and P. Wolański†
Technical University of Warsaw, Warsaw, Poland

Abstract

The structure and propagation mechanism of flames in dust-air and gaseous fuel-dust-air mixtures near their flammability limits was studied. Experiments were conducted mainly in a 0.035×0.088 m^2 rectangular duct 1.2 m long. Interferometric cinematography as well as frame, streak, and compensation photography were used for this study. Temperature measurements with a resistance thermometer were also made. Dusts of coal or lignite were used and in hybrid mixtures technical methane was used as the gaseous fuel. It was found that, under the same test conditions, flames in dust and dust-gas have a quite different structure than that of a gaseous fuel; the flame front thickness is about five times larger and the flame surface is about twice the size. Eddies within the flame are created due to the nonuniformity of heat release, a consequence of imperfect dust dispersion. Initially, the dust particles in the flame front burn individually and only later can the gas flame be seen inside the main flame structure. The average flame temperature is similar to the temperature of the limiting gas flames, but is more uniform across the duct cross section. The total combustion time of the dust is of the order of tenths of seconds, while the total flame thickness is about 0.5 m. The flame propagation velocity is due to its larger frontal area, as well as to the intensification of the heat and the mass exchange process within the dusty atmosphere.

Introduction

Combustion of gas mixtures near the lean flammability limit is now well documented. The concentration limits, structure, and mechanisms controlling gas flame propagation as well as the suppression mechanism are relatively well understood. The near limit behavior of

Presented at the 10th ICDERS, Berkeley, California, August 4-9, 1985. Copyright © 1986 by R. Klemens and P. Wolański. Published by the American Institute of Aeronautics and Astronautics, Inc., with permission.
*Assistant Professor, Instytut Techniki Cieplnej, Politechnika Warszawska.
† Associate Professor, Instytut Techniki Cieplnej, Politechnika Warszawska.

169

dust and hybrid flames (hybrid-consisting combustible dust and gas) has not been as thoroughly studied as those of gaseous flames. Until the late 1970s, the limiting concentration of the dust-air mixtures has not been determined properly. Recently Wolanski (1979) and Hertzberg et al. (1979) independently developed a technique for observing the minimum explosive concentration of dust mixtures, while Buksowicz et al. (1982) and Buksowicz and Wolanski (1983) were the first to evaluate the mechanism of flame propagation.

The aim of this work is to characterize the structure of dust and hybrid flames near the lean flammability limit, in comparison to that of lean gas flames, and to describe the mechanism of flame propagation in dust flames near the lean flammability limit.

Apparatus

Experiments were conducted mainly in a 0.035×0.088 m^2 rectangular duct (Fig 1). In some experiments, ducts with cross-sectional dimensions of 0.035×0.050 m^2 and 0.035×0.035 m^2 were also used. The lower end of the duct was always open. The vibrational dust feeder was placed at the top together with the system supplying the air and gas. To observe the process, two square windows of dimensions of 0.08×0.08 m^2 were installed at the upper part of the duct. At the same distance, a resistance thermometer with a very small time constant was also installed. The ignition system, placed at a lower end of the duct, ignited the mixture by an electrical discharge of a few joules of energy, but did not influence the flame propagation in the duct.

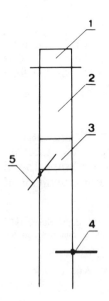

Fig. 1. Experimental set-up: (1) dust feeder, (2) duct, (3) observation window, (4) ignition source, (5) resistance thermometer.

Flow visualization was a main reason for the use of the rectangular tube instead of a 0.05×0.05 m² "standard tube." The flame structure was studied with a Mach-Zehnder interferometer and a high-speed framing camera. Flame spreading velocities were evaluated from streak photographs, while the flame structure was studied from direct frame and compensation photographs. The temperature of the flame was measured in different parts of a duct, mainly between the duct axis and a duct wall.

Experiments were carried out for two types of dust: coal and lignite. For both, the particles were smaller than 75 μm. In hybrid mixtures, technical methane was used as the gas fuel. The duct was cleaned of combustion products with fresh air before each run. When the gas-air mixture was used, the mixture was passed through the duct for at least a few minutes to assure the uniform concentration of the mixture. In all experiments, the mixture concentration was only slightly higher than the limiting concentration and was tested before each set of runs. Finally, the experimental studies were performed in a tube 0.16 m in diameter and 3 m in length.

Experiments

For purposes of comparison, the flame propagations in gas, hybrid, gas-dust and dust-gas, and dust mixtures were studied under the same conditions. In the gas-dust-air mixture, most of the fuel was gaseous and only low concentrations of dust were used. In that mixture the concentration of combustible gas was slightly less than the limiting concentration, but the addition of dust supported flame propagation in the mixture. In the dust-gas-air mixture, the fuel consisted mostly of dust and only low concentrations of the gaseous fuel were used. The dust concentration in the mixture was slightly below the limiting concentration, but the addition of gaseous fuel made combustion possible. The concentration of fuel in mixtures was always very close to the lean combustion limit. The dust-air mixture was called simply a dust mixture.

Figures 2-5 show the variation of the flame temperature measured at the duct axis 25, 6, and 2 mm from the duct wall. On the basis of these measurements, the temperature profiles for different flames were evaluated. The temperature profiles for gas, hybrid, and dust flames are shown in Fig. 6. It can be seen that for limiting gas flames, temperature profiles in our experimental duct are very similar to the temperature profiles of the methane-air lean mixture in a standard tube (Jarosinski et al, 1982). The highest flame temperature is near the duct wall, while at the center the temperature is about 150 K lower. The gas flame with a concentration higher than the lean limit (Fig. 6b) has the maximum flame temperature on the duct axis. For the limiting

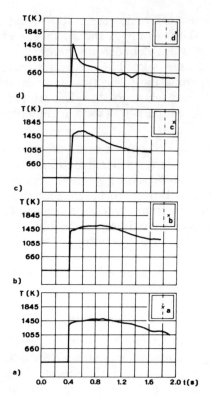

Fig. 2. Temperature distribution in the gas-air (5.1% CH$_4$) flame at various distances from the duct wall: (a) on the duct axis; (b) 25 mm from the wall; (c) 6 mm from the wall; (d) 2 mm from the wall.

gas-dust flames (Fig. 6c), the temperature profiles are similar to those of the limiting gas flame (Fig. 6a). In the gas-dust flames, however, one can see small temperature fluctuations (Fig. 3). As the dust concentration (in the limiting mixture) increases, the temperature fluctuations increase (Fig. 4). The average temperature remains close to the temperature of the lean gas flame, and the temperature is more uniform across the duct. In the dust flames, the temperature fluctuations are even higher (Fig. 5) and temperature differences as large as 1000 K may occur within the flame. This occurs because combustion products and a fresh mixture pass successively across the probe, a characteristic feature for the turbulent flame. For dust flames these fluctuations are most intense near the duct wall. The maximum temperature of the dust flame is slightly below that of similar gas flames (Fig. 6c).

The temperature measurements in the flame front may be used to evaluate the thickness of the flame front, defined as the distance between the crossing of the initial temperature level and the maximum

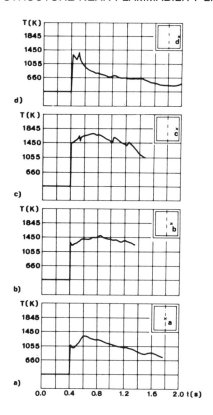

Fig. 3. Temperature distribution in the gas-lignite dust-air (4.2% CH_4) flame at various distances from the duct wall ($C < 0.019$ kg/m^3): (a) on the duct axis; (b) 25 mm from the wall; (c) 6 mm from the wall; (d) 2 mm from the wall.

Table 1
Flame parameters for different mixtures in a 0.035×0.088 m² flat duct

Mixture	flame front thickness, mm	flame velocity m/s	Re
Methane-air (5.1% CH_4)	2.8	0.26	910
Methane-air (6.6% CH_4)	2.3	0.58	2000
Gas-dust	2.6	0.27	940
Dust-gas	11.7	0.45	1550
Dust	11.4	0.57	1960

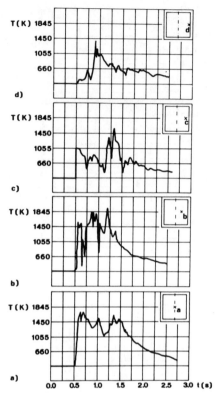

Fig. 4. Temperature distribution in the lignite dust-gas-air (4.2% CH_4) flame at various distances from the duct wall (C = 0.066 kg/m^3): (a) on the duct axis; (b) 25 mm from the wall; (c) 6 mm from the wall; (d) 2 mm from the wall.

flame temperature level in the flame front by the line tangent to the maximum temperature gradient. The results are presented in Table 1. The flame front thicknesses for dust and dust-gas flames is five times greater than those of gas and gas-dust flames.

The flame front velocity was evaluated from streak pictures. Typical streak records of gas-dust, dust-gas, and dust flames are shown in Fig. 7 and the flame front velocity measurements listed in Table 1. In this table the Reynolds numbers, calculated in the same way as specified by Zeldovitch et al. (1980) and Oppenheim and Ghoniem (1983) are given as well. The highest Reynolds numbers are for the gas flame with 6.6% CH_4 and the dust flames, while for the limiting gas and gas-dust flames Reynolds numbers are smaller. Although these calculations suggest that the flames should have a laminar structure, temperature measurements and compensation pictures (Figs. 8-11) indicate a different structure for the dust-gas and dust flames. Limiting gas and

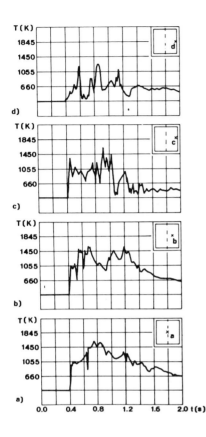

Fig. 5. Temperature distribution in the lignite dust-air flame at various distances from the duct wall ($C = 0.103$ kg/m^3): (a) on the duct axis; (b) 25 mm from the wall; (c) 6 mm from the wall; (d) 2 mm from the wall.

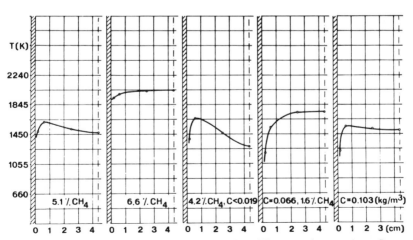

Fig. 6. Temperature profiles of flames for different mixtures in a flat duct (T=Tmax).

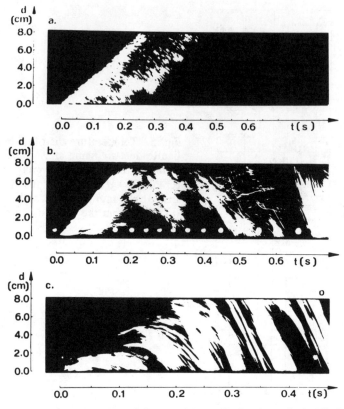

Fig. 7. Streak photographs of flames: (a) gas-lignite dust-air (4.2% CH_4, $C < 0.019$ kg/m³); (b) lignite dust-gas-air (1.6% CH_4, $C = 0.066$ kg/m³); (c) lignite dust-air ($C = 0.103$ kg/m³).

gas-dust flames have a purely laminar structure; even the 6.6% CH_4 gas flame with the highest Reynolds number exhibits a laminar structure, as can be very clearly seen in the streak record of a flame containing a very small amount of dust traces. The gas-dust flame also shows a structure similar to the classical structure of the gas flame described by Zeldovitch et al. (1980).

On the other hand, dust-gas and dust flames exhibit very clear eddies within the flame; sometimes the flame front breaks into a cellular structure (Figs. 10 and 11). In some cases, the front is laminar, but farther behind the appearance of eddies can be clearly observed (Fig. 12). In all flames, at larger distances from the flame front, the flow becomes laminar, despite the significant increase in the flow velocity, mainly because of the better uniformity of the products as well as to a significant increase in viscosity.

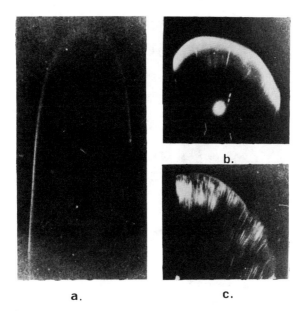

Fig. 8. Compensation photographs of gas-air flames: (a) 5.1% CH_4; (b) 6.6% CH_4; (c) 6.6% CH_4.

Fig. 9. Compensation photographs of gas-lignite dust-air flames (4.2% CH_4, $C < 0.019$ kg/m^3).

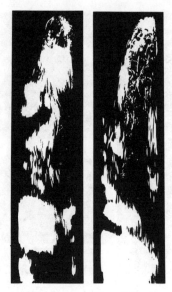

Fig. 10. Compensation photographs of lignite dust-gas-air flames (1.6% CH$_4$) C = 0.066 kg/m^3.

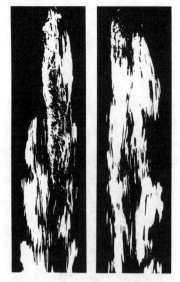

Fig. 11. Compensation photographs of lignite dust-air flames (C = 0.103 kg/m^3).

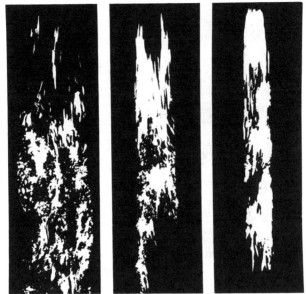

Fig. 12. Compensation photographs of lignite dust flames in ducts of different diameters: (a) 80 mm; (b) 50 mm; (c) 35 mm.

For a detailed examination of the processes occurring in the flame front, interferometric pictures of a narrow section of the combustion zone were taken. The interferometer was pointed at the center of a duct and its range of view was limited to 35 mm. The camera was arranged so a direct picture of burning particles as well as an interferometric picture of the flame front could be simultaneously recorded. Figure 13 shows two typical pictures from two different runs, recorded at a frequency of 200 frames per second. From these pictures, the location of the region of the flame where dust particles are ignited can be clearly seen. At the leading edge of the flame, dust particles are ignited 2-3 mm behind the front, while at the flame those distances are larger (5-10 mm). Dust particles and agglomerations burn individually at the flame front and there is no evidence of a gas flame. Interferometric pictures also show that the smallest eddy structures within the flame front have dimensions of 1-3 cm.

The influence of the duct width on the flame structure was also studied, demonstrating that in ducts of smaller width, the limiting concentration of dust as well as the limiting flame velocity were larger. While the Reynolds numbers do not change significantly, smaller fluctuations and less evidence of eddies was observed in smaller ducts.

Figure 14 shows eight frames of the leading portion of the limiting dust-air flame in the 0.16 m diameter tube. Small eddies are clearly

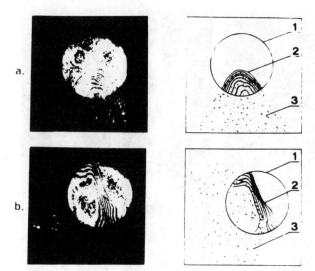

Fig. 13. Interferometric-direct photographs of lignite dust flames (1-the interferometric field of view, 2-front of the hot combustion products, 3-burning particles): a) the front of the eddy; b) side part of the eddy.

deforming the flame front and behind the flame front eddies of a much larger scale are observed. Visual observation of such flames suggest the spinning nature of such flames. In the flame front, individual burning dust particles and agglomerations can also be observed. However, those flames are in a purely turbulent region and the Reynolds number is about 6000.

Discussion

Gas-dust flames propagate in a way similar to gas flames. In this case, the process is controlled by free convection, which is in agreement with the model of Babkin et al. (1982). According to it, hot products created by combustion are lifted from the lower part of the duct by buoyancy. The velocity of such lifting hot gases, calculated by the Davis-Taylor relation, is in good agreement with measured flame velocities. The flame structure is also similar to the structure described by Zeldovitch et al. (1980), for gas flames propagating with a Reynolds number of about 2000.

Quite different structures are observed in the dust-gas and dust flames. The flame front thickness is about five times larger than that of a gas flame, while the flame surface is doubled and eddies are observed within the flame. The average flame temperature is similar, but the flame propagation velocity is twice that of the gas mixtures.

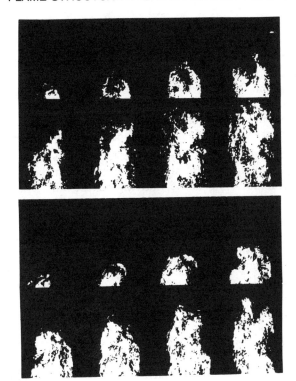

Fig. 14. Direct photographs of dust flame in a glass tube 0.16 m in diameter. (frequency 24 frames per second).

As the total combustion time of dust is large (0.2-0.5 seconds), the flame is relatively long.

The flame front in the dust-gas and dust mixtures is usually irregular in shape and can be divided into two subfronts (cells). Because of the flame induced flow the flame influences the motion of the particles ahead of the flame front. The aerodynamic interaction can change the velocity of particles from a free-falling velocity to a velocity very close to that of the flame propagation. Numerical calculation shows that the 50 μm particle can reach the gas velocity after having traveled a distance of a few millimeters. This is evident from an examination of streak records on which the burning particles move at nearly the velocity of the flame front. This process increases the time of interaction between the dust particles and the hot flame front gases and facilitates ignition. The interferometric pictures show that the dust particles are ignited at relatively small distances behind the leading edge, while on the side of the flame front this distance is a few times larger. This is due to the differences between the relative particle velocity in that region.

Eddies in the dust and dust-gas flames have different intensities in different flame regions. In the same type of mixture, they can occupy a large portion of the flame or be present in only one section of the flame. Their appearance is random and cannot be predicted. This is mainly due to the fact that they are created by nonuniformity of heat release within the flame caused by imperfect dust dispersion. Measurements of nonuniformity of the dust dispersion process show the variation in local concentration to be at least ±20% from the mean.

On the basis of the experiments, the following simplified model of flame propagation in dust-air mixture near the lean flammability limit can be constructed. The flame, moving in the duct, induces a flow ahead of the flame front. This flow interacts aerodynamically with the falling dust particles and dust agglomerations and results in nearly equal dust and gas velocities. As a consequence, the time of interaction between the hot gases and particles increases and facilitates particle ignition. Initially, dust particles and agglomerations burn "individually" and only later can the gas flame be seen inside the main flame structure. Imperfect dust dispersion produces nonuniformity in a heat release process and also leads to the creation of eddies. The eddies intensify the heat and mass exchange, influence the flame shape and area, and lead to a higher propagation velocity. The improvement in the uniformity by mixing and increasing viscosity as a result of a rise in temperature leads to the laminarization of the flow at the flame tail. Because of the relatively long burning time of the dust particles, the total flame thickness can reach 0.5 m.

Conclusions

It has been found that near the lean flammability limit the dust and dust-gas flames have a different structure and mechanism of propagation than purely gaseous flames. The former have the following characteristic features:

1. The thickness of the dust flame front is about five times larger than that of the gas flames and usually is about 11-12 mm thick.

2. The combustion of dust particles in the flame front initially occurs as the individual combustion of dust particles and dust agglomerations. Evidence of the presence of a gas flame can be found only at longer distances from the flame front.

3. In the flame front, eddies can be observed even when the Reynolds number is less than 2000. They are created due to the nonuniformity of heat release caused by the imperfect dust dispersion in the mixture. At larger distances from the flame front they vanish and the flow becomes laminar.

4. The flame-spreading velocities in dust and dust-gas mixtures are about two times larger than in limiting gas mixtures. This is due to a larger flame front area as well as to the intensification of the heat and mass exchange processes within the dust and dust-gas flames.

These conclusions cannot be extended to all conditions or to all dust mixtures. They can be applied only to similar dusts and to similar burning conditions. More study is needed for different dusts and different conditions before general conclusions about dust mixtures burning near the lean flammability limit can be established.

References

Babkin, V. S. et al. (1982) Effect of tubes diameter on the propagation limits of homogeneous gas flames. *Combust. Explos. Shock Waves* (USSR) **18**(2) (in Russian).

Buksowicz, W., Klemens, R. and Wolanski, P. (1982) An investigation of the structure of dust-air flame. *Second Int. Specialists Meeting of The Combustion Institute of Oxidation*, Budapest, Hungary.

Buksowicz, W. and Wolanski, P. (1983) Flame propagation in dust-air mixtures at minimum explosive concentration. Shock Waves, Explosions and Detonations: *Progress in Astronautics and Aeronautics* (edited by J. R. Bowen, N. Manson, A. K. Oppenheim, and R. I. Soloukhin), Vol 86, pp. 141-425. AIAA, New York.

Hertzberg, M., Cashdollar, K. L. and Opferman, J. J. (1979) The flammability of coal dust-air mixtures. RI 8360, U.S. Bureau of Mines.

Jarosinski, J., Strehlow, R. A. and Azarbarzin, A. (1982) The mechanism of lean limit extinguishment of an upward and downward propagation flame in a standard flammability tube. *19th Symposium (International) on Combustion*, pp. 1549-1557. The Combustion Institute, Pittsburgh, PA.

Oppenheim, A. K. and Ghoniem A. F. (1983) Aerodynamic features of turbulent flames. *AIAA* Paper, 83-0470, 10 pages. AIAA, New York.

Wolanski, P. (1979) Explosion hazards of agricultural dust. *Proceedings of International Symposium on Grain Dust*, pp. 422-426. Kansas State College Press, Manhattan, KS.

Zeldovich, Ya. B., Istratov, A. G., Kidin, N. I. and Librovitch, V. B. (1980) Flame propagation in tubes: hydrodynamics and stability. *Comb. Sci. and Tech.* **24**, 1-13.

Turbulent Burning Velocity Measurements for Dust/Air Mixtures in a Constant Volume Spherical Bomb

F.I. Tezok,* C.W. Kauffman,† M. Sichel,‡ and J.A. Nicholls‡
The University of Michigan, Ann Arbor, Michigan

Abstract

The propagation of turbulent dust/air flames has been
studied experimentally. Measurements of the burning velocity have been made using the constant volume bomb technique
in a 0.95-m³ spherical jet stirred reactor. A measured
dust/air mixture is fed to the reactor, and the inlet jets
provide a uniform mixture and turbulence intensity throughout the volume. Simultaneous with the central ignition of
the cloud, the bomb inlets and exits are isolated and the
flame propagation and pressure histories of the resulting
combustion process are recorded. Burning velocities of 45-
100 cm/s and 70-330 cm/s were obtained for mixed grain dust
and cornstarch, respectively, for controlled moisture contents (< 12%), particle size distributions (< 212 μm), and
turbulence intensities (1.4-4.2 m/s) for dust concentrations in the range of 50-1300 g/m³. The maximum pressure
and rate of pressure rise data showed general agreement
with other bomb data. The ratio of turbulent to laminar
burning velocity (obtained by interpolation of the burning
data to zero turbulence intensity) was found to correlate
well with both the ratio of rms turbulent velocity to laminar burning velocity and Reynolds number. The data show
a linear relationship for the quantities mentioned. Flame
thickness determined by an optical probe ranged from 15-80
cm, increasing with both turbulence intensity and dust con-

Presented at the 10th ICDERS, Berkeley, California, August
4-9, 1985. Copyright © 1986 by the American Institute of Aeronautics
and Astronautics, Inc. All rights reserved.
*Graduate Student, Department of Aerospace Engineering. Cur-
rent Address: General Dynamics of Turkey, Inc., 3 Tahran Cad.,
Ankara, Turkey.

†Lecturer, Department of Aerospace Engineering.

‡Professor, Department of Aerospace Engineering.

184

centration. The measurements and correlations strongly
suggest that the propagation mechanism of the flame for
high turbulence levels is dominated by turbulent exchange.
     The explosive nature of many organic dust/air mixtures
has been verified using testing apparatus primarily developed by the U.S. Bureau of Mines approximately a half centu-
ry ago. Some recent modifications of this equipment have
given rise to slightly different test results. The major
results of this testing are the ignition sensitivity and
the explosion severity which reflect such quantities as minimum ignition energy, ignition temperature, minimum explosive concentration, maximum pressure, and maximum rate of
pressure rise, as compared with Pittsburgh seam number
eight coal (Kauffman, C. W., 1982; Bartknecht, W., 1981;
and Field, P., 1982).
     The maximum rate of pressure rise is not a fundamental
physical variable, although it may be related to the burning velocity, and indeed it reflects an apparatus dependence. Uniformity of concentration and turbulence intensity for the dust/air mixture in the testing apparatus would
help isolate different mechanisms affecting dust flame propagation in order to aid in the fundamental understanding
of such flames. The burning velocity for a given dust at
a given concentration is certainly going to be affected by
the turbulence intensity and length scale as well as the
local uniformity of the dust/air mixture. It is these
quantities that cannot be quantitatively controlled in the
traditional explosion testing apparatus. In order to find
a solution to these problems, a new apparatus was developed
and measurements were made to determine the burning velocity for mixed grain dust and cornstarch as a function of
dust concentration, dust moisture content, and mixture turbulence intensity.

Experimental Setup

     A schematic diagram of the experimental apparatus, the
premixed turbulent combustion bomb (PTCB), is shown in Fig.
1. It is a spherical steel vessel with a diameter of 122
cm and a design pressure of 10 atm. Instrumentation consists of a pressure transducer, an ion or thermocouple rake
located along a radius, a highly collimated photodiode view-
ing a path perpendicular to a radius, and a U.S. Bureau of
Mines developed optical dust concentration probe. A pair
of windows are located at the opposite ends of a diameter,
and an exploding wire ignition source is located at the
center of the bomb. Symmetrically located around the
sphere are six inlet ports and eight exhaust ports.

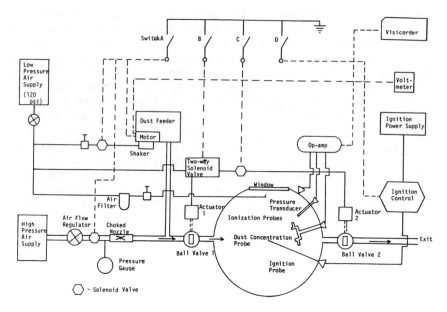

Fig. 1   Schematic diagram of the experimental apparatus.

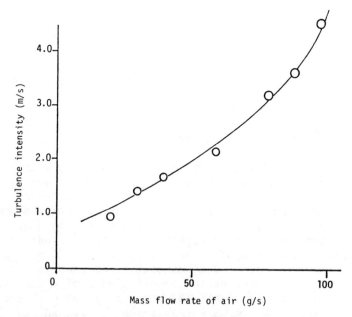

Fig. 2   Turbulence level calibration for different mass flow rates of air (no dust) at the center of the sphere.

The airflow rate is controlled through a choked nozzle. Dust is carefully metered into the airstream through an auger in order to establish a well mixed homogeneous flow condition at the inlet ports. A flow divider for the inlets is located on the bottom of the sphere, and the outlets are connected to a collector located on the top that directs the mixture to an exhaust stack. The pressure drop across the system is less than 2 psi for all flow conditions, thereby keeping the pressure inside the sphere very close to atmospheric before ignition. The bomb thus acts as a continuous jet stirred reactor, where the turbulent mixing keeps the dust suspended and well mixed in the air. During the operation of the bomb, quick-acting valves close all the inlets and exhausts simultaneous with ignition, creating a constant volume combustion situation with little decay of the existing turbulence.

The turbulence level inside the bomb is determined by the air mass flow through it which varies from 0.020 to 0.092 kg/s. The turbulence level inside the sphere was measured using a hot wire for the no-dust condition, as the dust would destroy the hot wire. The location and orientation of the hot wire was changed radially and azimuthally during these measurements. Figure 2 shows the change of the turbulence intensity at the center of the sphere for different air mass flow rates, whereas Fig. 3 shows the variation of turbulence level with radius. The azimuthal variation was insignificant. The turbulence level u' was determined using the isotropic turbulence assumption (Semenov 1965; Checkel 1981) where the measured value of rms was related to u' through $u' = \sqrt{(\sigma^2 + \varepsilon^2)}/2$; σ and ε are respectively the measured rms and mean values of the signal from the hot wire anemometer. The turbulence level associated with a given mass flow of air was set to be the rms obtained at the center of the sphere (Fig. 2), which was found to vary between 1.50 and 4.50 m/s for the different air mass flow rates employed. The upper limit on the airflow for a given dust concentration is established by the dust feeder.

The uniformity of the dust concentration inside the sphere was also checked using the Bureau of Mines optical dust probe (Cashdollar 1981), where the presence of dust in the sampling volume attenuates the signal recorded from a photodiode under the illumination from a constant light source from a fixed path length. The dust probe was first calibrated in a small closed wind tunnel by injecting known amounts of dust into the known volume to get a steady concentration reading from the probe. The probe output was proportional to the reference light beam attenuation and

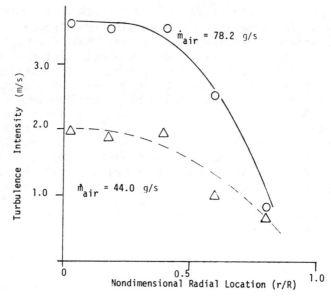

Fig. 3 Distribution of turbulence intensity inside the PTCB (no dust) for different air mass flow rates.

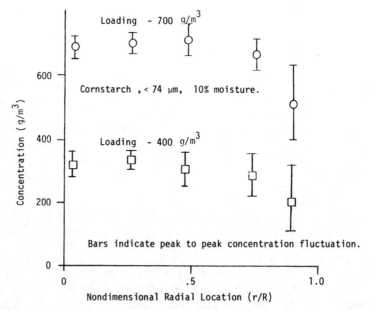

Fig. 4 Distribution of dust concentration inside the sphere for two different settings (u' = 4.2 m/s).

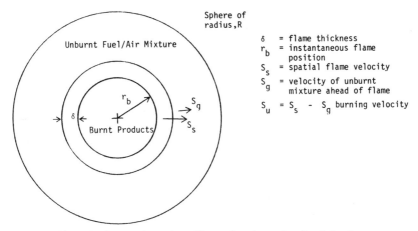

Fig. 5   Outward moving flame in the spherical bomb.

the calibration curve obtained using the above procedure
was used to control the dust concentration inside the spherical bomb. The local dust concentration was found to be
highly fluctuating but very uniform throughout the sphere
and in good agreement with the expected average concentration level from the auger and airflow setting (Fig. 4).
Thus, as opposed to many testing apparatus employed in evaluating the explosibility of dust/air mixtures, in the PTCB
the mixture and the turbulence are uniform and steady and
they can be controlled.

Calculation of the Turbulent Burning Velocity

By definition, the burning velocity is the difference
between the spatial flame velocity and the velocity of the
unburned mixture ahead of the flame (Fig. 5). The following assumptions have been made in order to get an expression for the turbulent burning velocity S_u; the flame remains smooth and spherical with no convective rise of the
unburned gases; no pressure gradients exist within the
PTCB, the flame zone has no thickness, and the unburned gas
is compressed isentropically (Rallis et al. 1965; Bradley
and Mitcheson 1976). The unburned gas conservation equations in a spherical vessel of radius R then become

$$\frac{dm_u}{dt} = -4\pi r_b^2 \rho_u S_u, \tag{1}$$

$$m_u = (\frac{4}{3})\pi(R^3 - r_b^3)\rho_u \tag{2}$$

and

$$P[\rho_u]^{-\gamma_u} = \text{const} \tag{3}$$

Combining Eqs. (1), (2), and (3) gives

$$S_u = \frac{dr_b}{dt} - \left(\frac{R^3 - r_b^3}{3r_b^2 \gamma_u P}\right) \frac{dP}{dt} \tag{4}$$

Even for thick flames this unburned gas equation remains the same (Tezok, 1985).

Results

Extensive combustion tests were conducted using mixed grain dust, corn and soybean (12% moisture, 74-212 μm) and cornstarch (4% and 10% moisture < 74 μm). Data typically obtained are shown in Fig. 6. The progress of the combustion process is shown on the pressure transducer as well as sequentially on the ion gauges. These data give the rate of pressure rise, the maximum pressure, and the trajectory of the flame front. The conventionally reported parameters for dust explosions such as $(dP/dt)_{max}$ and P_{max} were obtained from these data and they are in general agreement with those obtained using the traditional testing equipment which for cornstarch gives a K_{ST} = 71 bar ms^{-1} (K_{ST} = $V^{1/3}dP/dt_{max}$).

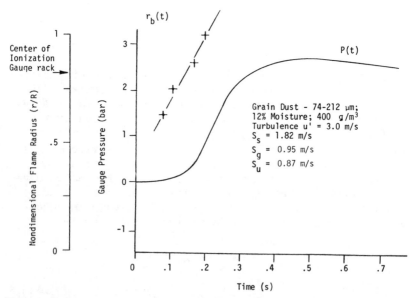

Fig. 6 Variation of flame position and pressure with time in the PTCB.

The turbulent burning velocities were calculated using Eq. (4) with the average value of dr_b/dt, and the values of P and dP/dt at the time the flame was located at the center of the ion gauge rake. Four ionization gauges were positioned radially at two inch intervals starting from the surface of the sphere. The burning velocities for cornstarch and mixed grain dust as a function of the variables considered here are given in Figs. 7 and 8. It is recognized that a dust concentration of ~ 700 g/m^3 gives a maximum burning velocity for both types of dust. An increasing level of turbulence increases the burning velocity significantly. It is also observed that burning velocities for cornstarch are higher than those for mixed grain dust, and an increase in the moisture content decreases the flame velocity. All of these trends are in agreement with the qualitative "explosibility" results produced in Hartmann bomb tests (Bartknecht 1982; Field 1982). Another item of interest is the less-sensitive dependence of the burning velocity to the dust concentration on the rich site.

For gaseous combustible mixtures, the ratio of turbulent to laminar burning velocity is found to correlate with the ratio of turbulence intensity over laminar burning velocity or the Reynolds number (Abdel-Gayed and Bradley 1976). The burning velocity for the dust/air mixtures may be extrapolated to zero turbulence intensity (Fig. 9) in

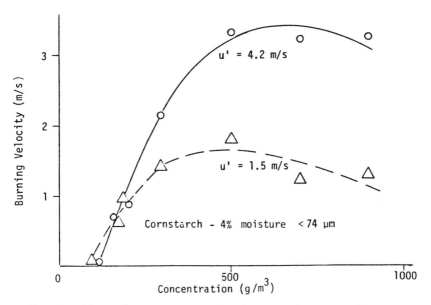

Fig. 7 Effect of turbulence on burning velocity for different dust concentrations.

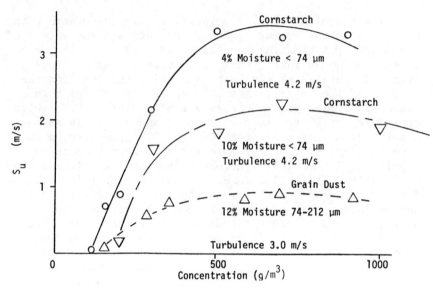

Fig. 8 Effect of moisture on burning velocity.

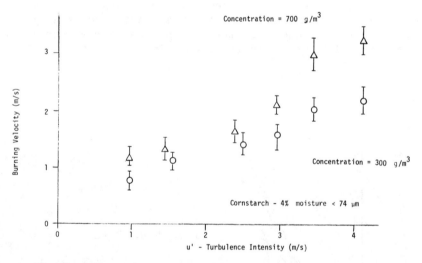

Fig. 9 Effect of turbulence intensity on burning velocity.

order to obtain the laminar burning velocity. The results
for the two turbulent flame velocity correlations are given
in Figs. 10 and 11, and it may be noted that both work reasonably well.

One of the critical assumptions made in deriving the
combined expressions for the burning velocity is that the

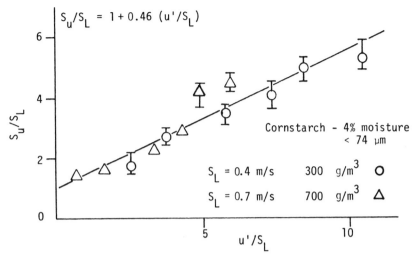

Fig. 10   Correlation of turbulent burning velocity with turbulence intensity.

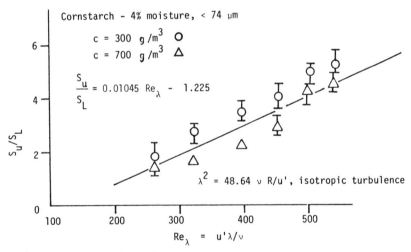

Fig. 11   Correlation of turbulent burning velocity with Reynolds number.

flame has a small thickness. The flame radiation profile which is the duration of the output of a photodiode above a preset trigger level measured normal to the flame front should be indicative of the thickness, and these data are presented in Fig. 12. It may be recognized that increasing dust concentrations and levels of turbulence give a thickening of the flame, and the value soon becomes comparable to

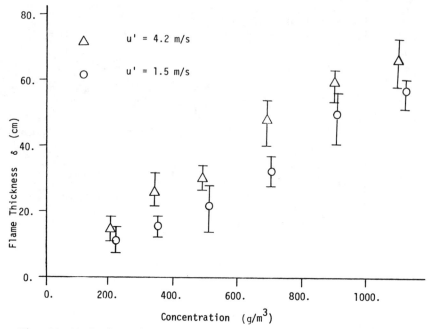

Fig. 12 Variation of flame thickness with concentration and turbulence level.

the characteristic bomb dimension. Corrections may be made and these are reported elsewhere (Tezok 1985).

Conclusions

Through the use of a premixed turbulent combustion bomb, it has been possible to measure the turbulent burning velocity in dust/air mixtures where the turbulence intensity and mixture ratio are quite uniform, steady, and controllable. The effects of the dust composition, moisture content, and concentration and of the mixture turbulence intensity on the burning velocity have been observed. The classical gaseous turbulent burning velocity correlations appear to be applicable to turbulent dust flames.

Acknowledgment

This work was supported by the Department of Health and Human Services, National Institute of Occupational Safety and Health, under grant number OHO1122-05.

References

Abdel-Gayed, R. and Bradley, D. (1976) Dependence of turbulent burning velocity on turbulent Reynolds number and ratio of laminar burning velocity to rms turbulent velocity. 16th Symposium (International) on Combustion, pp. 1725-1735, The Combustion Institute, Pittsburgh, PA.

Bartknecht, W. (1981) Explosions, Springer-Verlag, New York.

Bradley, D. J. and Mitcheson, A. (1976) Mathematical solutions for explosions in spherical vessels. Combust. Flame 26:2, 201-217.

Cashdollar, K. L., Liebman, I. and Conto, R. S. (1981) The Bureau of Mines optical dust probes. R.I. 8542, U.S. Bureau of Mines, Washington, D.C.

Checkel, D. M. (1981) Hot wire anemometer signal analysis for turbulence with no mean flow (private communication). University of Alberta, Edmonton, Alta. Canada

Field, P. (1982) Dust Explosions, Elsevier Scientific Publishing Company, New York.

Kauffman, C. W. (1982) Agricultural dust explosions in grain handling facilities. Fuel Air Explosions, SM Study 16, University of Waterloo Press, Waterloo, Ont. Canada.

Rallis, C. J., Garforth, A. M. and Steinz, J. A. (1965) Laminar burning velocity of acetylene-air mixtures by constant volume method: Dependence on mixture composition, pressure and temperature. Combust. Flame, 9, 345-356.

Semenov, E. S. (1965) Measurement of turbulence characteristics in a closed volume with artificial turbulence. Combust. Explos. Shock Waves (USSR) 1(2), 57-62.

Tezok, F. I. (1985) Flame propagation through turbulent combustible dust/air mixtures, Ph.D. Thesis, The University of Michigan, Ann Arbor, Mich.

Flame Propagation Due to Layered Combustible Dusts

S.R. Srinath,* C.W. Kauffman,† J.A. Nicholls,‡ and M. Sichel‡

The University of Michigan, Ann Arbor, Michigan

Abstract

An experimental study of flame propagation in a tube fueled by combustible dust layers was conducted to improve the understanding of the industrial and agricultural dust explosion problem. Toward this end, a 36.58-m-long tube closed at the ignition end and open at the opposite end, having an internal diameter of 29.85 cm and a length-to-diameter ratio of 120 to 1, was constructed. Dust was deposited throughout the length of the tube in the form of a layer using a motorized cart, except for the first 3.66 m where it was deposited in a V-channel. The V-channel dust was dispersed using air jets and ignited using a hydrogen and oxygen detonation tube, resulting in a primary explosion. The convective flow due to the pressure waves induced by the expanding burned gases caused the layered dust in the remainder of the tube to be distributed across the tube. This effect resulted in a secondary explosion, a rapidly accelerating flame propagating throughout the rest of the tube. The flame trajectory and time histories of pressure, gas velocity, gas temperature, and dispersed dust concentration were recorded at different tube locations. Tests were conducted with agricultural dusts to study the effect of composition, size, moisture content, and layer thickness, as well as the primary explosion strength on the flame propagation process. Flame velocities of about 600 m/s and overpressures of about 6 atm were measured under certain conditions. The process was studied theoretically using a simplified description of the flame as a discontinuity, characterized by a specified normal burning velocity

---

Presented at the 10th ICDERS, Berkeley, California, August 4-9, 1985. Copyright © 1985 by the American Institute of Aeronautics and Astronautics, Inc. All rights reserved.

*Research Fellow, Department of Aerospace Engineering.

†Research Scientist, Department of Aerospace Engineering.

‡Professor, Department of Aerospace Engineering.

196

and density expansion ratio. Numerically calculated results, obtained by modifying an available code to include
entrainment of the layered dust and the effects of turbulence, were found to agree favorably with experiment.

Nomenclature

a	=	speed of sound
C	=	dispersed dust concentration
C_s	=	stoichiometric dust concentration
C_o	=	nominal secondary dust concentration
D	=	diameter of FAT
e	=	energy per unit volume
E	=	total energy per unit volume
FAT	=	Flame Acceleration Tube
γ	=	ratio of specific heats
L	=	length of FAT
m	=	mass flux
P	=	static pressure
q	=	heat transfer rate
R	=	particular gas constant
ρ	=	density
S	=	normal burning velocity
t	=	time
t_o	=	L/a_o
T	=	gas temperature
V	=	gas velocity
V_f	=	flame velocity
x	=	distance from closed end of FAT
x_f	=	distance of flame from closed end of FAT
χ	=	expansion ratio ρ_u/ρ_b

Subscripts

o	=	ambient
L	=	laminar
t	=	turbulent
u	=	unburned
b	=	burned

Introduction

Dust explosions have frequently occurred in coal mines,
agricultural handling and processing facilities, and other
industrial facilities where combustible dusts are present in

confined volumes with typically large length-to-diameter ratios. Since the recognition of the importance of dust in
these explosions (Faraday and Lyell 1945; Price and Brown
1922), much research has been conducted to study dust combustion, dust flame propagation, and the explosive properties of combustible dusts.

In one type of experiment, different dusts are dispersed
in a constant volume apparatus and ignited (Dorsett et al.
1960; Ishihama and Enomoto 1973), and the resultant explosions are studied (Hartmann et al. 1950; Enomoto 1977). In
industrial and agricultural facilities, however, dust is
normally present in the form of layers. Accidental ignition of some dispersed dust or natural gas and the subsequent expansion of burned gases cause pressure waves to be
propagated outward. The convective flow due to these pressure waves, in turn, causes the layered dust from the floor
and walls to be distributed in the facility. This effect
is highly favorable to further flame acceleration accompanied by substantial overpressures resulting in extensive
damage and injury. This phenomenon has been studied in the
second type of experiment, where dust is distributed on
shelves and on the floor of long galleries and ignited at
one end (Rae 1973; Richmond and Liebman 1975).

Much of the available data pertains to coal dust, and
very limited data are available for agricultural dusts.
Further, there is a lack of data from direct studies of the
effect of parameters such as dust layer thickness, etc., on
entrainment of the layered dust and hence on explosion development, as dust entrainment studies were conducted primarily in nonreacting systems. Some results from a study
of propagating flames fueled by combustible dust layers
and associated pressure waves are presented in this paper.
Results from a theoretical study of the process are also
presented and discussed.

Experimental Facility

A schematic of the experimental apparatus, the Flame
Acceleration Tube (FAT), is given in Fig. 1. It consists
of a 36.58-m-long tube with an internal diameter of 29.85
cm, resulting in a length-to-diameter ratio of about 120
to 1. The ignition source, a smaller diameter detonation
tube filled with a stoichiometric mixture of hydrogen and
oxygen to a maximum pressure of 1 atm abs, is attached to
the closed end and separated by a thin Mylar diaphragm.
The other end of the tube is either open or terminates with
an exhaust section consisting of a smooth elbow and a vented tube section. The following instrumentation is mounted

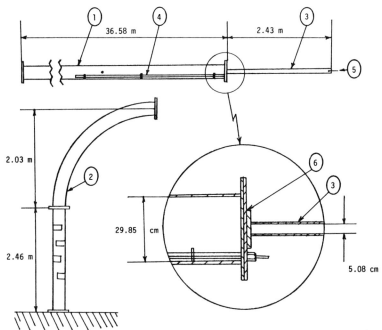

Fig. 1 Flame acceleration tube: (1) main tube; (2) exhaust elbow;
(3) detonation tube; (4) V-channel; (5) glow plug; (6) Mylar
diaphragm.

at different locations along the tube: static and total
pressure probes, thermocouples, and light-sensitive photodiodes. A dust concentration probe (Cashdollar et al.
1981) and a fast radiometer (Wolanski and Waudby-Smith,
1980) are mounted at midsection of the tube. The outputs
from the various instruments were originally recorded on a
24-channel analog oscillograph using miniature galvanome-
ters. Currently, a 32-channel analog-to-digital converter
module mounted in a CAMAC crate connected to an IBM 9000
laboratory computer is used for recording data.
     In the first 3.66 m of the tube, the dust is deposited
in a V-channel from which it can be dispersed using air
jets. The mass loading of the dust in the V-channel is the
primary dust concentration. In the rest of the tube, the
dust is deposited in the form of a layer using a motorized
cart at a given mass loading per unit length. This mass
loading is referred to as the nominal secondary dust explosion. This, in turn, results in a flame preceded by compression waves propagating through the remainder of the
tube. The progress of this wave, called the secondary explosion, is monitored. The flame trajectory and flame
velocity are measured using the output of the light-

sensitive photodiodes which detect the arrival of the leading edge of the luminous zone of the flame. The time history of static pressure is measured using the pressure
transducers, gas velocity using the dynamic pressure read-
ings, gas temperature using the thermocouples, and the temperature of the burning particles using the radiometer.
The concentration of the dust actually dispersed by the
convective flow, referred to as the dispersed dust concentration as opposed to the nominal secondary dust concentration, is measured using the dust concentration probe at
midsection of the tube.

Experimental Results

    Typical data from a test conducted with mixed grain
dust with a maximum particle diameter of 75 μm is given in
Figs. 2 through 5. The conditions for this test were as
follows — the dust moisture content was 0.2% by mass, the
oxidizer air dew point was -23°C, the nominal secondary
dust concentration was 366 g/m³, the V-channel mass loading
was 500 g/m^3, and the initiator detonation tube pressure
was 1 atm abs. The exhaust elbow was attached to the facility during this test.
    In Fig. 2 the static pressure is given as a function of
time at different tube locations. The pressure is normalized with atmospheric pressure, the distance with the diameter of the FAT, and the time with the time taken by an
acoustic wave to traverse the length of the tube. In Fig.
3 the gas velocity normalized with the ambient speed of
sound is presented in similar form, while Fig. 4 contains
the gas temperature history normalized with the room temperature. An x-t diagram for the flame front is also given
in these figures. Figure 5 contains the time history of
the dispersed dust concentration at midsection of the tube,
normalized with the nominal secondary dust concentration.
The first disturbance to arrive at any location is the
shock wave resulting from the rupture of the initiator diaphragm, as can be seen in Fig. 2. This results in an induced convective flow towards the open end of the tube, as
shown in Fig. 3, which in turn causes some of the layered
dust to be dispersed, as seen in Fig. 5. The compression
of the air ahead of the flame can be seen by noting the
gradual increase in static pressure at a given location.
This increase is more rapid at downstream locations due to
the coalescing of successive compression waves. This effect also results in an increase in gas velocity towards

the tube exit, which in turn results in an increase in the
dispersed dust concentration. Once the flame arrives at
any location, the gas temperature rises to about 1500 K
and remains approximately constant. It should be noted
that the concentration probe data is useless after the arrival of the flame. The flame exits from the open end resulting in an expansion wave propagating at about 700 m/s
(the speed of sound in the burnt gases) towards the closed
end of the FAT, which "quenches" the combustion process.
  A brief summary of the results presented earlier (Kauffman et al. 1984) is given here for continuity. It has
been shown that the primary explosion process occurring

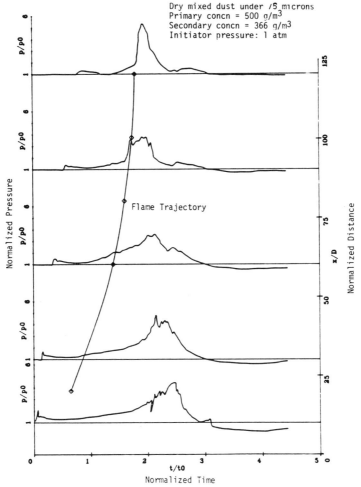

Fig. 2 Experimental pressure histories at different locations for
dry mixed dust.

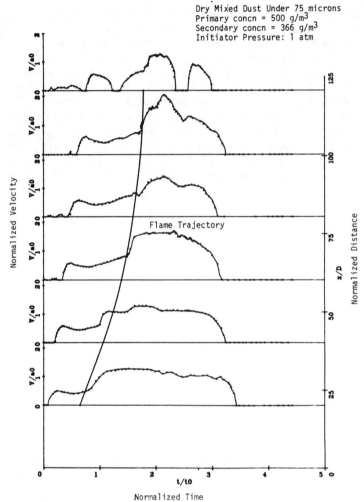

Fig. 3 Experimental gas velocity histories at different locations for dry mixed dust.

due to the ignition of dispersed V-channel dust does not dominate the secondary explosion process occurring due to the layered dust in the rest of the tube. Among the dusts tested, mixed grain dust and cornstarch were found to be more reactive than wheat dust and oil shale dust, resulting in flame velocities of about 600 m/s and overpressures of about 6 atm under dry conditions. Increasing the dust moisture content to 9% by mass suppressed dust entrainment and reduced the above values to 120 m/s and 1.3 atm, respectively. Contrary to existing data, larger particles (over 425 μm in diameter) were found to sustain a secondary

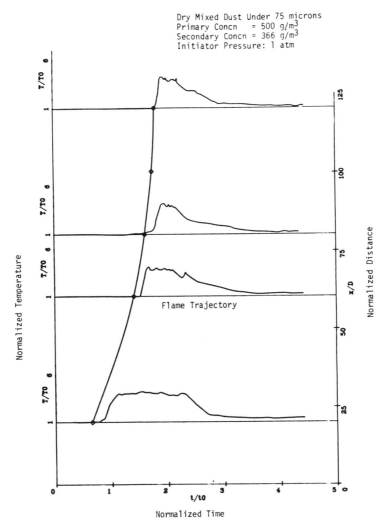

Fig. 4 Experimental temperature histories at different locations
for dry mixed dust.

explosion. Nominal secondary dust concentration limits for
flame propagation were determined under different conditions using mixed grain dust.

In tests conducted in the FAT, the fuel is distributed
in the form of a layer rather than as suspended dust.
Since the thickness of the layer affects the dust dispersion and hence the local concentration of the dust/air mixture, it has an effect on the secondary explosion process.
This effect was studied as follows. Mixed dust with a max-

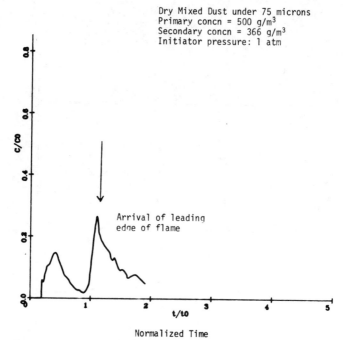

Fig. 5   Dust concentration-time curve at x/D = 40.2 for dry mixed
dust.

imum particle diameter of 75 μm and a moisture content of
9% by mass was deposited in the FAT at a given concentration in three different widths with attendant thicknesses:
a 12.5-mm-wide layer designated the thick layer, a 90-mm-
wide layer designated the medium layer, and a 700-mm-wide
layer designated the thin layer. The nominal dust concentration in the secondary section of the FAT was varied
from 100 to 1500 g/m^3. As the layer widths were held constant, the layer thicknesses increased as the dust concentration was increased. To give an idea of the layer thicknesses involved, at a nominal secondary dust concentration
of 366 g/m^3, the layer thicknesses were 3.81 mm, 0.53 mm,
and 0.10 mm for the thick, medium, and thin layers, respectively. In these tests the V-channel was filled with
dust at a mass loading equal to the nominal secondary dust
concentration. The initiator tube was at a pressure of 1
atm abs. The dew point of the air in the tube was -7°C.
The facility did not have the exhaust elbow for these
tests.
     The maximum pressure that was obtained in a test is
given in Fig. 6 as a function of the nominal secondary dust

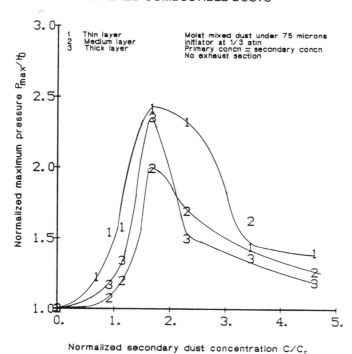

Fig. 6 Variation of maximum pressure with secondary dust concentration for different layer thicknesses.

concentration for the three different layers. The pressure
is normalized with room pressure and the concentration with
the theoretically calculated (Srinath 1985) stoichiometric
concentration of 217 g/m^3. The plotted maximum pressures
typically occurred near the open end of the FAT. For the
range of mixed grain dust concentrations for which the
flame propagated throughout the length of the FAT, an accelerating flame was realized. It can be seen that bell-
shaped curves were obtained with a maximum occurring at a
fuel rich nominal secondary dust concentration. This is
because of partial entrainment of the layered dust and incomplete combustion of the entrained dust. It is found
that the thin layer produced the highest overpressures at
all concentrations. At lower concentrations, the thick
layer produced higher overpressures than the medium layer,
while this order was reversed for higher concentrations.
The same behavior can be seen for the maximum flame velocity obtained in a test normalized with the ambient speed of
sound, given in Fig. 7, as a function of normalized secondary dust concentration for the three different layers.
Further, in the dust concentration limit data for flame

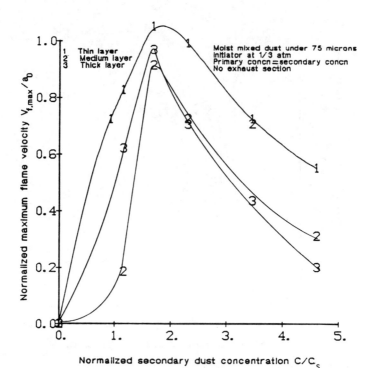

Fig. 7 Variation of maximum flame velocity with secondary dust concentration for different layer thicknesses.

propagation throughout the length of the FAT presented before (Kauffman et al. 1984), it was shown that a lean limit of 200 g/m^3 was obtained for the thin layer, 366 g/m^3 for the medium layer, and 250 g/m^3 for the thick layer.

Considering the above data, it is evident that the conditions most conducive to flame propagation are provided by the thin, thick, and medium layers, in that order, at lower dust concentrations. This could perhaps be attributed to a change in the mechanism by which the dust is entrained from the tube walls. Experimental observations have shown (Singer et al. 1972) that the dust is dispersed into the gas stream by two mechanisms; an erosion process by which dust is lifted up from the entire bed surface, and denudation whereby aggregates and clumps are fragmented from the leading edge and finally dispersed at midstream. The thick layer in the FAT tests is dispersed primarily by the denudation process, while the thin layer is dispersed by the faster and more efficient surface erosion process (Hwang et al. 1974). But for the medium layer, both the thickness and surface area are intermediate, and this may have resulted in conditions least conducive to flame propagation.

Once the layer thickness increases above a threshold value, the boundary layer denudation increases for the medium layer, and by virtue of its larger surface area, it might become more conducive to flame propagation than the thick layer.

Other tests were conducted with small particle mixed dust deposited in a thin layer at the conditions described above, with the exception of the initiator tube pressure, which was increased to 1 atm abs. It has been shown that this results in an increase in the primary explosion strength (Kauffman et al. 1984). It was found that increasing the initiator tube pressure resulted in increased pressure, gas velocity and flame velocity in the secondary explosion, while the temperatures remained about the same. This can be seen by the maximum pressure, temperature, gas velocity and flame velocity data presented in Table 1, at two different nominal secondary dust concentrations. These results were compared to the effect of increasing the V-channel dust concentration, which also caused an increase in the primary explosion strength (Kauffman et al. 1984). It was found that increasing the initiator detonation tube pressure had a more pronounced effect on the resultant flame velocities and overpressures than increasing the V-channel concentration, even though both resulted in similar increases in the primary explosion strength (Srinath 1985). This could be attributed to the stronger shock produced by the initiator at higher pressures, resulting in higher associated convective velocities and therefore more extensive entrainment of the layered dust. It has been shown earlier (Kauffman et al. 1984) that an increase in the primary explosion strength also has the effect of widening the nominal secondary dust concentration limits for flame propagation.

Theoretical Results

A mathematical study of a propagating flame and its induced aerodynamics in a dust has been conducted by Chi and Perlee (1974). A numerical code was developed at the U.S. Bureau of Mines, Pittsburgh, for this purpose, and it is described in the above reference. This code has now been modified to include entrainment of the layered secondary dust due to the convective flow and the effect of turbulence on the normal burning velocity through phenomenological expressions. The modified code is used to solve the one-dimensional, compressible flow equations governing a flame propagating through the FAT, a tube of finite length closed at the ignition end and open at the opposite end.

S.R. SRINATH ET AL.

Table 1 Effect of primary explosion strength on secondary explosion

Dust concentration g/m^3	Variable	Initiator pressure atm	Numerical value
250	P_{max}/P_o	1/3	1.57
		1	2.70
250	T_{max}/T_o	1/3	4.36
		1	4.40
250	V_{max}/a_o	1/3	1.02
		1	1.60
250	V_{fmax}/a_o	1/3	0.83
		1	1.39
366	P_{max}/P_o	1/3	2.36
		1	2.98
366	T_{max}/T_o	1/3	4.86
		1	4.81
366	V_{max}/a_o	1/3	1.29
		1	1.64
366	V_{fmax}/a_o	1/3	1.08
		1	1.73

The following assumptions are made:

1) The flame is a thin internal moving boundary across which the flow fields are discontinuous.

2) The flame is quasi-one-dimensional, i.e., gas properties and state variables represent averages over the cross-sectional area and vary only in the longitudinal direction.

3) The gas is homogeneous and continuous except across the flame front.

4) The gas is in thermodynamic equilibrium.

5) The gas obeys the ideal gas law.

6) Body forces, molecular transport phenomena, radiative transport, and viscous forces are neglected.

7) The flame is completely characterized by the specified expansion ratio x and the burning velocity S_t.

With the above assumptions, the governing equations can be written as follows:

$$D\rho/Dt + \rho\ \delta V/\delta x = 0 \qquad \text{(mass)} \qquad (1)$$

$$DV/Dt + 1/\rho\ \delta P/\delta x = 0 \qquad \text{(momentum)} \qquad (2)$$

$$D/Dt\,(e + V^2/2) = q - V/\rho\ [\delta p/\delta x] - P/\rho\,(\delta V/\delta x) \quad \text{(energy)} \quad (3)$$

$$\text{where} \quad D/Dt = \delta/\delta t + V\ \delta/\delta x \qquad (4)$$

The equation of state used for system closure is

$$P = \rho RT = (\gamma-1)\rho e \qquad (5)$$

Using the following transformations,

$$m = \rho V \qquad \text{(mass flux)}, \qquad (6)$$

and

$$E = \rho\,(1/2V^2 + e) \qquad \text{(total energy per unit volume)} \quad (7)$$

The governing equations can be rewritten as follows:

$$\delta\rho/\delta t = \delta m/\delta x \qquad (8)$$

$$\delta m/\delta x = -\ \delta/\delta x\,[m^2/\rho + P] \qquad (9)$$

$$\delta E/\delta t = -\ \delta/\delta x\,[m/\rho\,[E+P]] + \rho q \qquad (10)$$

and

$$P = (\gamma-1)\,(E - m^2/2\rho) \qquad (11)$$

The transformed equations (8) to (11) can be shown to be in conservative form (Courant, Friedrichs and Lewey 1967). This property guarantees satisfaction of the Rankine-Hugoniot relations across a normal shock. Therefore the finite difference forms of the above equations capture shocks implicitly and give the correct jump conditions across a shock front.

The boundary conditions are given by:
1) At the closed end $V(0,t) = 0$.
2) At the open end $P(L,t) = 0$.
3) Across the flame, the density, velocity, and pressure relations are given by the following equations:

$$\rho_u = \chi \rho_b$$

$$V_u = V_b + (\chi-1)S_t,$$

and
$$P_u = P_b[1 + \gamma_b \chi(\chi-1)S_t^2/a_b]$$

The initial conditions are given by:
1) $V(0,0) = 0$, $T = T_b$, $P = P_b$ on the burned side.

2) $V(0,0) = (\chi-1)S_t$, $T=T_0$, $P=P_0$ on the unburned side.

3) Elsewhere all flow fields are set equal to their ambient values.

The equations (8) to (11), the boundary conditions, and the initial conditions are nondimensionalized using appropriate constants. The nondimensional form of the equations are omitted here since they are similar to the dimensional form, but for the appearance of a dimensionless constant γ_0 in the momentum and energy equations.

The flame is completely characterized by the specified density expansion ratio χ and the turbulent normal burning velocity S_t. This avoids the need for speculation regarding the one-dimensional nature of the transport mechanism or the functional nature of the exothermic reactions involved. However, neither χ nor S_t is contained within the governing equations. Therefore the flame is treated as an internal moving boundary. The partial differential equations describing the burned and unburned regions are solved separately using a modified Lax-Wendroff algorithm and their solutions are matched at the flame front with the appropriate boundary conditions. The detailed governing equations, boundary conditions, and solution procedure can be found in Chi and Perlee (1974).

Numerical experiments were conducted to describe a test conducted with dry mixed dust having a maximum diameter of 75 μm at a nominal secondary dust concentration of 366 g/m³. As discussed above, it is necessary to specify the normal burning velocity S_t and the density expansion ratio χ for the numerical integration. The turbulent normal burning velocity S_t, defined as the velocity with which the flame front propagates relative to the unburned mixture ahead, is given in Fig. 8 as a function of location in the FAT for a number of conditions. It should be noted that S_t is the normal burning velocity and is different from the flame velocity which is measured in laboratory coordinates. It can be seen that S_t increases continuously as the flame propa-

1 Moist mixed dust over 425 microns,
   1 atm initiator, with exhaust section
2 Moist.mixed dust under 75 microns,
   1/3 atm initiator, without exhaust section
3 Moist mixed dust under 75 microns,
   1 atm initiator, without exhaust section
4 Moist mixed dust under 75 microns,
   1 atm initiator, with exhaust section
5 Dry mixed dust under 75 microns,
   1 atm initiator, with exhaust section
6 Dry corn starch under 75 microns,
   1 atm initiator, with exhaust section

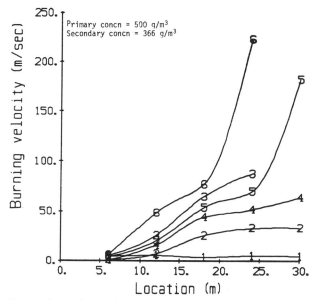

Fig. 8 Variation of burning velocity with location for different conditions.

gates from the closed end to the open end of the FAT. This is caused by an increase in the local turbulence intensity due to the increase in the mean gas velocity. Such large burning velocity values have been noted by other authors under similar conditions. For coal dust propagation in a 366-m-long gallery of 5.6-m^2 internal cross-section (Rae 1973), it has been reported that "the front of the combustion zone may have an appreciably greater velocity (up to some hundreds of meters per second) than the preceding air." Also, burning velocities of the order of 120 m/s have been estimated (Clark and Smoot 1985) for similar tests conducted with coal dust elsewhere (Richmond and Liebman 1975). It should be noted, however, that the S_t values presented

in Fig. 8 are the computed differences between two large
quantities and should hence be treated with some caution,
especially towards the open end of the FAT. All the values
given above are much smaller than the turbulent burning
velocities of the order of 3.5 m/s measured by other researchers in constant volume combustion bombs (Tezok 1985).
    Guided by the above results, the following linear vari-
ation of the burning velocity with location in the FAT was
used:

$$S_t/S_L = 14.29 + 1.07 \ x_f/D \qquad (12)$$

The above values for the slope and intercept reproduced the
experimental results best. A value of 0.7 m/s was used for
the laminar burning velocity (Tezok 1985). It was found
that by using a constant laminar burning velocity of 0.7
m/s, or a constant turbulent burning velocity of 10 m/s,
the experimental results could not be reproduced. In both
cases the flame did accelerate due to the expansion of the
burned gases and the subsequent increase in the convective
flow velocity ahead of the flame, but the flame acceleration predicted was much less rapid than what was observed
in the experiments and the predicted overpressures were
significantly lower than those obtained in the experiment.
    For determining the expansion ratio χ, it was neces-
sary to estimate the entrained dust concentration just a-
head of the flame. The dispersed dust concentration at the
instant of flame arrival, which can be determined from the
data in Fig. 5, was plotted as a function of the gas velocity at the same location under a number of different conditions. The following linearized form of the experimentally
determined relationship was used:

$$C/C_o = 0.137 + 1.13 \ V/a_o \qquad (13)$$

Using the above determined values for the entrained dust
concentration, the corresponding expansion ratio χ was theoretically calculated (Srinath 1985).
    Pressure histories at different locations obtained
from the numerical calculations are given in Fig. 9. Time
is normalized with t_o, the time taken by an acoustic wave
to traverse the length of the FAT, and distance with D,
the diameter of the FAT. The flame trajectory is also
given. Because of limitations in the code, numerical results are available only until the flame reaches to within
0.72 m of the open end. The given theoretical pressures
at the exit end actually correspond to 35.64 m. The pressure at 36 m, the open end, is set equal to 1 atm at all
times by the boundary condition. Results from every fourth

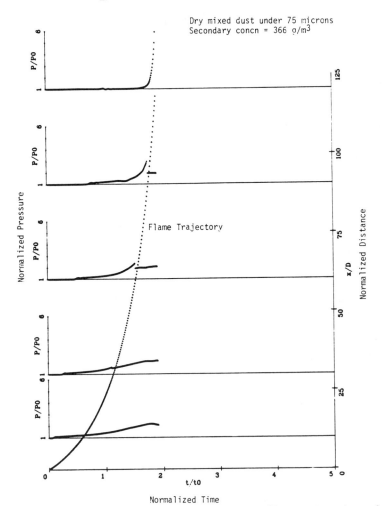

Fig. 9 Theoretical pressure histories at different locations for dry mixed dust.

iteration are plotted to avoid smearing of the data.

The numerical results presented in Fig. 9 can be compared to the corresponding experimental results given in Fig. 2 for a test conducted under the same conditions. On comparing the two figures, it is evident that the disturbance due to the rupture of the initiator diaphragm, seen in the experiment, is not predicted by the theory since the primary explosion is not included in the numerical simulation. After the initiator effects subside, the theoretical and experimental pressure histories are found to be in good agreement until the arrival of the flame front at each

location. The predicted drop in pressure across the flame
front at $x/D = 60.3$ and $x/D = 90.5$ is not seen in the experimental results. This is due to the longer flame zones
realized in reality, unlike the thin flame assumed, causing
pressure changes across the flame to occur over a longer
time. The experiment and theoretical flame trajectories
also match closely in the latter half of the tube, by which
time the initiator effects would have subsided. The theoretically predicted maximum pressure of 5.89 atm and maximum flame velocity of 607 m/s are in good agreement with
the corresponding experimental values of 5.39 atm and 673
m/s. The temperatures are overpredicted by about 100 to
300 K, while the profiles match experimentally determined
curves. The gas velocities at the instant of flame arrival
are also found to be in reasonable agreement with experiment (Srinath 1985).

A sensitivity analysis was performed to determine the
dependence of the flow parameters on the slopes of the nondimensional equations (12) and (13) which determine the
normal burning velocity and the entrained dust concentration ahead of the flame, respectively. It was found that
a 20% change in the slope of the normal burning velocity
equation (12) causes the predicted overpressures to double,
while a corresponding change in the slope of the dispersed
dust concentration equation (13) had a smaller effect on
the overpressures predicted. This result is in agreement
with findings from similar work elsewhere (Clark and Smoot
1984), suggesting that the increase in burning velocity
due to turbulence might also have a strong influence on the
flame propagation process studied.

Conclusions

Experimental evidence has been presented to show that,
for propagating flames in ducts supported by layered dust
as opposed to dispersed dust, the thickness of the dust
layer is an important parameter that governs the entrainment of dust and thus has an effect on the resultant overpressures and flame velocities. The results suggest that
at lower dust concentrations there might be a change in the
mechanism of dust entrainment as the thickness of the layer
is increased at a given mass loading. The primary explosion strength also has a strong effect on the amount of
dust dispersed initially and thus affects the development
of the secondary explosion. Further, an increase in the
primary explosion strength results in the widening of the
nominal secondary dust concentration limits for flame
propagation. Numerical simulation using a simplified de-

scription of the flame propagation process has been shown to give results in reasonable agreement with experiments, especially quantities of importance such as maximum over-pressure, flame velocity, etc. The increase in burning velocity due to an increase in the turbulence intensity has been shown to also have a strong influence on the flame propagation process.

Acknowledgment

The authors would like to thank the National Institute for Occupational Safety and Health for funding this re-search program; F. J. Liu and R. J. Ettema for their assis-tance with the experimental studies; and the U.S. Bureau of Mines, Pittsburgh, for providing the source listing for the numerical code.

References

Cashdollar, K.L., Liebman, I. and Conti, R.S. (1981) Three bureau of mines optical dust probes. Bureau of Mines Report of Investigation 8542, U.S. Bureau of Mines, Pittsburgh, Pa.

Chi, D.N. and Perlee, H.E. (1974) Mathematical study of a propagating flame and its induced aerodynamics in a coal mine passageway. Bureau of Mines Report of Investigation 7908, U.S. Bureau of Mines, Pittsburgh, Pa.

Clark, D.P. and Smoot L.D. (1985) Model of accelerating coal flames. Combust. Flame 62, 3, 255-269.

Courant, R., Friedrichs, K. and Lewy, H. (1967) On the partial difference equations of mathematical physics. IBM J. 2, 215-234.

Dorsett, H.G., Jr., Jacobson, M., Nagy, J. and Williams, R.(1960) Laboratory equipment and test procedure for evaluating explosibility of dusts. Bureau of Mines Report of Investigation 5624, U.S. Bureau of Mines, Pittsburgh, Pa.

Enomoto, H. (1977) Explosion characteristics of agricultural dust clouds. Proceedings of the International Symposium on Grain Dust Explosions, pp. 163-170. National Academy of Sciences, Washington, D.C.

Faraday, M. and Lyell, C. (1845) Report to the Home Secretary on the explosion at the Haswell colliery on 28 September, 1844. Philos. Mag. 26, 16-30.

Hartmann, J., Jacobson, M. and Williams, R. P. (1950) Laboratory explosion study of American coals. Bureau of Mines Report 4725, U.S. Bureau of Mines, Pittsburgh, Pa.

Hwang, C.C., Singer, J.M. and Hartz, T.N. (1974) Dispersion of dust in a channel by a turbulent gas stream. Bureau of Mines Report of Investigation 7854, U.S. Bureau of Mines, Pittsburgh, Pa.

Ishihama, W. and Enomoto, H. (1973) New experimental method for studies of dust explosions. Combust. Flame 21, 2, 177-186.

Kauffman, C.W., Srinath, S.R., Tezok, F.I., Nicholls, J.A. and Sichel, M. (1985) Turbulent and accelerating dust flames. 20th Symposium (International) on Combustion, pp. 1701-1708, The Combustion Institute, Pittsburgh, Pa.

Price, D.J. and Brown, H.H. (1922) Dust Explosions. National Fire Protection Association, Boston, Mass.

Rae, D. (1973) Initiation of weak coal-dust explosions in long galleries and the importance of the time dependence of the explosion pressure. 14th Symposium (International) on Combustion, pp. 1225-1236. The Combustion Institute, Pittsburgh, Pa.

Richmond, J.K. and Liebman, I. (1975) A physical description of coal mine explosions. 15th Symposium (International) on Combustion, pp. 115-126. The Combustion Institute, Pittsburgh, Pa.

Singer, J.M., Cook, E. and Grumer, J. (1972) Dispersal of coal and rock dust deposits. Bureau of Mines Report of Investigation 7642, U.S. Bureau of Mines, Pittsburgh, Pa.

Srinath, S.R. (1985) Flame propagation due to layered combustible dusts. Ph.D. Thesis, The University of Michigan, Ann Arbor, Mich.

Tezok, F.I. (1985) Flame propagation through turbulent combustible dust/air mixtures. Ph.D. Thesis, The University of Michigan, Ann Arbor, Mich.

Wolanski, P. and Waudby-Smith, P. (1980) Four wavelength pyrometer: Operation and calibration. Departmental communication, Department of Aerospace Engineering, The University of Michigan, Ann Arbor, Mich.

Stability of a Droplet Vaporizing in a Hot Atmosphere

F.J. Higuera* and A. Liñán†

Universidad Politécnica de Madrid, Madrid, Spain

Abstract

A linearized stability analysis is carried out for an evaporating liquid droplet in a hot atmosphere with a view to understand the stabilizing effects of surface tension together with the unstabilizing effects of expansion, due to the phase change, and of the motion in the liquid induced by tangential viscous stress on the interface. The analysis is carried out for both small and order-one gas Reynolds number based on the unperturbed Stefan flow. Two different surfaces bound the stability domain; on one of them, the growth rate of the perturbations is imaginary and on the other, vanishes.

Introduction

The subject of quasisteady droplet vaporization has received considerable attention in the literature. Under very general conditions, the mass vaporization rate per unit area of the interface of a droplet evaporating in a stagnant hot atmosphere is proportional to the inverse of the droplet radius. Therefore, the area of the interface, or the square of its diameter, decreases at a constant rate. This behavior was first observed by Sreznersky in 1882, and its explanation, given by Langmuir in 1918, is based on the fact that the mass flux is directly related to the heat flux coming to the interface from the gas, which is proportional to the inverse of the droplet radius. Later the "d-square law" was found to hold true for the consumption of a droplet of fuel burning in an oxidizing atmosphere.

The "d-square law" results from assuming that the droplet temperature is constant equal to the boiling temperature, so that no heat is transferred to the interior of the droplet. In addition the quasisteady assumption is used for

---

Presented at the 10th ICDERS, Berkeley, California, August 4-9, 1985. Copyright © 1986 by The American Institute of Aeronautics and and Astronautics, Inc. All rights reserved.

*Assistant Professor, Department of Fluid Mechanics, E.T.S.I. Aeronáuticos.

†Professor, Department of Fluid Mechanics, E.T.S.I.Aeronáuticos.

the gas phase. These assumptions have not been used in more
refined analyses, see for example the review by Sirignano
(1983), but the literature on the stability of the droplet
vaporization process is very scarce.

When describing the liquid oscillations, one can take
advantage of the fact that the gas-to-liquid density and
viscosity ratios are small to neglect, in first approximation, the effect of the gas phase stresses on the droplet
surface. This has been done in analysing the free oscillations of droplets and bubbles, first considered, in a way
or another, by Kelvin (1890) and Rayleigh (1894) for the
inviscid case; they found and undamped oscillatory motion.
However viscous effects lead to the damping of these oscillations as it as been shown first by Lamb (1932), who considered the small viscosity case. The more viscous cases
were treated by Chandrasekhar (1959) and Reid (1960). An
attempt to generalize the results of the small viscosity
case to account for the possibly destabilizing effect of
the vaporization, when this can be considered as a small
perturbation, is presently under study and will be published
elsewhere.

In this paper we shall analyze another aspect of the
linear stability of the vaporization of a droplet in a hot
atmosphere, involved with a slow response of the system associated with the nonuniform temperature distribution within
the droplet, found in the initial stages of the droplet vaporization. Small perturbations in the local vaporization
rate can induce instabilities which disappear altogether
when the liquid temperature distribution becomes uniform.

Instabilities of planar vaporizing interfaces, associated with nonuniform temperature distributions within the
liquid have been described in the literature. For example,
Palmer (1976) analyzed the stability of the vaporization of
a superheated liquid, pointing out the role of the shear
stress exerted on the liquid surface by the gas as a possibly relevant destabilizing factor, among others. The effect on the interface of pressure perturbations in the gas
due to changes in the local vaporization rate and due to
the motion of the gas was pointed out by Hickman (1952;
1972) for this same problem, and similar effects had been
found before by Landau (1944) for a flame propagating in a
combustible gas mixture. In all these cases of unperturbed
planar interfaces the viscosity does not play an important
role in the determination of the velocity and pressure perturbations. Transport effects are, however, essential in
the droplet vaporization problem, because the heat has to
arrive at the interface by conduction against the radial
outward flow induced by the vaporization. The pressure per-

turbations now take a different form; however, they still have an important effect on the stability when the deformation of the interface is to be accounted for in the analysis.

Here the limit of small gas-to-liquid density and viscosity ratios will be used. At the interface, we assume a local thermodynamic equilibrium condition, leading to the Clausius-Clapeyron relation, but, in many cases, when vaporization occurs, the interface temperature is almost constant, close to the boiling temperature, because the ratio cT_b/L of the specific internal energy at the boiling temperature to the specific latent heat is moderately small.

We begin by describing the spherically symmetrical, unperturbed, vaporization process that in the limiting case indicated above includes three stages.

In a first stage the droplet is heated without significant vaporization until the surface temperature reaches a value close to the boiling temperature T_b; in this stage the stability analyses for a nonvaporizing droplet are applicable.

In a second stage the surface temperature is close to T_b, but a fraction of the heat reaching the droplet surface is transported to the interior of the droplet to uniformize the temperature distribution. The vaporization mass flux increases during the second stage from a negligible value to the quasisteady value corresponding to the third stage, that cover most of the lifetime of the droplet, when the liquid temperature is nearly uniform.

In the stability analysis two time scales appear. There is a fast response associated to the motion in the gas phase and a slow one associated to the liquid. When trying to describe the instabilities associated with the fast response, we found that, within the framework of the present model, the effects of the perturbations in the heat flux entering the droplet during the second heating-vaporization stage are negligible, so that this analysis is essentially the same for the second and third stages. It turns out that there are not instabilities associated to the fast response of the gas, and for this reason the stability analysis presented in this paper is restricted only to the slow time scale. The quasisteady state approximation is used in the analysis that follows, because the times involved are long compared with the gas-phase response time.

Two main parameters are found to control the stability properties during the heating-vaporization period. One is the Reynolds number β, based on the droplet radius and the typical gas phase vaporization flow; the other parameter,

Σ, is the ratio of the pressure jump due to surface tension and the typical gas viscous stresses.

The analysis is first carried out for low ambient temperatures, when the vaporization rate is small enough to make the Reynolds number small, and for large values of Σ, so that the deformations of the interface can be neglected: The motion induced in the liquid by the viscous drag of the gas through the interface has a destabilizing effect, opposite to that of the motion induced by changes in surface tension, due to interface temperature perturbations, which is stabilizing.

For small droplet radius Σ becomes of order unity and the deformation of the interface has to be accounted for. The analysis of this case is given for small and order one gas Reynolds numbers β. Two stability limits are found; on one of them the growth rate of the perturbations vanishes and on the other, which provides an upper bound for the heat flux coming from the gas to the droplet, it is imaginary, resulting in an oscillatory behavior of the perturbations.

The same analysis is applicable to a droplet of fuel burning in an oxidizing atmosphere when the stoichiometric ratio is large or the oxidizer concentration in the gas is small, so that the combustion takes place far from the droplet.

The Unperturbed State

The following description of the unperturbed droplet vaporization process is justified by the small value of the ratio of gas-to-liquid densities, that allows us to use the quasisteady approximation for the gas phase processes, and by the small value of the parameter cT_b/L, that makes the vapor pressure very sensitively dependent on the interface temperature. The effect of surface tension on phase equilibrium is neglected, retaining only its influence on the mechanical equilibrium of the interface.

In the first of the three stages mentioned in the Introduction, the vaporization rate is negligible because the interface temperature is not yet close to the boiling temperature T_b. The heat coming from the gas phase is used to heat the droplet until, after a fairly well-defined heating time, the interface temperature reaches a value that differs from T_b by a small amount of order cT_b^2/L and the vaporization begins; the temperature ceases to increase at the interface but not in the interior of the droplet, where it is lower. In a second transient period of order $t_c = \rho_L cR_o^2/k_L$, after the initiation of the vaporization, the temperature in the bulk of the liquid is raising up to the boiling temperature. Here, ρ_L, c, and k_L are the liquid density, specific heat, and thermal conductivity, respectively, and R_o is the initial droplet radius. At the end of this second stage, the liquid temperature is nearly uniform and does not change in the rest of the vaporization. The droplet lifetime is $t_v \sim \rho_L R_o/m$, in terms of the characteristic mass flux at the interface $m \sim k_g(T_\infty - T_b)/R_o L$; k_g is the gas thermal conductivity and T_∞ the temperature far from the droplet. The droplet radius does not change appreciably in the second transient step because $t_v/t_c \sim (k_L/c)/(k_g/c_p) L/c_p(T_\infty - T_b)$ is usually large, for example $t_v/t_c \sim 8$ for a droplet of water in air at $T_\infty = 1000°K$.

In the first heating period, there is only pure conduction in the gas phase at distances of the order of the droplet radius, and the heat flux coming from the gas to the interface is given by $k_g(T_\infty - T_s)/R_o$, where T_s is the unknown instantaneous interface temperature. In the liquid, the temperature distribution is the solution of the problem

$$\frac{\partial \theta}{\partial \tau} = \frac{1}{r^2} \frac{\partial}{\partial r} \left(r^2 \frac{\partial \theta}{\partial r} \right) \tag{1}$$

$$\tau = 0: \quad \theta = 1 \tag{2}$$

$$r = 0: \quad \frac{\partial \theta}{\partial r} = 0 \tag{3}$$

$$r = 1: \quad \frac{\partial \theta}{\partial r} = -\psi_1 \tag{4}$$

where $\theta = (T-T_b)/(T_i-T_b)$ is the nondimensional liquid temperature; r is measured with the droplet radius as a unit; $\tau = = t/t_c$, and $\psi_1 = (k_g/k_L)(T_\infty - T_b)/(T_b - T_i)$. A small term $-\theta k_g/k_L$ has been omitted in the right-hand side of Eq. (4) because $k_g \ll k_L$. The solution can be written in the form of a infinite series (Carslaw and Jaeger 1959) and is valid up to the instant $\tau = \tau^*$ in which $\theta(r=1) = 0$ and the vaporization begins. τ^* as a function of ψ_1 is plotted in Fig. 1a.

In the second step, the temperature and the mass fraction of the vapor in the gas are given by

$$T_{go} = T_b + (T_\infty - T_b) \frac{\exp\left[-\beta(1/r-1)\right]}{e^\beta - 1} \tag{5}$$

$$Y_o = 1-e^{-\beta Le/r} \tag{6}$$

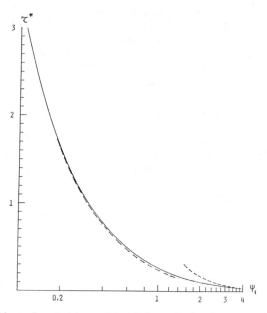

Fig. 1.a. Nondimensional time of begining of the droplet vaporization as a function of ψ_1. The dashed lines are the asymptotic expressions $\tau^* = \pi/4\psi_1^2$ and $\tau^* = 1/3\psi_1-1/15$ for large and small values of ψ_1.

where the subscript o is used for the basic solution and $\beta = mc_pR/k_g$ is the nondimensional vaporization rate, which is a function of τ, to be calculated as part of the solution. Le is the Lewis number of the vapor in the gas. In the liquid, the heat conduction equation (1) is still valid but the conditions (2-4) change to

$$\tau = \tau^*: \quad \theta = \theta^*(r) \tag{7}$$

$$r = 0: \quad \frac{\partial \theta}{\partial r} = 0 \tag{8}$$

$$r = 1: \quad \theta = 0 \tag{9}$$

where $\theta^*(r)$ is the temperature distribution inside the droplet at the end of the first heating step. In order to calculate $\beta(\tau)$, the additional boundary condition

$$r = 1: \quad \frac{\psi_1\beta}{e^{\beta}-1} = \psi_2\beta - \frac{\partial \theta}{\partial r} \tag{10}$$

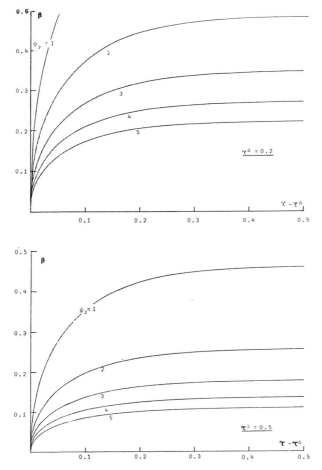

Fig. 1.b. Time evolution of the nondimensional vaporization mass flux.

is needed, where $\psi_2 = \left[(k_g/c_p)/(k_L/c)\right] \times L/c(T_b - T_i)$. Equation (10) is an energy balance through the interface. Again, the solution of Eq. (1) and Eqs. (7-9) can be written in the form of an infinite series, and the value of $\beta(\tau)$ resulting from Eq. (10) is plotted in Fig. 1b. For a detailed account of the short transition between the first and second steps see Liñán and Rodríguez (1985).

When $\tau \to \infty$, θ goes to zero and all the heat coming from the gas is used in the vaporization. The final, asymptotic value of β is

$$\beta = \beta_e = \ln\left[1 + c_p(T_\infty - T_b)/L\right] \qquad (11)$$

and this expression is also valid in the last step if the instantaneous droplet radius is used in the definition of β. The total mass flux $4\pi R^2 m$ is proportional to the droplet radius $R(t)$, whose decay follows the "R-square law"

$$\left(\frac{R(t)}{R_o}\right)^2 - 1 = 2 \frac{L\beta_e}{c_p(T_\infty - T_b)} \frac{t}{t_v} \tag{12}$$

with

$$t_v = \frac{\rho_L c_p R_o^2 / k_g}{c_p(T_\infty - T_b)/L}$$

Formulation of the Linear Stability Problem

The heat flux neccesary for vaporizing the liquid has to arrive from the gas counteracting the effect of the convection due to the vaporization; therefore, in the gas, the ratio of convection to conduction, measured by β, cannot be large. The characteristic Reynolds number is β/Pr and $Pr \sim 1$, so that viscosity is always important in the motion of the gas, and when the basic spherical solution is slightly perturbated, the pressure and velocity variations p_g' and u_g' are in the relation $p_g' \sim \mu_g u_g'/R$, where μ_g is the gas viscosity coefficient. On the other hand, the equilibrium of tangential stresses at the interface leads to the condition $u_L'/u_g' \sim \mu_g/\mu_L$. The ratio of viscosity coefficients is very small and the parameter $\varepsilon = \sqrt{\rho_g/\rho_L} \, \mu_L/\mu_g$ will be assumed to be of order unity. Two main simplifications result from the small value of the gas-to-liquid density ratio and the foregoing estimates. First, the time derivatives can be neglected in the gas conservation equations because they are of order $\sqrt{\rho_g/\rho_L}$ compared to the others terms in these equations. Second, the interface is nearly a fluid surface from the point of view of the liquid.

In order to estimate the characteristic time for the evolution of the perturbations, t_o, the relation $k_L T_L'/R \sim L\rho_g u_g'$, where T_L' is the temperature perturbation in the liquid, must be taken into account. This relation comes from the energy balance through the interface. The heat flux entering the liquid is always important in this balance, the basic spherical configuration being stable when it is negligible, as will be seen later. However, the liquid Prandtl number Pr_L will be assumed to be large, as is often the case so that the heat conduction in the bulk of the liquid is too slow to affect the evolution of the perturbations, except in a thin thermal layer close to the interface, and the energy conservation equation for the liquid leads to the relation $T'_L \sim t_o u'_L (\partial T_{Lo}/\partial r)$. Carrying this to the previous relations results finally in

$$t_o \mu_L / \rho_L R^2 \sim \text{Pr}\varepsilon^2/\beta_e \tag{13}$$

The ratio $t_c/t_o \sim \beta_e \text{Pr}_L/\text{Pr}\varepsilon^2$ is large, and therefore the change in the basic unperturbated solution is negligible in the stability analysis and the normal mode method for steady basic solutions is applicable.

As a further simplification, density variations, which are due to temperature changes, are neglected in both fluids. This allows us to write the momentum equations linearized for small perturbations in terms of the pressure and the radial components of the vorticity ω_{gr} and the velocity u_{gr}, multiplying them by $\vec{x} = (r,0,0)$ after the operator $\nabla\times$ has been applied a certain number of times and the continuity equation $\nabla \cdot \vec{v} = 0$ has been used. The result in nondimensional form is

$$\frac{\partial}{\partial t}(ru_{Lr}) = -r\frac{\partial P_L}{\partial r} + \nabla^2(ru_{Lr}) \tag{14}$$

$$\frac{\partial}{\partial t}(r\omega_{Lr}) = \nabla^2(r\omega_{Lr}) \tag{15}$$

$$\frac{\partial}{\partial t}(\nabla^2 ru_{Lr}) = \nabla^2(\nabla^2 ru_{Lr}) \tag{16}$$

for the liquid, and

$$\frac{\beta/P_r}{r^3}\left[r\frac{\partial ru_{gr}}{\partial r} - 3ru_{gr}\right] = -r\frac{\partial P_g}{\partial r} + \nabla^2(ru_{gr}) \tag{17}$$

$$\frac{\beta/P_r}{r^3}\left[r\frac{\partial r\omega_{gr}}{\partial r} + r\omega_{gr}\right] = \nabla^2(r\omega_{gr}) \tag{18}$$

$$\frac{\beta/P_r}{r^3}\left[r\frac{\partial \nabla^2 ru_{gr}}{\partial r} - \nabla^2 ru_{gr}\right] = \nabla^2(\nabla^2 ru_{gr}) \tag{19}$$

for the gas. The unprimed variables u_{gr}, P_g, u_{Lr}, and P_L are the velocity and pressure perturbations in each fluid nondimensionalized with the factors m/ρ_g, $\mu_g m/\rho_g R$, $(m/\rho_L)\sqrt{\rho_L/\rho_g}$, and $(\mu_L m/\rho_L R)\sqrt{\rho_L/\rho_g}$, respectively. The time has been referred to $\rho_L R^2/\mu_L$, and r to the instantaneous droplet radius. Terms of order $\sqrt{\rho_g/\rho_L}$ corresponding to nonsteady effects in the gas and to recession of the interface in the unperturbed solution have been omitted. $r^2\omega_r$ and $r^2 u_r$ are the defining scalars for the toroidal and poloidal components of the velocity when the spherical harmonic decomposition is used (Chandrasekhar 1961).

The linearized energy conservation equations for the gas and liquid phases are

$$\alpha\,\beta\,e^{-\beta(1/r-1)}\,\frac{ru_{gr}}{r^3}+\frac{\beta}{r^2}\,\frac{\partial T_g}{\partial r}=\nabla^2 T_g \tag{20}$$

$$\frac{\partial T_L}{\partial t}+\frac{\beta}{Pr\varepsilon}\,u_{Lr}\,\frac{\partial T_{Lo}}{\partial r}=\frac{1}{Pr_L}\,\nabla^2 T_L \tag{21}$$

where

$$\alpha=\frac{c_p(T_\infty-T_b)}{L}\,\frac{\beta}{e^\beta-1} \tag{22}$$

and the temperatures of the gas and liquid have been non-dimensionalized by multiplying them by c_p/L and $c/L\delta$, respectively, with $\delta=(k_g/c_p)/(k_L/c)$. Notice the difference with the previous section.

The mass conservation equation for the vaporizing species in the gas can be obtained from Eq. (20) by replacing β by βLe and α by $-\beta Le$.

As the unperturbed solution is quasisteady and normal modes are used, the time dependence is exponential for every variable. In addition, the dependence on the angular variables is accounted for with the spherical harmonic decomposition, and the gas velocity and interface deformation, for example, are written as the real parts of $u_{gr}(r,\theta,\psi,t)=$ $=u_{gr}(r)\exp(\Omega t)Y_1^n(\theta,\psi)$ and $r_s-1=X\exp(\Omega t)Y_1^n(\theta,\psi)$, respectively, where the deformation has been referred to the droplet radius. The same name is used for a variable and for its radial part. l and n are integers defining the angular mode. As a consequence of the symmetry of the basic solution, n does not appear in the results and Ω, which is a complex number in general, depends only on l and the parameters. The equations for the radial functions result from Eqs. (14-21), replacing $\partial/\partial t$ by Ω and ∇^2 by $1/r^2 d/dr(r^2 d/dr)-l(l+1)/r^2$.

In addition to the regularity conditions at the origin and at infinity, the solutions of these equations are subject to the following continuity and conservation conditions at the interface, written for the radial part of the variables at $r = 1$.

$$(\beta/\epsilon Pr)u_{L_r} = \Omega X \tag{23}$$

$$\epsilon\left[P_L - 2\,\frac{du_{L_r}}{dr}\right] = P_g + 2\,\frac{\beta}{Pr}\,(u_{g_r} - X) - 2\left(\frac{du_{g_r}}{dr} + 6X\right)$$

$$+ (1-1)(1+2)\Sigma X + M(Y - \beta LeX) \tag{24}$$

$$\frac{d}{dr}\,\frac{r^2 u_{g_r}}{1(1+1)} + X = 0 \tag{25}$$

$$r^2 \omega_{g_r} = 0 \tag{26}$$

$$\epsilon\left[\frac{d^2}{dr^2} + 1(1+1) - 2\right]\frac{ru_{L_r}}{1(1+1)} = \left[\frac{d^2}{dr^2} + 1(1+1) - 2\right]\frac{ru_{g_r}}{1(1+1)} - 4X$$

$$+ M(Y - \beta LeX) \tag{27}$$

$$\frac{d}{dr}\,(\omega_{L_r} - \omega_{g_r}) = 0 \tag{28}$$

$$T_g + \alpha X = 0 \tag{29a}$$

$$T_L + \frac{\partial T_{Lo}}{\partial r}\,X = A(Y - \beta LeX) \tag{29b}$$

$$\beta u_{g_r} + \frac{dT_L}{dr} = \frac{dT_g}{dr} + \alpha\beta X \tag{30}$$

$$\beta Le\,e^{-\beta Le}\,u_{g_r} - \beta LeY + \frac{dY}{dr} = 0 \tag{31}$$

where terms of order $\sqrt{\rho_g/\rho_L}$ have been neglected, as well as a small term $X \partial T_{Lo}/\partial\tau$ in the left-hand side of Eq. (30). Equation (23) is the mass conservation condition. Equation (24) is the momentum conservation condition, the convective flux being negligible in the liquid. $\Sigma = \rho_g\sigma/m\mu_g$ is the non-dimensional surface tension coefficient and $M = (R_g T_b^2/LY_o) \times \rho_g(d\sigma/dT)/m\mu_g < 0$, where R_g is the vapor constant, is a

Marangoni number. The conditions of continuity of tangential velocities at the interface and the balance of tangential stresses lead to Eqs. (25,26) and (27,28), taking into account the continuity equation. The last term in Eq. (27) is the tangential stress due to the change of surface tension with the temperature, which, in turn, is related to the local concentration of the vapor at the interface through the equilibrium condition. $A = R_g c T_b^2 / L^2 Y_o(1) \delta$, appearing in the linearized equilibrium condition (29b), gives the magnitude of the interface temperature changes on the liquid-vapor saturation curve. Notice that these temperature changes are neglected in Eq. (29a) because they are small compared to the temperature perturbations in the gas. Finally Eqs. (30) and (31) come from energy and mass balance through the interface.

The equations set (15), (18), (26), and (28), with the regularity conditions at the center of the droplet and at infinity, has the trivial solution $\omega_{gr} = \omega_{Lr} = 0$. Only the other equations are going to be considered from here on.

Perturbations Without Deformation of the Interface

The normal stress due to the surface tension is proportional to $1/R$, whereas the normal viscous stress of the gas on the interface varies like $1/R^2$ because it arises from a velocity gradient and the velocity itself varies like $1/R$. Σ is a measure of the relative importance of both effects and it is proportional to the droplet radius ($m \sim 1/R$); therefore, the larger the droplet size, the more difficult it is to deform the interface by the action of perturbations due to local changes in the vaporization rate. Here we consider the limit $\Sigma \gg 1$, which is appropriate for all but very small droplets, neglecting the deformation of the interface. In addition, it will be assumed that $\beta_e \ll 1$ (always with $\beta_e Pr_L \gg 1$), so that viscosity dominates the motion of both fluids and the left-hand sides of Eqs. (14-19) can be neglected. The velocity and pressure perturbations in the gas and liquid phases can be written as

$$u_{gr} = \frac{-c_1}{2(2l-1)r^l} + \frac{b_1}{r^{l+2}} \qquad\qquad P_g = -\frac{c_1}{(l+1)r^{l+1}}$$

$$u_{Lr} = \frac{B_1 r^{l+1}}{2(2l+3)} + C_1 r^{l-1} \qquad\qquad P_L = \frac{B_1 r^l}{l}$$

$$(32)$$

whereas the temperature and concentration perturbations are given by

$$T_g = \frac{\alpha\beta}{r^l}\left(1 - \frac{1}{r}\right)\left(\frac{c_1}{4l(2l-1)} - \frac{b_1}{2(l+1)r}\right) \tag{33}$$

$$Y = \frac{\beta Le}{l+1}\left(-\frac{c_1}{2(2l-1)} + b_1\right)\frac{1}{r^{l+1}} + O(\beta^2) \tag{34}$$

$$T_L = -\frac{\beta}{\epsilon Pr\Omega}\frac{\partial T_{Lo}}{\partial r}\left(\frac{B_1 r^{l+1}}{2(2l+3)} + C_1 r^{l-1}\right) + AY(1)\exp\left[\sqrt{\Omega Pr_L}(r-1)\right] \tag{35}$$

where use has been made of Eqs. (29) and (31). The last term in Eq. (35) is due to the temperature changes at the interface and is associated with the thin thermal layer close to it. Carrying these results to conditions (23), (25), (27), and (30) at the interface, the system

$$\frac{B_1}{2(2l+3)} + C_1 = 0 \tag{36a}$$

$$c_1\frac{l-2}{2(2l-1)} - b_1 l = 0 \tag{36b}$$

$$\epsilon\left[B_1\frac{l(l+2)}{2l+3} + C_1 2(l^2-1)\right] = -c_1\frac{l^2-1}{2l-1} + b_1 2l(l+2) + MY(1)l(l+1) \tag{36c}$$

$$-\frac{\beta}{\epsilon Pr\Omega}\frac{\partial T_{Lo}}{\partial r}\bigg|_1\left[B_1\frac{l-1}{2(2l+3)} + C_1(l-3)\right] + A\sqrt{\Omega Pr_L}\,Y(1)$$

$$+ \beta\left(-\frac{c_1}{2(2l-1)} + b_1\right) = 0 \tag{36d}$$

is obtained. The last equation comes from Eq. (30); the perturbation in the heat flux from the gas to the interface drops out because it is of order β compared to the others terms in this equation. The system (36) leads to the dispersion relation

$$\Omega = \frac{(\alpha-\beta)(3+M\beta Le)l(l+1)}{\epsilon^2 Pr(2l+1)(l+1+ALe\sqrt{\Omega Pr_L})}, \qquad l \geqslant 2 \tag{37}$$

where use has been made of the result $\partial T_{Lo}/\partial r\big|_1 = \alpha-\beta$. When $cT_b/L \to 0$, both A and M vanish and the interface temperature does not change. In this case, the basic configuration is unstable. The reason is that below the interface regions

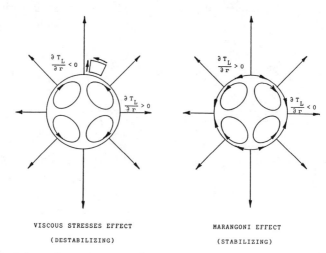

(DESTABILIZING)   (STABILIZING)

Fig. 2. Schematic representation of the viscous stresses and Marangoni effects in the case $\Sigma \gg 1$, $\beta_e \ll 1$.

where the mass flux increases ———— that is, where $Y_1^n(\theta,\psi)$ < 0 if it is chosen $c_1 > 0$, because $u_{gr}(r=1) = -c_1 Y_1^n/1(2l-1)$ ———— the radial velocity in the liquid is directed toward the center of the droplet (for $r < 1$, the radial velocity is dominated by the component $C_1 r^{1-1} Y_1^n(\theta,\psi)$ and $C_1 = 3(c_1/\varepsilon)/2(4l^2-1) > 0$). The temperature of every liquid particle does not change due to the perturbations because the heat conduction is negligible in the bulk of the liquid, and therefore the heat flux entering the droplet decreases at these regions of the interface and increases where $u_{gr}(r=1) < 0$. There is more energy available for vaporization where initially $u_{gr} > 0$ than where $u_{gr} < 0$, and, as shown in the result (37) with $A = M = 0$, the differences in the local vaporization rate increase even more, (see Fig. 2a).

The interface temperature changes, when cT_b/L is different from zero, are reflected in the denominator of Eq. (37). Their effect is not able by itself to stabilize the spherical configuration, but the temperature changes also lead to changes in surface tension whose effect is the term proportional to M in Eq. (37), which is stabilizing for the usual case $M < 0$. At the interface regions where the vaporization rate increases, $u_{gr} > 0$, the vapor mass fraction Y and the temperature also increase, leading to a decrease in surface tension; then the liquid is dragged along the interface toward the colder regions, being replaced by cold liquid coming from the interior of the droplet, (see Fig. 2b). This

liquid motion due to Marangoni effects increases the heat
flux toward the liquid side at these hot interface regions.

The basic configuration will be stable when the stabilizing Marangoni effects prevail over destabilizing effects
due to viscous stresses.

The previously discused effects are proportional to
$\partial T_{Lo}/\partial r|_1$, and, therefore, the growth rate of the perturbations decreases with time due to the time evolution of the
basic state. For values of $\beta_e Pr_L \sim 1$ the response time of the
perturbations equals the characteristic time of the basic
solution and the previous normal mode analysis ceases to be
valid. However $\beta_e Pr_L \sim 1$ can be understood as a qualitative
criterion for stability, taking into account that in this
case there is not time enough for the growth of the perturbations.

Perturbations with Deformation of the Interface

When the droplet is small enough, Σ is not large compared to unity and the effect of surface tension is not able to
keep the interface spherical, so that its deformation has to
be accounted for in the analysis. Notice that $\Sigma \sim 1$ for a water droplet of a few microns with $\beta \sim 1$.

We begin by analysing the effects of the surface deformation for $\Sigma \sim 1$ in the case of small values of β_e (with
$\beta_e Pr_L \gg 1$). Afterward the analysis is extended to order one
gas Reynolds number but for simplicity we will consider
here the limiting case $cT_b/L \to 0$, so that $A = M = 0$ and therefore, in particular, the Marangoni effects are excluded. The
velocity and pressure perturbations are still given by Eq.
(32), but new terms, proportional to the interface deformation X, have to be included in the right-hand sides of the
conditions (36) at the interface; these terms are $(\epsilon Pr/\beta)\Omega X$
in Eq. (36a), $-l(l+1)X$ in Eq. (36b), and $-4l(l+1)X$ in Eq.
(36c). An additional relation, to determining X, comes from
the normal momentum conservation condition through the interface (24), which can be written as

$$-\epsilon\left[B_1 \frac{l^2-l-3}{l(2l+3)} + 2(l-1)C_1\right] = -c_1 \frac{l^2+3l-1}{(l+1)(2l-1)} + 2(l+2)b_1 - 12X$$

$$+ (l-1)(l+2)\Sigma X + MY(l) \qquad (36e)$$

The deformation of the interface induces a temperature perturbation in the gas phase given by $T_g = -\alpha X/r^{l+1}$, which is
much greater than that in Eq. (33), and a new term $-\alpha(l+1)X$
appears in the right-hand side of Eq. (36d). The modified

set of equations, with Eq. (36e) included, leads to the dispersion relation

$$A\Omega_1^2 + B\Omega_1 + C = 0 \tag{38}$$

where

$$A = 2\,\frac{l^2+4l+3}{l^2}\,\frac{1+ALe\sqrt{Pr_L}\,\Omega/(l+1)}{\alpha/\beta-1}$$

$$B = \left\{\frac{(2l+1)(l+2)}{l}\,\Sigma + 4\,\frac{l^3+2l^2-4l-2}{l(l-1)}\right\}\frac{1+ALe\sqrt{Pr_L}\,\Omega/(l+1)}{\alpha/\beta-1}$$

$$+\,2\,\frac{2l^3-8l-9-M\beta Le(l^2+3l+3)}{l(l+1)}$$

$$C = 2\,\frac{2l^3-5l+12-M\beta Le(l^2+2l-4)}{l-1} - (3+M\beta Le)(l+2)X$$

and $\quad \Omega_1 = \epsilon^2 Pr\Omega/\beta$.

The basic state is seen to be stable when the temperature of the liquid is uniform, $\alpha = \beta$. In this case $c_1 = 0$ and the normal stress of the gas on the interface has a restoring effect, like the surface tension. When $(\alpha-\beta)/\beta \sim 1/Pr_L \ll 1$ the time evolution of the perturbations become very slow and the normal mode analysis is not applicable.

The modes $l=1$ are associated with translations of the droplet without deformation of the interface. Only Marangoni effects can limit the growth of these perturbations, when $-M\beta Le > 9$. The discussion that follows for the other modes is restricted to the case $A = 0$, so that perturbations in the interface temperature appear only through their effect on the surface tension. Eq. (38) becomes a quadratic expression and the stability limits can be obtained by equating to zero the coefficients of the different powers of Ω_1. The limit with $\Omega_1 = 0$ is given by

$$\Sigma = \frac{2}{3}\,\frac{2l^3-5l+12-(l^2+2l-4)M\beta Le}{(l-1)(l+2)(1+M\beta Le/3)} \tag{39}$$

The minimum value of the right-hand side for l integer is obtained for $l=2$. These modes are the first to lose stability and the basic state is unstable for $\Sigma > (3 - 2M\beta Le/3)/(1 + M\beta Le/3)$. When $-M\beta Le > 3$ the stability limit dissappears and every perturbation decreases with time.

The stability limit with Ω_1 imaginary is obtained from
(38) by equating to zero the coefficient of the linear term
in Ω_1. With $\alpha/\beta > 1$, the result is meaningful only for $l=2$,
because greater values of l lead to $\Sigma < 0$ on the stability
limit. There are five modes with $l=2$, and for these the
stability limit is

$$\frac{\alpha-\beta}{\beta} = \frac{2}{3} \frac{6+5\Sigma}{1+13\overline{M}\beta Le/9} \qquad (40)$$

which gives an upper bound to the nondimensional heat flux
coming from the gas, α. This limit moves to infinity and
disappears   for $-\overline{M}\beta Le = 9/13$; only the previous limit, with
$\Omega_1 = 0$, remains for $9/13 < -\overline{M}\beta Le < 3$.

The results (39) and (40) are plotted in Fig. 3, as
boundaries of the stability domain of the basic solution.

As can be easily seen from the previous equations, the
velocity and pressure perturbations in the liquid vanish
when $\Omega_1 = 0$. The perturbations of pressure and normal viscous
stress of the gas on the interface are positive in the regions where $\chi < 0$, and negative where $\chi > 0$, whereas the ef-

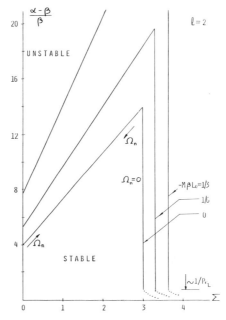

Fig. 3. Stability region for $1 \gg \beta_e \gg 1/\mathrm{Pr_L}$ and several values of
$-\overline{M}\beta Le$. The arrows indicate the direction of increase of the frequency on the stability limit. When $(\alpha - \beta) \sim 1/\mathrm{Pr_L}$ the normal mode
analysis fails and this is represented by the dotted lines.

fect of the surface tension is opposite. Therefore the equilibrium is possible only for a value of Σ, given by Eq. (39).
It is also easy to compute the small perturbations in the
pressure and normal stress of both fluids on the interface
close to this stability limit, leading to the result

$$\Omega_1 = \frac{(\alpha/\beta-1)12\Sigma'}{15-\alpha/\beta} \quad \text{for} \quad \Sigma' = (\Sigma-3) \ll 1 \quad \text{and} \quad M = 0 \qquad (41)$$

so that the perturbations grow if Σ is greater than 3, at
least while $0 < (\alpha/\beta-1) < 14$, which is the range of interest
(see Fig. 3).

The second stability limit, with Ω_1 imaginary, is associated with the existence of a term quadratic in Ω_1 in Eq.
(38). This derives from the fact that the velocity perturbations are proportional to both the interface deformation
and its velocity, whereas the perturbation in the heat flux
entering the liquid, which, together with the heat flux coming from the gas, determines the local vaporization rate, is
proportional to the integral of the velocity, generating a
term $1/\Omega_1$ in Eq. (36d).

The analysis can be extended to see how both stability
limits change when β_e grows to values of order unity. According to the estimates in the section on formulation, the
time derivatives can no longer be neglected in the momentum
equations for the liquid, Eqs. (14-16), but the solutions
of these equations can be written in terms of modified Bes-
sel functions. The left-hand sides of Eqs. (17-19) must also
be retained; the solutions of these equations must be obtained numerically. Carrying them and the solutions of Eqs.
(20,21) to the boundary conditions (23-30), an homogeneous
system of linear equations results; the compatibility condition provides the dispersion relation between Ω and l, with
the seven parameters α, β, Pr, Σ, ε, A, and M. This will not
be written for the sake of brevity. The following results
correspond to the limiting case of constant interface temperature, so that $A = M = 0$.

The stability limit with Ω imaginary is obtained by putting $\Omega = i\Omega_n$ in this dispersion relation. The result for Pr =
$= 0.7$, $\varepsilon = 1$, and several values of β is plotted in Fig. 4.
Also plotted in this figure is the stability limit with $\Omega=0$,
which does not depends on α. Like in the case $\beta_e \ll 1$, the
perturbations in the liquid vanish on this stability limit,
but now we find also perturbations in the flux of momentum
with the same direction as the perturbations in the pressure
and normal viscous stress of the gas on the interface; such
perturbations have to be balanced by the surface tension.

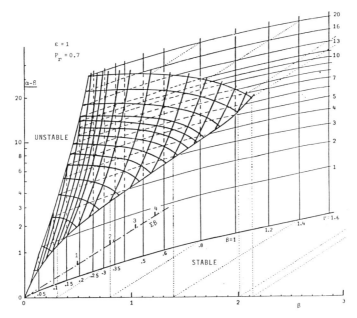

Fig. 4. Stability limits for $\Sigma \sim 1$. On the vertical axis $\alpha-\beta$ is the instantaneous heat flux entering the liquid. The vertical surface is the stability limit with $\Omega = 0$ and the stability region is ahead this vertical surface and below the inclined one. It is $\Sigma\beta = \sigma Rc_p\rho_g / \mu_g k_g$.

The two stability limits enclose the region, extending to infinity in some parts, where the basic state is stable. The plot in Fig. 4 corresponds to $l = 2$, as this is the critical value of l for the most unstable perturbations. There are five different angular functions for $l = 2$ ($n = -2$ to 2), the degeneracy being associated with the symmetry of the unperturbated solution. Two maxima and two minima appear in the deformation of the interface, so that, when unstable, the droplet is likely to slipt into two parts.

There is a region of thickness $1/Pr_L$ around the surface $\Omega = 0$ in Fig. 4 (and also around the vertical line in Fig. 3), in which $\Omega \sim 1/Pr_L$ and the normal mode method fails. This region widens when $(\alpha-\beta)$ decreases and for $(\alpha-\beta) \sim 1/Pr_L$, is $\Omega \sim 1/Pr_L$ everywhere. In the stability analysis, the unsteady effects become negligible in the liquid momentum equations, and the problem can be reformulated in terms of $T_L(r,t)$ and $X(t)$ only, but the generality of the normal mode method is lost. The situation is similar, but simpler, for $\beta \sim 1/Pr_L$. If $\alpha = \beta$ there is not heat flux toward the liquid from the interface and the spherical solution is stable, as mentioned before.

Concluding Remarks

The analysis carried out in this paper shows the existence of instabilities of the droplet vaporization process in a stagnant atmosphere that may appear during the early stage of the vaporization, when the temperature within the droplet is not yet uniform. These instabilities are due to the motion of the liquid toward the interior of the droplet in those regions where the surface temperature, and therefore the vaporization rate, is higher; the liquid motion is induced by the gas phase viscous stress.

The destabilizing effects of the viscous stresses are stronger for $\Sigma \gg 1$, when the effects of the surface tension forces are capable to maintain the droplet spherical, counteracting the deforming effects of viscous stresses, so as to impede the deformation of the droplet. This is the case for large droplet radius.

The analysis of the strong vaporization case, involving values of $\beta \sim 1$, should be extended to include the Marangoni effects that have been found to be stabilizing in the simplified analysis carried for $\beta_e \ll 1$.

It should also be noticed that the characteristic growth time of these instabilities, associated with the temperature nonuniformity within the droplet, becomes very large when the temperature gradients in the liquid tend to zero, in the last long stage of the droplet history, and these instabilities cease to be important.

During the droplet lifetime the representative point of the quasisteady state in the parameter space in Fig. 4 changes. Early during the second stage $\beta \ll 1$, while α and $\Sigma\beta = = \sigma R c_p \rho_g / \mu_g k_g$ are finite, so that the spherically symmetrical state appears as unstable. However, the characteristic time of the perturbations is βPr_L times shorter than the characteristic time t_c for change in β; so that at early times when β is small the perturbations grow very slowly unless $\beta Pr_L \gg 1$. For liquids with nor very large values of Pr_L, β will grow, taking the system to the stable region in Fig. 4, before the perturbations grow significantly.

As mentioned in the Introduction, there are no instabilities associated to nonsteady effects in the gas phase. Although the analysis is not presented in the text, it is easy to understand this result when $\beta_e \ll 1$ and radial convection effects are negligible. In this case the temperature perturbations in the gas are due mainly to deformations of the interface, which are very small, so that the changes in the heat flux toward the interface are not strong enough to modify the local mass vaporization rate; the resulting linearized problem is the same as that for a constant-volume bub-

ble in a viscous fluid, which leads to a stable behavior,
(Lamb 1932, Chandrasekhar 1961, Reid 1969). Numerical results for the corresponding eigenvalue problem show the same
decaying behavior of the perturbations even for large values
of the Reynolds number, when $\beta_e \sim 1$.

The reader should remember that the analysis given here
does not include the effect of the motion of the droplet
with respect to the enviroment, based on the assumption that
the corresponding Reynolds number is small compared with
unity.

The analysis can be extended to account for the linear
transient response of the liquid phase in the burning of a
fuel droplet whose vapor reacts with an oxidizer present in
the atmosphere, leading to a diffusion controlled flame in
the gas phase. The results (see Higuera 1985) are very similar to the ones of this paper and both coincide exactly, as
mentioned in the Introduction, in the limiting cases when
the flame sheet lies far from the droplet, and also when
$\beta_e \ll 1$, or when the Lewis number of the fuel vapor and the
oxidizer are equal to each other, or one of them is equal
to the unity.

Acknowledgment

Partial support for this research by the Spanish CAICYT
under Project 2291083 is acknowledged.

References

Carslaw, H.S. and Jaeger, J.C. (1959) Conduction of Heat in Solids,
2nd ed. Oxford University Press, New York.

Chandrasekhar, S. (1959) The oscillations of a viscous liquid globe.
Proc. London Math. Soc. 9(3), 141-149.

Chandrasekhar, S. (1961) Hydrodynamic and Hydromagnetic Stability.
Clarendon Press, Oxford (reprinted by Dover, New York, 1981).

Hickman, K.C.D. (1952) Surface behaviour in the pot still. Indust.
Engng. Chem. 44(10), 1892-1902.

Hickman, K.C.D. (1972) Torpid phenomena in pump oils. J. Vac. Sci.
Techn. 9(10), 960-976.

Higuera, F.J. (1985) Estabilidad de la Combustión de Propulsantes
Condensados, Ph.D. Thesis, Universidad Politécnica de Madrid,
Spain.

Lord Kelvin (1980) Oscillations of a liquid sphere. Mathematical and
Physical Papers, Vol. 3, pp. 384-386. Clay and Sons, London.

Lamb, H. (1932) Hydrodynamics, 6th ed. Cambridge University Press, Cambridge, England (reprinted by Dover, New York, 1945).

Landau, L.D. (1944) On the theory of slow combustion. Acta Phys. Chem. 19(1), 77-85.

Langmuir, I. (1918) The evaporation of small spheres. Phys. Rev. 12(5), 368-370.

Liñán, A. and Rodríguez, M. (1985) Droplet vaporization, ignition and combustion. Proc. 1st Colloque on Combustion in Thermal Engines, Madrid, Spain, pp. 10/0-10/22.

Palmer, H.J. (1976) The hydrodynamic stability of a rapidly evaporating liquid at reduced pressure. J. Fluid Mech. 75(3), 487-511.

Lord Rayleigh (1894) The Theory of Sound, 2nd ed., Vol. 2, p. 371. MacMillan, London (reprinted by Dover, New York, 1945).

Reid, W.H. (1960) The oscillations of a viscous liquid drop. J. Appl. Math. 18(2), 86-89.

Droplet Ignition in Mixed Convection

R.H. Rangel* and A.C. Fernandez-Pello†

University of California, Berkeley, California

Abstract

A theoretical model is developed of the gas phase
thermal ignition of a fuel droplet in a mixed convective
hot oxidizing flow. The droplet is assumed to have reached
equilibrium vaporization and to have a constant surface
temperature. In the gas, a finite rate chemical reaction
with a large activation energy is considered. The analysis
makes use of the boundary layer approximation to describe
the gas flow around the droplet and of first order matched
asymptotic expansions to define ignition. An explicit
expression is derived for the critical Damköhler number for
ignition as a function of the angular location along the
droplet periphery. The results show that ignition will
occur further downstream from the forward stagnation point
as the intensity of the convective flow and activation
energy increase and as the initial temperature and oxygen
concentration of the oxidizer gas decrease. This implies
that ignition will occur first downstream from the forward
stagnation point -- most likely near the point of boundary
layer separation. However, since the analysis is not
applicable in the droplet wake, it is not possible to predict whether or not ignition will first occur in this
region. As the Damköhler number is increased, ignition
will occur simultaneously over a larger portion of the fuel
droplet and, finally, when the Damköhler number equals its
critical value for ignition at the forward stagnation
point, the whole droplet ignites at once.

Presented at the 10th ICDERS, Berkeley, California, August
4-9, 1985. Copyright © 1985 by the American Institute of
Aeronautics and Astronautics, Inc. All rights reserved.
*Research Assistant, currently Postdoctoral Fellow,
Department of Mechanical Engineering, University of California,
Irvine, CA .
†Associate Professor, Department of Mechanical Engineering.

Nomenclature

A = pre-exponential constant Eq. (1)

A_p = flow parameter Eq. (11)

a = radial distance from symmetrical axis

c_p = specific heat

D = diffusion coefficient

Da = parameter defined in Eq. (10)

E = activation energy

f = scaled stream function

g = acceleration of gravity

Gr = Grashof number, $= gR^3 \rho_\infty^2 \tau / \mu_\infty^2$

h = function defined in Eq. (11)

h_{fg} = heat of vaporization of fuel

p = pressure

Pr = Prandtl number, $= c_p \mu_\infty / \lambda_\infty$

Q = heat of combustion of the fuel

R = droplet radius

R = universal gas constant

Re = Reynolds number, $= 3 \rho_\infty u_\infty R / 2\mu_\infty$

Sc = Schmidt number, $= \mu_\infty / \rho_\infty^2 D_\infty$

$\tilde{T}$ = temperature

$\tilde{Ta}$ = parameter, $= c_p E/QR$

T = dimensionless temperature, $= c_p \tilde{T}/Q$

u = velocity in x direction

v = velocity in y direction

W_i = molecular weight of species i

x = coordinate along cylinder surface

y = coordinate normal to cylinder surface

$\tilde{Y}_i$ = mass fraction of species i

Y_i = dimensionless mass fraction, $= \tilde{Y}_i / \gamma_i$

α = parameter, $= [c_p (T_\infty - T_w) + h_{fg}] / Q$

Δ = Damköhler number for ignition, Eq. (18a)

γ_i = mass ratio between species i and fuel

ϵ = small parameter, $= T_w^2 / Ta$

η = transformation coordinate

θ = perturbation temperature introduced in Eq. (13)

λ = thermal conductivity

μ = viscosity

ν_i = stoichiometric coefficient of species i

ξ = transformed normal coordinate, Eq. (11)

ρ   =   density
σ   =   angular coordinate, = x/R
τ   =   parameter, = $(T_\infty - T_w)/T_\infty$
χ   =   stretched coordinate, Eq. (14)
ψ   =   stream function, Eq. (2)

Subscripts

c   =   critical
f   =   frozen flow
F   =   fuel
i   =   species i
0   =   oxydizer
P   =   products
w   =   droplet surface
∞   =   free stream

Introduction

In most practical combustors, the fuel droplets react
under the combined influence of both natural and forced
convection. The forced flow condition is the result of the
relative velocity between the droplet and the ambient gas,
which is the result of the droplet injection or of turbulence, in the combustion chamber. The natural convection
is caused by the density differences in the gas between the
droplet surface and the ambient. The ignition of the
droplet is affected by the convective characteristics of
the flow near the droplet surface. These characteristics
not only determine the critical conditions for ignition,
but also the location along the droplet surface where ignition will occur. The gas-phase ignition of the fuel
droplet will occur when the residence and chemical times
are of the same order of magnitude. However, because of
the competing effects of the chemical and residence times,
ignition can occur at different locations along the droplet
periphery. At the forward stagnation point, the boundary
layer is thinnest and thus the region of high temperature
is closest to the droplet surface. Since the rate of the
chemical reaction is very sensitive to temperature, the
chemical time will be shortest at the forward stagnation
point. The residence time, however, is also the shortest
due to the large velocity and temperature gradients present
in this region. Downstream from the forward stagnation
point, the boundary layer thickens and, as a consequence,
both the chemical and residence times increase. The relative variation of these times depend on the characteristics
of the convective flow. Finally, in the wake region near

the rear stagnation point, the residence time is very
large, but the chemical time is also very large due to the
fuel accumulation in this region. From these phenomenological arguments, it is seen that a detailed analysis of the
droplet ignition process is necessary if accurate information about the critical conditions for ignition is
required. The objective of the present work is to develop
an analysis of the thermal convective ignition of a fuel
droplet capable of predicting the critical fluid mechanical
and thermal conditions at which ignition of the gas phase
will occur.

Previous theoretical studies of the thermal ignition
of a fuel droplet have either considered the case of a
still environment (Law, 1975, Linan and Rodriguez, 1985) or
have analyzed the convective ignition of the droplet at its
stagnation point (Krishnamurthy, 1976; Fernandez-Pello and
Law, 1982). All of these authors used matched asymptotic
expansion for large activation energies to determine the
critical Damkohler number for ignition. The analytical
methods established in the above-cited works will be
followed in the present work. The boundary-layer approximation will be used to describe the flow around the sphere
and large activation energy asymptotics will be used to
define ignition. The critical Damkohler number for ignition will be deduced. From its expression, the location
along the droplet periphery at which ignition of the gas
phase occurs will be identified. However, the model cannot
describe the ignition process in the wake region of the
droplet, since the analysis does not apply after the location of boundary layer separation.

Analysis

The physical problem to be analyzed and the coordinate
system used in the analysis are shown in Fig. 1. The analysis considers the mixed convection, laminar, boundary
layer ignition of a spherical fuel droplet in a hot oxidizing environment. An oxidizer with a velocity U_∞, temperature T_∞, and concentration $Y_{O\infty}$, flows over a fuel
droplet forming a thin boundary layer near the liquid surface. The large difference between ambient density and the
density near the fuel surface generates a buoyant flow that
reinforces the forced flow. (The gravity vector is assumed
to be in the direction of the flow.) The Reynolds and
Grashof numbers of the droplet are assumed to be large
enough so that the boundary layer approximation holds and
the thickness of the boundary layer is small in comparison

with the droplet diameter. Heat transfer from the hot
environment to the droplet causes the liquid to evaporate.
It is assumed that the fuel has reached equilibrium vaporization and that the surface droplet temperature is
constant, T_v. The vaporized fuel is convected and diffused
away from the surface toward the hot ambient where ignition
may occur if the residence time is of the order of the chemical time. The gas-phase activation energy of the fuel is
assumed to be large. Since the initiation of the chemical
reaction is very sensitive to temperature, it is expected
that ignition will occur at the outer edge of the boundary
layer where the temperature is higher. Because of the competing effects of the residence and chemical times, ignition may occur first in the downstream region of the
droplet, where the residence times are larger (lower
velocities), or at the forward stagnation point (higher
temperature). The determinations of the critical conditions for ignition and of the ignition location as a
function of fuel and environmental conditions are the primary objectives of this work.
     The equations and boundary conditions that specify the
problem are the mass, momentum, energy, and species boundary layer equations for a reacting, axisymmetric, flow
past a vaporizing sphere, neglecting curvature. They are
given in Rangel and Fernandez-Pello (1984) and will not be
repeated here. Assuming that the gas phase reaction is of

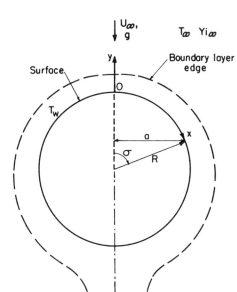

Fig. 1   Schematic of the
problem and coordinate system.

the Arrhenius type, the production term in the energy and species equations are

$$q'''/Q = m_i''' = - A \, W_i \nu_i \rho^2 T \tilde{Y}_F \tilde{Y}_0 \mathrm{Exp}(E/\bar{R}T) \tag{1}$$

where a one-step, second-order irreversible chemical reaction has been assumed to occur between the fuel and oxidizer. For simplicity of the analysis, motion within the liquid droplet is not considered in this work.

The mathematical analysis is simplified by introducing a stream function ψ satisfying the mass conservation equation:

$$a\rho u = \mu_\infty \frac{\partial(a\psi)}{\partial y} \,, \qquad a\rho v = - \mu \frac{\partial(a\psi)}{\partial x} \tag{2}$$

transformation independent variables

$$\sigma = x/R \,, \quad \eta = (\mathrm{Re}^4 + \mathrm{Gr}^2)^{1/8} \int_0^y (\rho/\rho_\infty R)\, dy \tag{3}$$

and functions

$$f = \psi/\sigma(\mathrm{Re}^4 + \mathrm{Gr}^2)^{1/8} \,, \quad T = C_p \tilde{T}/Q \,, \quad Y_i = \tilde{Y}_i/Y_i \tag{4}$$

where Re and Gr are the Reynolds and Grashof droplet numbers, respectively. The definition of the symbols in the above equations are given in the nomenclature. With these variables and assuming that ρT, $\rho\lambda$, $\rho\mu$, and $\rho^2 D$ are constant, the boundary-layer equations and their boundary conditions become

$$\left(\frac{\partial f}{\partial \eta}\right)^2 + \sigma \frac{\partial f}{\partial \eta} \frac{\partial^2 f}{\partial \sigma \partial \eta} - \left(f + \sigma \frac{\partial f}{\partial \sigma} + f\sigma \cot\sigma\right) \frac{\partial^2 f}{\partial \eta^2}$$

$$= \frac{\sin(2\sigma)}{2\sigma(1 + \phi^2)^{1/2}} + \frac{T - T_\infty}{T_w - T_\infty} \frac{\phi\sin\sigma}{\sigma(1 + \phi^2)^{1/2}} + \frac{\partial^3 f}{\partial \eta^3} \tag{5}$$

$$\sigma \frac{\partial f}{\partial \eta} \frac{\partial T}{\partial \sigma} - \left(f + \sigma \frac{\partial f}{\partial \sigma} + f\sigma \cot\sigma\right) \frac{\partial T}{\partial \eta}$$

$$= \frac{1}{\mathrm{Pr}} \frac{\partial^2 T}{\partial \eta^2} + D_a Y_F F_0 \exp\left(-\frac{Ta}{T}\right) \tag{6}$$

$$\sigma \frac{\partial f}{\partial \eta} \frac{\partial Y_i}{\partial \sigma} - (f + \sigma \frac{\partial f}{\partial \sigma} + f\sigma \cot\sigma) \frac{\partial Y_i}{\partial \eta}$$

$$= \frac{1}{Sc} \frac{\partial^2 Y_i}{\partial \eta^2} - D_a Y_F Y_0 \exp(-\frac{Ta}{T}) \qquad (7)$$

with the following boundary conditions.

At $\eta = 0$:

$$\frac{\partial f}{\partial \eta} = 0 \ ; \ f + \sigma \frac{\partial f}{\partial \sigma} + f\sigma \cot\sigma = \frac{Q}{Pr \ h_{fg}} \frac{\partial T}{\partial \eta} \ , \qquad T = T_w$$

$$(f + \sigma \frac{\partial f}{\partial \sigma} + f\sigma \cot\sigma)(1 - Y_F) = \frac{1}{Sc} \frac{\partial Y_F}{\partial \eta}$$

$$- (f + \sigma \frac{\partial f}{\partial \sigma} + f\sigma \cot\sigma) Y_i = \frac{1}{Sc} \frac{\partial Y_i}{\partial \eta} \qquad (\text{for } i \neq F) \qquad (8)$$

At $\eta = \infty$:

$$\frac{\partial f}{\partial \eta} = \frac{\sin\sigma}{\sigma(1 + \phi^2)^{1/4}} \ , \ T = T_\infty \ , \ Y_F = 0 \ , \ Y_0 = Y_{0\infty} \qquad (9)$$

where

$$Da = \frac{A W_0 \nu_0 R^2 \rho_\infty \rho_\infty}{(Re^4 + Gr^2)^{1/4} \bar{R} \bar{W} \mu_\infty} \qquad (10)$$

is the relevant Damköhler number for the problem.

It is convenient to introduce a new transverse independent variable defined (Krishnamurthy, 1976) as

$$\xi = 1 - \frac{\int_0^\eta e^{-h} d\eta'}{A_p} \qquad (11)$$

where

$$h = Pr \int_0^\eta (f + \sigma \frac{\partial f}{\partial \sigma} + f\sigma \cot\sigma) \, d\eta'$$

and

$$A_p = \int_0^\infty e^{-h} \, d\eta'$$

Assuming unity Lewis number, the energy and species equations (6) and (7) may now be written as

$$\frac{1}{Pr} \; (\frac{\partial \xi}{\partial \eta})^2 \frac{\partial^2 J}{\partial \xi^2} - \sigma \; (\frac{\partial f}{\partial \eta}) \frac{\partial J}{\partial \sigma}$$

$$+ DaY_F Y_0 \; \exp(- \frac{Ta}{T}) = 0 \; , \; \text{for } J = T, Y_i \qquad (12)$$

The onset of ignition in the gas phase will be analyzed under the assumption that the activation energy is large. The development of the analysis will follow the guidelines established by Krishnamurthy (1976) and Fernandez-Pello and Law (1982) in their studies of the stagnation point ignition of a condensed fuel using matched asymptotic expansions for large activation energies. Since the mathematical procedure is similar to those of the above referenced works, it is presented only briefly here.

Before the establishment of the ignition reaction, the gas flow is assumed to be chemically frozen. Frozen flow conditions are obtained by letting Ta → ∞ in Eqs. (12). The solution corresponds to fuel vaporization without chemical reaction (Rangel and Fernandez-Pello, 1984). If the physical conditions are sufficient for the ignition of the gas phase, the chemical reaction will be initiated near the outer edge of the boundary layer where the temperature is higher. This occurs even though the fuel concentration is small in this region, because of the stronger dependence of the reaction rate on temperature, particularly for large activation energy. At the onset of ignition the temperature and concentration profiles will be perturbed by a small quantity of order ε from the corresponding profiles in the frozen flow. Hence, an inner solution is assumed of the form (Krishnamurthy, 1976)

$$T = T_\infty + \varepsilon(\theta - x)$$

$$Y_0 = Y_{0\infty} + O(\varepsilon)$$

$$Y_F = \varepsilon(x - \alpha\theta) \qquad (13)$$

where θ is the perturbed temperature, x the stretched transverse independent variable

$$\chi = \frac{(-\partial T_f)}{\partial \xi} \Bigg|_{\xi=0} \frac{\xi}{\varepsilon} \qquad (14)$$

and the small parameter is ε and the parameter α are

$$\varepsilon = Tw^2/Ta \quad , \quad \alpha = (c_p(T_\infty - T_w) + h_{fg})/Q \tag{15}$$

Substitution of Eqs. (14) and (15) into Eq. (12) gives the following equation for the perturbed temperature

$$\chi^2 \frac{\partial^2 \theta}{\partial x^2} - \Omega \frac{\partial \theta}{\partial \sigma} + \Delta(\chi - \alpha\theta) \exp(\theta - \chi) = 0 \tag{16}$$

where

$$\Delta = \frac{Pr\, Y_{0\infty} Da\, \exp(-T_a/T_\infty)}{\alpha H^2} \quad , \quad \Omega = \frac{Pr\, \sin\sigma}{H^2(1 + \phi^2)^{1/4}}$$

and

$$H = Pr(f(\infty) + \sigma \frac{\partial f(\infty)}{\partial \sigma} + f(\infty)\, \sigma\cot\sigma)$$

The asymptotic solution for $f(\infty)$ is of the form

$$f(\infty) = \frac{\dfrac{\sin\sigma}{\sigma} \eta_\infty + C_1 + \phi^{1/2} C_2}{(1 + \phi^2)^{1/4}}$$

where C_1 and C_2 are constants. The parameter η_∞ is the value of η at the edge of the boundary layer, which has been taken here as the location where the velocity equals 99% of its free stream value.

From Eq. (16) it is seen that for natural convection ($\phi \to \infty$, $\Omega = 0$) the convective term vanishes. For forced convection ($\phi = 0$), Ω is small except near the location of boundary layer separation (see Fig. 3), and thus the convective term can be neglected in most of the region of validity of the analysis. For intermediate value of ϕ (mixed convection), Ω has a small but non-zero value. The effect of the convective term on the ignition process should be, however, small in comparison with the effects of the diffusion and reaction terms (Law, 1984) since these are the dominant mechanisms in the ignition process. Hence, this term is neglected in a first-order approximation. The resulting equation for the perturbed temperature is

$$\chi^2 \frac{\partial^2 \theta}{\partial x^2} + \Delta(\chi - \alpha\theta) \exp(\theta - \chi) = 0 \tag{17}$$

where

$$\Delta = \frac{A\tilde{Y}_{O\infty}P_\infty\mu_\infty \, e^{-Ta/T_\infty}}{\alpha \, (\frac{\partial T_f}{\partial \xi})\Big|_{\xi=0}) \, \rho_\infty \overline{Pr} \, \overline{RW}v_\infty^2} \tag{18a}$$

is the relevant Damköhler number for the ignition problem
and

$$v_\infty = (Re^4 + Gr^2)^{1/8} \frac{\mu_\infty}{\rho_\infty R} \, (f(\infty) + \sigma \frac{\partial f(\infty)}{\partial \sigma} + f(\infty)\sigma \, \cot\sigma)$$

$$\tag{18b}$$

is the radial velocity at the outer edge of the boundary
layer.
    Equation (16) must be solved in conjunction with the
boundary conditions $\theta(0) = 0$ and $\partial\theta/\partial\chi = 0$. This last
boundary condition is obtained through the matching of the
inner and outer solutions of the problem (Law and Law,
1979).

Results and Conclusions

    Equation (16) has been solved by Linan (1974) for
several values of the parameter α. The solution provides
the Damkohler number for ignition Δ_c as a function of α.
Those solutions are shown in Fig. 2 where the characteristic S-shaped ignition curves can be seen. For values
of Δ below the critical value Δ_c, ignition is not possible.

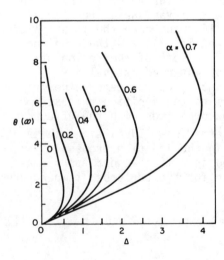

Fig. 2   Lower portions of the
ignition extinction curves for
several values of α.

The influence of the flow parameters on the onset of ignition was analyzed by Fernandez-Pello and Law (1982) at the
stagnation point of a fuel particle. In this work, the
concern is the influence of the local flow conditions along
the droplet surface on the occurrence of ignition. For
given physical properties and flow conditions, the desire
is to establish the location of the ignition point along
the droplet periphery. From the definition of the
Damköhler number [Eq. 18a)], it is apparent that only one
parameter is a function of the angular coordinate σ: the
entrainment velocity V_∞. This parameter has been obtained
from a numerical solution of the problem of mixed convective vaporization of a spherical fuel. In Eqs. (5) and
(6), the exponential reaction term is dropped by letting
$Ta \rightarrow \infty$, which corresponds to the frozen flow, and derivatives in σ are replaced by backward differences. The
resulting ordinary differential equations are solved using
a quasilinearization and iteration technique at each radial
location. The edge of the boundary layer is taken as the
location where the velocity equals 99% of its free stream
value. Figure 3 is a plot of v_∞, normalized by its value
at the stagnation point, vs. the angular coordinate σ for a
fuel with $Ja = 1$ vaporizing in air in free=/convection
mode. Near the rear stagnation point of the droplet, a
rapidly converging flow results in the rapid thickening of
the boundary layer and the formation of a steady downward
plume. A boundary layer solution does not exist in that
region.
      The location of the ignition angular location can be
obtained by defining a σ-independent Damköhler number as

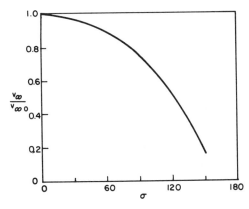

Fig. 3   Entrainment velocity v_∞ as a function of the angular
coordinate, σ.

$$\Delta_\sigma \; = \; \frac{AY_{O\infty} \; P_\infty \mu_\infty \; e^{-Ta/T_\infty}}{\alpha \; \rho_\infty \; Pr \; RW \; V_{\infty,0}^2} \tag{19}$$

where $v_{\infty,0}$ is the entrainment velocity at the front stagnation point. In terms of the Damköhler number defined in Eq. (18a), the σ-independent Damköhler number can be written as

$$\Delta_\sigma \; = \; (V_\infty^2 \; / \; V_{\infty,0}^2) \; \Delta \tag{20}$$

and the corresponding critical Damkohler number

$$\Delta_{\sigma,c} \; = \; (V_\infty^2 \; / \; V_{\infty,0}^2) \; \Delta_c \tag{21}$$

The ignition distance as given by the angular coordinate σ can now be obtained from Eq. (21) and with the aid of Figs. 2 and 3. Figure 4 shows the ignition angular location σ_c as a function of the critical Damkohler number $\Delta_{\sigma,c}$ for several values of the parameter α for the case of a vaporizing droplet in free convection. The parameter α is expected to be less than one because of the large values of the heat of combustion Q for most fuels. Figure 4 shows

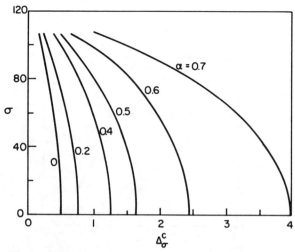

Fig. 4   Ignition location σ as a function of the critical Damkohler number $\Delta_{\sigma,c}$ for several values of α.

that the critical Damkohler number $\Delta_{\sigma,c}$ decreases with increasing distance away from the front stagnation point. It can be deduced that the ignition angular coordinate increases as the Grashof number and the activation energy increase or as the air temperature decreases. The analysis shows that ignition of the vaporizing fuel will be likely to start near the rear stagnation point where the analysis is not valid. In the case considered here of natural convection, the boundary layer exists up to very near the rear stagnation point and the analysis indicates that ignition will occur close to that point and the reaction will then propagate towards the front stagnation point. It may also be concluded from the analysis that, in the cases of forced or mixed convection, ignition is most likely to occur near the point of boundary layer separation. Although the analysis presented here is not valid in the wake region, it is reasonable to expect that fuel accumulation in the recirculating wake will make the chemical time too large in that region, consequently rendering igniton unlikely in spite of the fact that in the recirculating wake the residence time is large.

Acknowledgment

This work was supported by the U.S. Army Tank-Automotive Command under contract DAAE07-85-C-R025 and by the National Science Foundation under grant CPE-8502114. The support and encourgement of Professor A.K. Oppenheim is greatly appreciated.

References

Fernandez-Pello, A.C. and Law, C.K. (1982) On the mixed convective flame structure in the stagnation point of a fuel particle. Nineteenth Symposium (International) on Combustion, pp. 1037-1044, The Combustion Institute, Pittsburgh, PA.

Krishnamurthy, L. (1976) On gas phase ignition of diffusion flame in the stagnation point boundary layer. Acta Astron. 3, 935-942.

Law, C.K. (1975) Asymptotic theory for ignition and extinction in droplet burning. Combust. Flame 24, 89-98.

Law, C.K. and Law, H.K. (1979) Thermal ignition analysis in boundary-layer flows. J. Fluid Mech. 92, 97-108.

Law, C.K. (1984) Heat and mass transfer in combustion: Fundamental concepts and analytical techniques, ASME/JSME Thermal Engineering Joint Conference, Proceedings, Vol. 2, pp. 535-559, ASME, New York.

Linan, A. (1974) The asymptotic structure of counterflow diffusion flames for large activation energies. Acta. Astron. 1, 1007-1039.

Linan, A. and Rodriguez (1985) Droplet vaporization, ignition and combustion. Submitted for publication.

Rangel,R. H. and Fernandez-Pello, A. C. (1984) Mixed convection droplet combustion with internal circulation. Combust. Sci. Technol. 42, 47-66.

A Numerical Technique for the Solution of a Vaporizing Fuel Droplet

Gopal Patnaik*
Carnegie-Mellon University, Pittsburgh, Pennsylvania
and
William A. Sirignano†
University of California, Irvine, California
and
H.A. Dwyer‡ and B.R. Sanders§
Sandia National Laboratories, Livermore, California

Abstract

This study deals with the behavior of a fuel droplet in a hot gas stream. Models exist of single-droplet vaporization in a stagnant environment but are inadequate for a convecting droplet. A study by Prakash and Sirignano that included slip and internal circulation has shown the importance of convection. A fully numerical solution obtained by Dwyer and Sanders assumes an incompressible gas phase and thus its application to situations with high temperature or concentration gradients is difficult. The technique now being developed is designed to handle variable density, two-phase flows. In addition, the method is sufficiently general to handle arbitrary axisymmetric flows. This technique has been applied to several cases of single-droplet vaporization in a convecting environment. A detailed picture of the transport processes throughout the droplet lifetime has been obtained. Due to drag, the dominance of convection on transport processes gives way to diffusion. Most earlier models assume the dominance of one mode or the other. A parameter survey with this technique will determine the regimes in which simpler models can be used and yield correlations that can be included in droplet spray calculations.

Presented at the 10th ICDERS, Berkeley, California, August 4-9, 1985. Copyright © by the American Institute of Aeronautics and Astronautics, Inc. All rights reserved.
* Department of Mechanical Engineering
† School of Engineering
‡ Combustion Research Facility
§ Combustion Research Facility

Nomenclature

a   = droplet radius
C_v = specific heat, constant volume
C_p = specific heat, constant pressure
D   = diffusion coefficient
k   = thermal conductivity
P   = pressure
r   = radial coordinate
t   = time
T   = temperature
V_r = radial velocity
V_z = axial velocity
Y   = mass fraction of fuel
z   = axial coordinate
η   = nonorthogonal coordinate
μ   = viscosity
ν   = kinematic viscosity
ρ   = density
τ   = nondimensional time
ω   = vorticity
ξ   = nonorthogonal coordinate
ψ   = streamfunction

Subscripts

g   = denotes gas phase
ℓ   = denotes liquid phase

Introduction

        This study deals with the behavior of a fuel droplet
in a hot gas stream, a situation that occurs in many
instances, for example, fuel injection in gas turbine
engines.   Classical droplet vaporization theory developed
in the 1950's by Spalding (1953), Godsave (1954),
Goldsmith and Penner (1954), and Wise et al. (1955)
describes a single, isolated droplet vaporizing in a
stagnant environment with no relative motion between
droplet and gas.   This theory yields the well-known D^2-
law, i.e., the square of the droplet diameter decreases
linearly with time.   The diffusion limit model (Law 1976,
1977; Faeth 1969) developed in the 1970's includes
transient conduction into the drop and is the most
accurate representation of single-droplet evaporation in a
stagnant environment.   These models prove quite inadequate
when applied to convecting droplets.   Experimental
correlations (Ranz and Marshall 1952) exist that correct
the stagnant droplet results for convection, but have been

shown to be inadequate for the problem of interest (Sirignano 1983).

Droplet vaporization with slip and an internal circulation has been studied analytically and numerically by Prakash and Sirignano (1978). This detailed study demonstrated the importance of convection by using reasonable assumptions for the flowfield both inside and outside the drop.

This study has shown that internal circulation serves to shorten the heat diffusion time by an order of magnitude, but not to the point where the drop temperature can be considered spatially uniform. The need for a careful inclusion of hydrodynamics in any study of droplet vaporization with convection has been highlighted by this work.

A numerical study of droplet vaporization including slip has been carried out by Dwyer and Sanders (1984). The study considered the coupled problem of transient heat, mass, and momentum transfer to a vaporizing fuel droplet. The full form of the incompressible Navier-Stokes equations was used, allowing details of the flow, including separation, to be considered. The results, while in qualitative agreement with the Prakash and Sirignano (1978) study, do show enough difference to warrant closer investigation to understand the limitations of both studies. Dwyer and Sanders have shown that the various heat, mass, and momentum transfer processes occurring at the drop surface are tightly coupled, making it very difficult to obtain simple global correlations for overall droplet behavior.

The applicability of the various models of a droplet in a quiescent ambience to a drop with convection is seriously limited by the radical alteration in behavior caused by slip and internal circulation. Thus, it is quite unlikely that the experimental correlations available in the literature can be used to extend the quiescent droplet results to the convecting droplet case. The Prakash and Sirignano model takes into account droplet convection in a realistic manner. Their results show a strong dependence on the flowfield. The work by Dwyer and Sanders shows some qualitative differences in the flowfield from that used by Prakash and Sirignano and further underline the need for a closer look at the hydrodynamics of the problem. The assumption of incompressible flow made in the Dwyer-Sanders study is suspect in light of the large density gradients expected in the gas boundary layer. These density gradients arise due to steep temperature and concentration gradients across the thin gas boundary layer.

The numerical technique developed for the present study has been developed specifically to handle large density gradients. It was felt that a detailed approach with as few assumptions as possible is needed to evaluate other simpler techniques. The other techniques do have regimes in which they are valid, and a comprehensive method has to be used to identify them.

Modifications in individual droplet behavior will occur whenever sprays are dense enough for interaction between droplets to occur. This is certainly true for many sprays of practical interest. Chigier (1981) has cited experimental evidence that droplets rarely burn alone, pointing out the importance of droplet interaction. Unfortunately, the geometry in real sprays is so complex and subject to uncertainty that calculations involving the entire spray are at present unfeasible. If statistical approaches are to succeed, information is still needed about droplet interaction.

Two and three burning droplets have been considered by several authors (Twardus and Brzustowski 1977; Labowsky 1978); however, none of these studies includes droplet slip or transient heating. Work involving heat and mass transfer to an array of nonvaporizing droplets and spheres with convection has been done by Tal and Sirignano (1982). While none of these studies includes all phenomena, they indicate strong interaction if droplet centers are less than two diameters apart. These studies serve to demonstrate the need for detailed investigation in the area of droplet interaction.

The numerical technique developed has been designed for use on arbitrary geometries. An arbitrary computational grid may be specified, thus enabling the use of this technique to the more complex geometries required in multiple droplet studies.

Solution Technique

A second-order, transient finite-difference code in primitive variables has been developed to handle variable density, two-phase flows. Two momentum equations, equations for energy and species conservation, and the continuity equation are the equations governing the gas phase. An equation for pressure is obtained by combining the momentum equations with the continuity equation as in the ICE method (Westbrook 1978). Gas density is given by the equation of state for an ideal gas. The equations for the gas phase are given below:

$$\frac{\partial \hat{Q}}{\partial t} + \frac{\partial \hat{E}}{\partial r} + \frac{\partial \hat{F}}{\partial z} = \frac{\partial \hat{R}}{\partial r} + \frac{\partial \hat{S}}{\partial z} + \hat{H}$$

where

$$\hat{Q} = \begin{bmatrix} \rho r \\ \rho rV_r \\ \rho rV_z \\ \rho rC_v T \\ \rho rY \end{bmatrix} \qquad \hat{E} = \begin{bmatrix} \rho rV_r \\ \rho rV_r V_r + rP \\ \rho rV_r V_z \\ \rho rC_p V_r T \\ \rho rV_r Y \end{bmatrix}$$

$$\hat{F} = \begin{bmatrix} \rho rV_z \\ \rho rV_z V_r \\ \rho rV_z V_z + rP \\ \rho rC_p V_z T \\ \rho rV_z Y \end{bmatrix} \qquad \hat{R} = \begin{bmatrix} 0 \\ 2r\mu\dfrac{\partial V_r}{\partial r} \\ r\mu\left(\dfrac{\partial V_r}{\partial z} + \dfrac{\partial V_z}{\partial r}\right) \\ rk\dfrac{\partial T}{\partial r} \\ rD\dfrac{\partial Y}{\partial r} \end{bmatrix}$$

$$\hat{S} = \begin{bmatrix} 0 \\ r\mu\left(\dfrac{\partial V_r}{\partial z} + \dfrac{\partial V_z}{\partial r}\right) \\ 2r\mu\dfrac{\partial V_z}{\partial z} \\ rk\dfrac{\partial T}{\partial z} \\ rD\dfrac{\partial Y}{\partial z} \end{bmatrix} \qquad \hat{H} = \begin{bmatrix} 0 \\ P - \dfrac{2\mu V_r}{r} \\ 0 \\ 0 \\ 0 \end{bmatrix}$$

All gas phase properties shown above except density are at freestream values.

The liquid phase can be considered incompressible, and the continuity and two momentum equations are replaced by two equations for streamfunction and vorticity. These two equations, along with the energy equation, are solved in the liquid phase. In the liquid phase,

$$\hat{Q} = \begin{bmatrix} \rho\, r\omega \\[8pt] 0 \\[8pt] \rho\, r C_p T \end{bmatrix} \qquad \hat{E} = \begin{bmatrix} \rho\, \dfrac{\partial \Psi}{\partial z}\, \omega \\[8pt] 0 \\[8pt] \rho C_p \dfrac{\partial \Psi}{\partial z}\, T \end{bmatrix} \qquad \hat{F} = \begin{bmatrix} -\rho\, \dfrac{\partial \Psi}{\partial r}\, \omega \\[8pt] 0 \\[8pt] -\rho C_p \dfrac{\partial \Psi}{\partial r}\, T \end{bmatrix}$$

$$\hat{R} = \begin{bmatrix} r\dfrac{\partial}{\partial r}(\mu\omega) \\[10pt] \dfrac{1}{r}\dfrac{\partial \Psi}{\partial r} \\[10pt] r k \dfrac{\partial T}{\partial r} \end{bmatrix} \qquad \hat{S} = \begin{bmatrix} r\dfrac{\partial}{\partial z}(\mu\omega) \\[10pt] \dfrac{1}{r}\dfrac{\partial \Psi}{\partial z} \\[10pt] r k \dfrac{\partial T}{\partial z} \end{bmatrix}$$

$$\hat{H} = \begin{bmatrix} \dfrac{\mu\omega}{r} + \rho\, r\omega \\[10pt] -\omega \\[10pt] 0 \end{bmatrix}$$

All properties referred to above are for the liquid phase and are considered constant.

Coupling at the interface is through special cells composed partly of liquid and gas. Momentum, mass, and energy balances on these cells yield velocities tangent and normal to the surface and surface temperature, respectively. The mass fraction at the droplet surface is given by the Clausius–Clapeyron relation.

The governing equations are transformed from cylindrical coordinates to arbitrary nonorthogonal ξ, η coordinates in which the Jacobian and other metrics of the transformation are calculated numerically (Viviand 1975; Vinokur 1974). Thus the code can easily be applied to any geometry merely by specifying a computational mesh that is best suited to it. Computations are performed on the transformed coordinates. The transformed equation is

$$\frac{\partial Q}{\partial \tau} + \frac{\partial E}{\partial \xi} + \frac{\partial F}{\partial \eta} = \frac{\partial R}{\partial \xi} + \frac{\partial S}{\partial \eta} + H$$

where $\hat{Q} = Q/J$; $E = (\hat{Q}\xi_\tau + \hat{E}\xi_r + \hat{F}\xi_z)/J$;

$F = (\hat{Q}\eta_\tau + \hat{E}\eta_r + \hat{F}\eta_z)/J$; $R = (\hat{R}\xi_r + \hat{S}\xi_z)/J$;

$S = (\hat{R}\eta_r + \hat{S}\eta_z)/J$; $\quad H = \hat{H}/J$; $\quad J = z\xi r_\eta - z_\eta r\xi$; $\quad$ and subscripts denote differentiation.

The momentum, energy, species, and vorticity equations are solved by an alternate direction implicit method. The pressure and streamfunction equations are solved efficiently by a relaxation technique. These equations are solved sequentially, and the whole set of equations is iterated until the set reaches convergence at each time step. This iterative procedure allows the solution of this tightly coupled system with large time steps with the achievement of convergence. Since the equations are solved sequentially, smaller computer facilities can be used.

Results

The results of detailed time-dependent calculations for four typical fuel droplets are presented in Table 1. Each drop, initially cold and without any internal

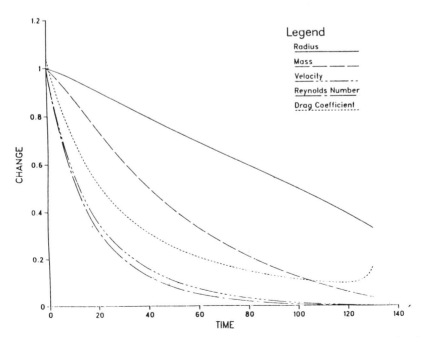

Figure 1 Droplet vaporization, Reynolds number: 100, velocity: 25 m/s.

circulation, is injected into a hot gas stream. In the base case, a 50 μm drop is injected at 25 m/s, thus giving an initial Reynolds number of 100 based on diameter. The four cases are summarized in Table 1.

Other physical parameters of interest which have been held constant in all cases follow: ambient temperature = 1000°K; ambient pressure = 10 atm; initial drop temperature = 500°K; viscosity ratio μ_ℓ/μ_g = 25; density ratio ρ_ℓ/ρ_g = 300; molecular weight of fuel = 192 kg|mole; boiling point of fuel = 573°K; Prandtl number of liquid = 10.

The transport processes initially are convection controlled but become diffusion dominated as the Reynolds number drops throughout the droplet lifetime. There is an initial period during which the Reynolds number falls drastically due to a sharp reduction in velocity while the drop diameter remains essentially unchanged.

Figure 1 presents the overall behavior of the first case during the droplet lifetime. Droplet radius, mass, velocity, and Reynolds number have been normalized by their initial values. The results are given in terms of a nondimensionalzied time scale defined by $\tau = tv|a^2$. During the heat-up phase, little change in radius is

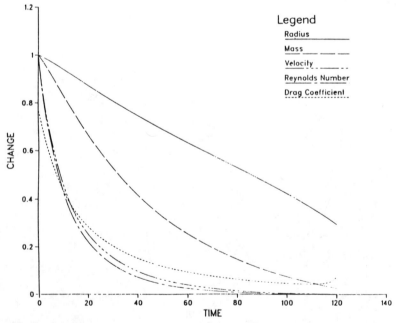

Figure 2 Droplet vaporization, Reynolds number: 200, velocity: 50 m/s.

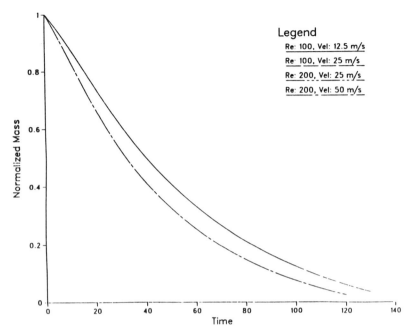

Figure 3   Droplet vaporization, change in mass,

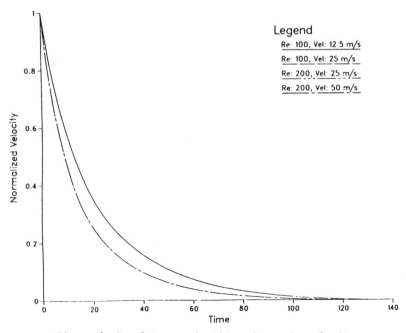

Figure 4   Droplet vaporization, change in velocity.

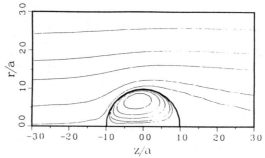

Figure 5   Stream function, time: 6.0.

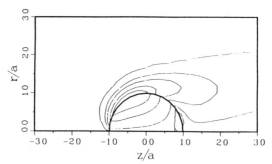

Figure 6   Vorticity, time: 6.0.

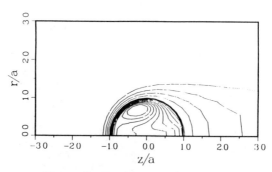

Figure 7   Isotherms, time: 6.0.

Table 1   Cases for calculation

	Diameter, μm	Initial velocity, m/s	Reynolds number
Case 1	50	25	100
Case 2	100	12.5	100
Case 3	50	50	200
Case 4	100	25	200

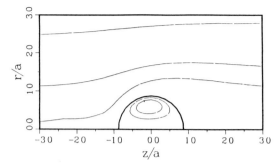

Figure 8   Stream function, time: 40.0.

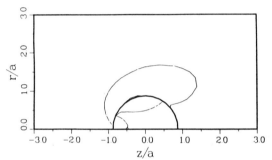

Figure 9   Vorticity, time: 40.0.

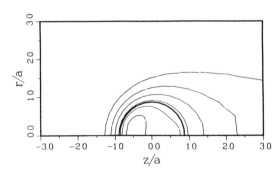

Figure 10   Isotherms, time: 40.0.

noticed until the droplet approaches its boiling point. The drag coefficient is seen to be lower than for a nonvaporizing droplet due to the blowing effect caused by the vaporizing fuel.   Figure 2 shows the results for the third case, with an initial Reynolds number of 200.   The same qualitative behavior is shown, but there is an enhancement of the vaporization rate and a consequent decrease in droplet lifetime.

Figures 3 and 4 give the change in mass and velocity, respectively, for all four cases. With the choice of axes as shown, the first two cases, with Reynolds number of 100, collapse onto one line while the last two fall on another. It should be noted that the actual time scale is not the same for drops differing in diameter. These two figures clearly indicate that the Reynolds number is an important similarity parameter. In spite of the complex coupling between heat, mass, and momentum transfer at the droplet surface, the relative velocity and diameter play no role in determining droplet history as long as the Reynolds number is held constant. This is as expected for the very low-speed flow in this problem.

Figures 5-7 show the streamlines, vorticity, and isotherms for the base case at an early time. The dominance of convection on the transport processes is easily seen. Figures 8-10 are for a later time when diffusion is the dominant mode. The gas phase streamfunction and vorticity have been calculated from velocities obtained by the computation.

Conclusions

A technique suitable for the investigation of droplet vaporization has been developed. The method is sufficiently general to be applicable to other axisymmetric geometries. A preliminary study of droplet vaporization has been completed.

There is a substantial decay in Reynolds number during the early part of the droplet lifetime. Transport processes change from being convection dominated to diffusion dominated while passing through a long period when both convection and diffusion play important roles. All earlier models, except the Dwyer-Sanders model, have assumed the dominance of one mode or the other. This study shows that the use of simpler models is limited only to certain regimes. For example, the simple one-dimensional diffusion model seems to be adequate only in the later stages of the droplet lifetime. A detailed parametric study will be useful in mapping out the various regimes so that simpler models can be used with confidence.

The study so far indicates that the Reynolds number is a similarity parameter. For a given fuel and initial temperatures, the droplet transfer number is a function of Reynolds number therefore. However, a study of different fuels is needed to confirm this. An exhaustive parameter survey may determine other useful nondimensional

parameters.   Simple correlations for practical use will
result from such a survey.

Present   correlations   for   drag   on   a   solid   sphere
appear inadequate when vaporization is present.   The tight
coupling between the transport processes brings in many
factors into the determination of drag and results in a
reduction of drag when vaporization is present.   A
detailed study is needed to isolate the various influences
on the drag coefficient.

References

Chigier, N. (1981) Energy, Combustion and Environment. McGraw-
     Hill, New York.

Dwyer, H.A. and Sanders, B.R. (1984) Comparative study of droplet
     heating and vaporization at high Reynolds and Peclet numbers.
     Prog. Astronautics and Aeronautics, 95, AIAA 464-483.

Faeth, G.M. and Lazar, R.A. (1969) Bipropellant droplet burning
     rates and lifetimes in a combustion gas environment. NASA CR-
     72622.

Godsave, G.A.E. (1954) Studies of the combustion of drops in a
     fuel spray: The burning of single drops of fuel. Fourth
     Symposium (International) on Combustion, pp. 810-830.

Goldsmith, M. and Penner, S.S. (1954) On the burning of single
     drops of fuel in an oxidizing atmosphere. Jet Propul. 24,
     245-251.

Labowsky, M. (1978) A formalism for calculating the evaporation
     rates of rapidly evaporating particles. Combust. Sci.
     Technol. 18, 145-151.

Law, C.K. (1976) Unsteady droplet combustion with droplet heating.
     Combust. Flame 26, 17-22.

Law, C.K. (1977) Unsteady droplet combustion with droplet heating
     - II. Combust. Flame 28, 175-186.

Prakash, S. and Sirignano, W.A. (1978) Liquid fuel heating with
     internal circulation. Int. J. Heat Mass Transfer 21, 885-895.

Ranz, W.E. and Marshall, W.R. (1952) Evaporation from drops. Chem.
     Eng. Prog. 48, 141-146 and 173-180.

Sirignano, W.A. (1983) Fuel droplet vaporization and spray combus-
     tion theory. Prog. Energy Combust. Sci. 9, 281-332.

Spalding D.B. (1953) The combustion of liquid fuels. Fourth
     Symposium (International) on Combustion, Combustion
     Institute, Pittsburgh, pp. 847-864.

Tal, R and Sirignano, W.A. (1982) Cylindrical cell model for the hydrodynamics of particle assemblages at intermediate Reynolds numbers. AIChE J. 28, 233–237.

Twardus, E.M. and Brzustowski, T.A. (1977) The interaction between two burning droplets. Arch. Procesow Spalania 8, 347–358.

Vinokur, M (1974) Conservation equations of gas-dynamics in curvilinear coordinate systems. J. Comput. Phys. 14, pp. 105–125.

Viviand, H. (1975) Formes conservatives des equations del dynamique des gaz. Rech. Aeros. 1, 65–68

Westbrook, C.K. (1978) A generalized ICE method for chemically reactive flows in combustion systems. J. Comput. Sci. 29, 67–80.

Wise, H., Lorell, H. and Wood, B.J. (1955) The effects of chemical and physical parameters on the burning rate of a liquid droplet. Fifth Symposium (International) on Combustion, Combustion Institute, Pittsburgh, pp. 132–141.

Interaction of Flame Spreading, Combustion, and Fracture of Single-Perforated Stick Propellants Under Dynamic Conditions

M.M. Athavale,* K.C. Hsieh,* W.H. Hsieh,* J.M. Char,* and K.K. Kuo†
The Pennsylvania State University, University Park, Pennsylvania

Abstract

Single-perforated long-stick propellants have been used in large-caliber gun systems. Flame spreading and combustion inside the perforation of a stick propellant involve many complicated processes. Pressure differential across the web causes grain deformation, which further interacts with the fluid dynamic processes inside the perforation. The propellant grain may rupture at some stage during flame spreading and combustion. It is important to determine the effect of pressurization rate, initial temperature, grain geometry, and pressure outside the stick propellant on grain fracture. In this research, theoretical analysis was performed using a coupled finite-difference and finite-element code that solves 1) one-dimensional transient flow of the gas phase, 2) transient flame spreading over the propellant surface, and 3) dynamic deformation of the axisymmetric propellant grain. Experimental investigations were conducted using a windowed test chamber. A single stick propellant was mounted in the test chamber, and the internal perforation was rapidly pressurized using either compressed nitrogen gas or combustion product gases from a driving motor. Test results indicate that the critical pressure differential for grain rupture increases monotonically with the internal pressurization rate, and under extremely

---

Presented at the 10th ICDERS, Berkeley, California, August 4-9, 1985. Copyright © 1985 by the American Institute of Aeronautics and Astronautics, Inc. All rights reserved.
*Graduate Student, Department of Mechanical Engineering.
†Distinguished Alumni Professor, Department of Mechanical Engineering.

rapid pressurization (~70,000 MPa/s or 10.15×10^6 psi/s), it could be substantially higher than that for the steady-state conditions. Recovered propellant samples showed that at low pressurization rates, longitudinal slits were formed, while at rapid pressurization (>35 MPa/s or 3.68×10^3 psi/s), the grains shattered into many small pieces. This implies that at high pressurization rates, propellant mass burning rate and chamber pressure could be affected substantially by grain fracture.

Nomenclature

A_p	= cross-sectional area of perforation, m^2
b	= covolume of Noble-Able equation of state, m^3/kg
c	= speed of sound, m/s
C_p	= constant pressure specific heat, J/kg-K
C_s	= specific heat of stick propellant, J/kg-K
D_v	= viscous drag force per unit area, N/m^2
$\mathcal{D}$	= binary diffusion coefficient, m^2/s
e_{ij}	= deviatoric strain
E	= total stored energy (internal plus kinetic), J/kg
f^k	= body force density in k^{th} direction, N/m^3
h	= specific enthalpy, J/kg
h_c	= convective heat-transfer coefficient, W/m^2-K
k	= thermal conductivity, W/m-K
K	= bulk modulus of solid propellant material, N/m^2
P_b	= perimeter of internal perforation, m
q_{rad}	= radiative heat flux absorbed by propellant, W/m^2
$Q_{s_{chem}}$	= surface heat release due to pyrolysis, J/kg
r_b	= propellant burning rate, m/s
r_i	= inner radius of perforation, m
R	= gas constant, J/kg-K
S	= surface, m^2
S_{ij}	= deviatoric stress, N/m^2
t	= time, s
T	= temperature, K
v_g	= gas velocity, m/s
V	= volume, m^3
x_k	= coordinate axis in k^{th} direction, m
$\ddot{x}_k$	= acceleration in k^{th} direction, m/s^2
Y_i	= mass fraction of the ith species
z	= axial coordinate, m
$\delta x_{k,m}$	= deformation gradient tensor

ε_{kk}	=	dilatory strain
π	=	virtual work, J
ρ	=	density, kg/m^3
σ^{km}	=	stress tensor, N/m^2
τ	=	viscous shear stress, N/m^2
τ^k	=	surface traction in k^{th} direction, N/m^2
τ_{zz}	=	normal viscous stress, N/m^2
$\dot{\omega}_i$	=	rate of production of ith species, kg/m^3-s

Subscript

g   = gas phase, internal perforation region

Introduction

Use of single-perforated long-stick propellant charges in large-caliber gun systems is becoming more attractive because of the following advantages offered over conventional granular charges:

1) Long-stick propellant grains are less mobile than granular propellants and hence stay near the breech end for a longer period of time and gasify with more localized burning under high-pressure conditions, as evidenced in studies by Robbins and Horst (1983).

2) The flow channels in stick propellant charges offer less flow resistance and reduce undesirable pressure overshoots and oscillations.

3) The regular geometry of stick propellants allows a higher charge density, charge design flexibility, and easy charge loading.

4) Higher charge density also allows the use of low-vulnerability ammunition (LOVA) propellants in this configuration.

A number of studies (Robbins et al. 1980; Robbins and Horst 1981, 1983; Robbins 1982; Robbins and Einstein 1984; Minor 1982; Gough 1983; Horst et al. 1983; Kuo et al. 1984) have been conducted on the combustion and performance of stick propellants. Studies by Robbins and Horst (1981), Robbins (1982), and Minor (1982) show that unslotted single-perforated stick propellants rupture during combustion. The geometry of the single-perforated unslotted long-stick charge leads to different combustion and flame-spreading rates in the stick perforation and the channels between grains. This results in a pressure differential across the web of the grain that may lead to fracture of the propellant. This phenomenon is important in the overall ballistic process because a fractured grain provides a higher burning surface area which, in turn, alters the total mass burning rate and performance of the gun. The excessive pressure buildup in the internal perforation region can also alter the flow channel area and affect the combustion process prior to rupture.

Robbins and Horst (1983) and Robbins (1982) carried out studies to determine the critical fracture pressure differential for several stick propellants. Their tests involved 1) quasisteady pressurization of the perforation of an unslotted stick propellant using compressed nitrogen gas, and 2) rapid pressurization of the perforation using product gases from combustion of black powder. In their experiments, the stick propellant sample was mounted on several supports. Transient pressure-time traces of internal pressure at several locations along the stick axis were obtained while the external pressure was kept at the ambient level. Although the study was useful in understanding the subject area, in actual gun chambers, the external pressure also changes rapidly. This factor could be important and must be taken into consideration in determination of the fracture criterion.

In the present research, an attempt has been made to determine the interrelationship between such phenomena as grain deformation, fracture, flame spreading, and combustion inside the perforation of a stick propellant. Effects of internal and external pressurization levels and pressurization rates on the above phenomena have been considered.

Experimental Approach

A test chamber was designed and fabricated to study flame spreading and combustion inside the stick perforation, and deformation and fracture of the propellant grain under dynamic loading. Figure 1 is a schematic drawing of the windowed test rig. A single stick propellant is mounted in the test chamber. A photograph of the stick propellant sample mounted on the sample holder is shown in Fig. 2.

The chamber has facilities to measure both transient pressures inside the perforation at several axial locations and pressures outside the stick propellant. The chamber has two long windows through which the phenomena of flame spreading, combustion, and fracture can be observed. When necessary, a light pipe can be mounted into the light pipe holder at the end of the propellant grain to detect the arrival of hot gases or a flame front inside the stick perforation. Pressure external to the stick propellant can be kept at a fixed level using compressed nitrogen gas, or controlled by dynamic pressurization by

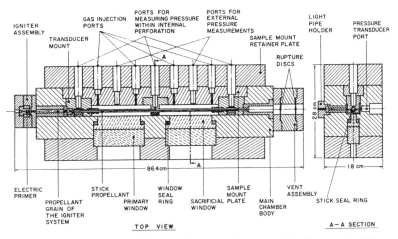

Fig. 1 Schematic diagram of test rig assembly for studying flame spreading and combustion of a single stick propellant.

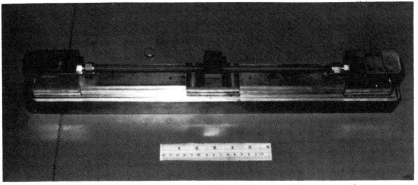

Fig. 2 Sample holder assembly with propellant sample.

opening a high-pressure nitrogen feedline at a preselected time.

The internal perforation of the stick propellant can be pressurized separately using either 1) compressed nitrogen gas at a low (quasi-steady) pressurization rate, 2) rapid injection of nitrogen gas at high pressurization rates, or 3) hot combustion gases generated from a driving motor. The third method can be used to pressurize and ignite the propellant in order to study flame spreading and combustion in the perforation. In the second method, a flying pin/rupture diaphragm replaces the igniter assembly. Rapid injection of high-pressure nitrogen gases into

M.M. ATHAVALE ET AL.

the internal perforation of a single stick propellant is achieved by rupturing a copper disk (0.5 mm thick) with a flying pin propelled by high-pressure combustion gases generated from a gas generator. In the third method, combustion product gases can provide extremely high pressurization rates, whereas the second method yields somewhat lower pressurization rates, depending on the maximum feedline pressure selected.

The data acquisition system contains a transient waveform recorder to store the pressure-time traces during dynamic pressurization. The time of grain fracture as well as the critical pressure differential across the propellant web were determined from these traces. A highspeed movie camera was used to obtain a record of the flame spreading and combustion as well as grain fracture.

Theoretical Approach

The overall phenomenon involves both the combustion process and the coupled grain deformation and fracture processes. The theoretical model consists of analyses for both fluid mechanics and solid mechanics. Major assumptions in the fluid-mechanics process are 1) fluid flow in the perforation can be considered one-dimensional; 2) body forces and bulk viscosity are negligible; 3) Soret and Dufour effects are negligible; and 4) Fick's law of diffusion is valid.

Governing Equations for the Gas Phase

Governing equations for the gas phase follow.
Continuity:

$$\frac{\partial(\rho_g A_p)}{\partial t} + \frac{\partial(\rho_g A_p v_g)}{\partial z} = r_b \rho_s P_b \tag{1}$$

Momentum:

$$\frac{\partial(\rho_g v_g A_p)}{\partial t} + \frac{\partial(\rho_g v_g^2 A_p)}{\partial z}$$

$$= - A_p \frac{\partial P}{\partial z} - D_{v_i} P_b + \underbrace{\frac{\partial(\tau_{zz} A_p)}{\partial z}}_{\text{small}} + \underbrace{\frac{\rho_s^2 r_b^2}{\rho_g} \frac{\partial A_p}{\partial z}}_{\text{small}} \tag{2}$$

The energy equation for the stored total energy (internal and kinetic) per unit mass, E_g, is

$$\frac{\partial(\rho_g A_p E_g)}{\partial t} + \frac{\partial(\rho_g A_p v_g E_g)}{\partial z} = \frac{\partial}{\partial z}\left(k_g A_p \frac{\partial T_g}{\partial z}\right) - \frac{\partial(A_p P v_g)}{\partial z}$$

small

$$- \frac{\partial}{\partial z}(\tau_{zz} A_p v_g) - \bar{h}_t(T_g - T_s)P_b$$

small

$$+ \rho_s r_b P_b \sum_{j=1}^{M} Y_j^* h_j(T_s) \tag{3}$$

The species continuity equation is

$$\frac{\partial(\rho_g Y_i)}{\partial t} + \frac{\partial(\rho_g v_g Y_i)}{\partial z} = \frac{\partial}{\partial z}\left[\rho_g D \frac{\partial y_i}{\partial z}\right] + \dot{\omega}_i \tag{4}$$

small

where the source term $\dot{\omega}_i$ consists of contributions from surface pyrolysis and gas-phase reactions. The flame model assumed follows the approach taken by Wu et al. (1983) and Kuo et al. (1984). The homogeneous propellant is assumed to pyrolyze into three groups of species: oxidizer, fuel, and first group of delayed-reaction species. In the pres= ent case, the species liberated from the propellant surface can be entrained by the gas flow in the perforation. The gas-phase mechanism involves from the oxidizer and fuel, which further yield final products. The first group of delayed-reaction species also gives rise to the final-product species, as shown in Kuo et al. (1984). Because the flow inside the perforation is highly turbulent, the species diffusion time is extremely short and the overall chemical reaction rate can be assumed to be kinetics controlled. The rate expressions for various species are the same as in Kuo et al. (1984).

Equation of state: The gas is assumed to obey either the Noble-Abel dense gas law,

$$P\left(1/\rho_g - b\right) = RT_g \tag{5a}$$

or the perfect gas law,

$$P = R\rho_g T_g \tag{5b}$$

The fluid-mechanics processes are solved using a modified version of the high-velocity transient (HVT) computer code, as discussed in Peretz et al. (1972), which was used to predict the ignition transient of rocket motors with a low port-to-throat area ratio.

Initial and Boundary Conditions for the Gas Phase

The initial conditions can be specified readily since the gas velocity is zero, and the pressure and temperature of gas are known initially. The nature of the governing equations for the gas phase is hyperbolic, since all eigenvalues are real. In addition to the physical boundary conditions, method of characteristics is employed to determine the flow properties at the boundaries of the gas phase. At the igniter end of the stick propellant grain, one of three conditions can occur: subsonic inflow, subsonic outflow, or supersonic outflow. For subsonic inflow, during the igniter discharge, the measured pressure immediately outside the perforation is used as the local pressure, and the stagnation temperature at boundary is assumed to be close to the flame temperature of the igniter gases.

The left-running characteristic line is used together with the above two conditions to determine the properties at the left boundary. The species at the left boundary can be considered solely as product species. In the case of subsonic outflow at the left end, only one physical condition is necessary; this can be specified using measured pressures outside the perforation. In this case, the left-running characteristic line and the particle-path line are used together with the physical condition for numerical computation. At the right end of the stick propellant, the right-running characteristic line and particle-path line are merging for subsonic outflow; hence, only one external boundary condition is needed at this location. This is again specified using measured pressure at the exit of the stick perforation. In the case of supersonic outflow at either end, no physical boundary conditions are necessary since flow properties at boundary conditions are necessary since flow properties at the boundaries are determined from the interior region.

Solid-Mechanics Equations

The solid-phase analysis involves calculation of solid-phase temperature distribution and deformation. The grain deformation analysis assumes that the solid propellant is linear viscoelastic in pure shear deformation, and elastic in bulk deformation. The elastic bulk behavior is assumed to follow

$$\sigma_i^i = 3K\varepsilon_i^i$$

(6)

where K is the bulk modulus. The shear behavior is expressed as

$$S_{ij} = \int_0^t G_1(t - t') \frac{\partial e_{ij}(t')}{\partial t'} dt'$$

(7)

where the relaxation modulus $G_1(t)$ is assumed to be

$$G_1(t) = G_\infty + (G_0 - G_\infty)e^{-\beta t}$$

(8)

As a closed-form solution to this dynamic problem is not possible, the finite-element code HONDO-II (Key et al. 1978) is used to compute the solid-phase deformation resulting from the pressure differential generated by combustion processes. The code utilizes the principle of virtual work to derive the equations of motion. This principle states that at all points along the path of motion of the body, the differential work $\delta\pi$ must vanish for all variations δx_k satisfying imposed motion boundary conditions. The differential virtual work $\delta\pi$ is defined as

$$\delta\pi = \int_V \rho \ddot{x}^k \delta x_k dv + \int_V \sigma^{km} \delta x_{k,m} dv$$

$$- \int_V \rho f^k \delta x_k dv - \int_S \tau^k \delta x_k ds$$

(9)

The HONDO-II code uses four-node bilinear isoparametric elements. The principle of virtual work is then written for each of the nodes as

$$\delta\pi = \sum_{n=1}^N \left(\int_{V_n} \rho \ddot{x}^k \theta_{\sim k}^\ell dv + \int_{V_n} \sigma^{km} \theta_{\sim k,m}^\ell dv \right.$$

$$\left. - \int_{V_n} \rho f^k \theta_{\sim k}^\ell dv - \int_{S_n} \tau^k \theta_{\sim k}^\ell ds \right) = 0$$

(10)

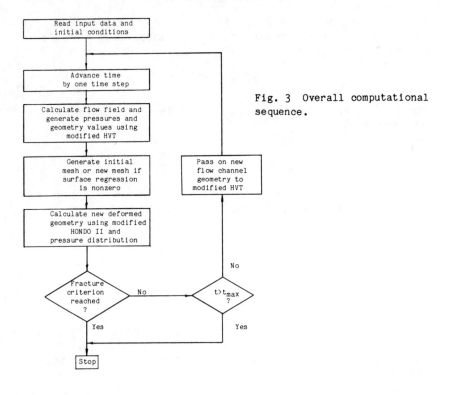

Fig. 3 Overall computational sequence.

where

$$\underset{\sim k}{\Theta}^{\ell} = [\phi_k^1, \phi_k^2, \phi_k^3, \phi_k^4] \Big|_{V_n} \qquad (11)$$

and ϕ_k^{ℓ} are the bilinear interpolation functions, and N is the total number of elements surrounding the node.

Heat Conduction Equation for the Solid Phase

The temperature distribution on the solid propellant can be solved from the heat conduction equation for the solid phase. Assuming that the conduction heat transfer in the axial direction is negligibly small compared to that in the radial direction, the one-dimensional transient heat equation becomes

$$\frac{\partial T_{pr}(t,r)}{\partial t} = \alpha_{pr} \frac{1}{r} \frac{\partial}{\partial r} \left(r \frac{\partial T_{pr}(t,r)}{\partial r} \right) \qquad (12)$$

The initial condition is

$$T_{pr}(o,r) = T_{pi} \tag{13}$$

The boundary conditions are

$$T_{pr}(t,r_o) = T_{pi} \tag{14}$$

and

$$-k_s \frac{\partial T_s}{\partial r}\bigg|_{r_i^+} = \bar{h}_c(T_g - T_s) + \rho_s r_b (Q_s)_{chem}$$

$$+ (\dot{q}_{rad})_{r_i} + \rho_s r_b (C_s - C_{pg})T_s \tag{15}$$

where $(\dot{q}_{rad})_{r_i}$ is the net radiation flux absorbed at the inner radius of the stick propellant. To obtain approximate surface temperature variation with respect to time, a third-order polynomial in r is assumed to represent the transient temperature profile in solid propellant. A quantity $\delta(t)$ (the thermal penetration depth) is defined, so that for $r \geq (r_i+\delta)$ the solid is at equilibrium temperature and there is no heat transfer. An integral method is used here to obtain the approximate solution of the heat equation. This method has proven to be quite accurate and economical.

Numerical Approach

Combustion Code

A generalized implicit scheme, based on central differences in spacewise derivatives, was chosen to solve the governing equations numerically. A weighting parameter θ is used to control the degree of implicitness of the numerical scheme. In the Crank-Nicolson scheme, θ is 0.5 and a value of 0.6 is used in the numerical simulations.

The nonlinear coefficients and inhomogeneous terms in the governing equations are linearized by following the procedure in the HVT program (Peretz et al. 1972). In spite of the fact that the theoretical model accounts for a distended (thick) flame zone, for simplicity, the following calculations were performed, assuming that the reaction occurs instantaneously in the local region.

Solid Mechanics Code

Using divergence theorem, one can reduce Eq. (9) to the equations given below. Motion of the body:

$$\sigma^{km}_{,m} + \rho f^k = \rho \ddot{x}^k \qquad \text{in } V \tag{16}$$

The traction boundary condition requires that

$$\sigma^{km} n_m = S^k(t) \qquad \text{on } S \tag{17}$$

The displacement boundary condition is

$$x^i(X^\alpha, t) = x^i(t) \qquad \text{on } S \tag{18}$$

After the elements, nodal velocities, and deformation gradients are initialized, Eq. (17) is used to calculate stresses in the first element from the applied surface traction load. The stress divergence, which is needed in Eq. (16), and the critical time step are then determined. An element-by-element process is used to generate the equations of motion for all nodes. Therefore, the resulting simultaneous equations in time are integrated using central difference expressions for velocities and displacements. Because a diagonal mass matrix is used, the scheme is explicit and therefore computationally very fast per time step. This integration procedure is conditionally stable with respect to time step size, and hence a very simple, but reliable, continuous monitor of the step size is used in the program.

The criterion for numerical stability used originally by Bertholf and Benzley and discussed by Key et al. (1978) is employed in HONDO-II. It accounts for the effects of bulk viscosity of viscoelastic material. The critical time step size is

$$\Delta t_{crit} = \delta \left/ \left\{ B_2 C + (B_1)^2 \delta \left| \frac{1}{\rho} \frac{\partial \rho}{\partial t} \right| \right. \right.$$
$$\left. + \sqrt{ \left(B_2 C + (B_1)^2 \delta \left| \frac{1}{\rho} \frac{\partial \rho}{\partial t} \right| \right)^2 + c^2 } \right\} \tag{19}$$

where B_1 and B_2 are constants, and δ is the minimum of the side lengths and diagonals of the mesh in question. The

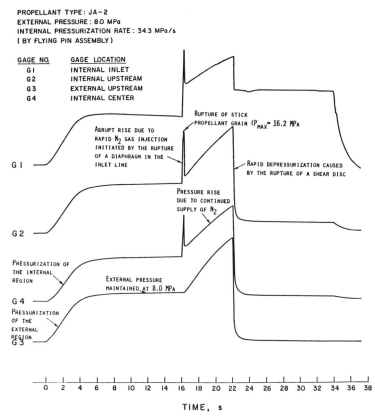

PROPELLANT TYPE : JA-2
EXTERNAL PRESSURE : 8.0 MPa
INTERNAL PRESSURIZATION RATE : 34.3 MPa/s
(BY FLYING PIN ASSEMBLY)

GAGE NO.   GAGE LOCATION
   G1   INTERNAL INLET
   G2   INTERNAL UPSTREAM
   G3   EXTERNAL UPSTREAM
   G4   INTERNAL CENTER

Fig. 4 Typical set of measured pressure-time traces of dynamic
pressurization of stick propellant by rapid nitrogen gas
injection (test No. DADP-5).

value used for each time step is 90% of the minimum Δt_{crit}
found for the entire mesh.

After the updated nodal velocities are obtained, the
displacements and current geometry are calculated. Strains
and surface traction loads at the current time step can
then be determined. Because the time step in HONDO-II is
determined internally, a check for matching the time with
that of the modified HVT program is made before the information of propellant deformation is transferred to the
combustion code. Pressure distribution along the internal
perforation and the external surface serves as the surface
loading for the solid-phase structural calculations, and
displacement boundary conditions are determined depending
on the location of the node.

Coupling of Combustion and Solid-Mechanics Codes

The two computer codes are modified and coupled to
analyze the combined processes of combustion and grain
deformation. The computations are done in the following
sequence:
1) The modified HVT code is used to calculate the
flow properties for a time step.
2) The generated pressure and new flow channel geom⁼
etry are passed on to the coupling subprogram.
3) If there is any combustion, the coupling subprogram generates a new mesh and passes on the values (together with pressure profiles) to the HONDO-II code.
4) The HONDO-II code is then used to calculate grain
deformation and stresses for the new time step.

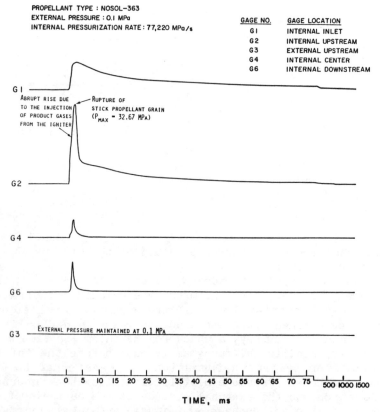

Fig. 5 Typical set of measured pressure-time traces of dynamic
pressurization of stick propellant by injection of combustion
product gases from igniter (test No. DADP-9).

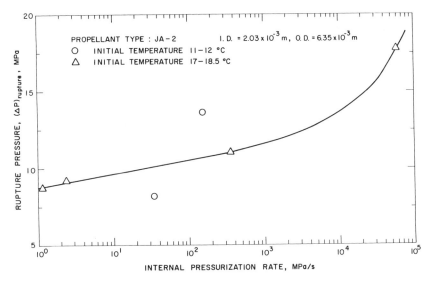

Fig. 6   Plot of rupture pressure $(\Delta P)_{rupture}$ versus pressurization rates.

5) The propellant geometry after deformation is then passed back to HVT for calculations of flow field for the next time step.

Figure 3 presents a schematic of the overall computational sequence.

Calculations are executed until a fracture criterion is reached. The fracture criterion is based upon attainment of a critical stress in the web of the stick propellant. When this point is reached, the calculations are stopped, because once the grain ruptures the above analysis is no longer applicable.

Discussion of Results

Several tests were conducted using the test chamber described earlier. The experiments involved pressurization of internal perforation of a single stick propellant at various rates at different external pressure levels. Pressurization was achieved with either nitrogen gas or combustion product gases from a driving motor. Two different types (NOSOL-363 and JA-2) of stick propellants were used in the tests. The pressure differential at which the grain fractured for each test was determined using the pressure-time traces.

Table 1 Dependence of critical rupture pressure
on pressurization rate of internal perforation
of NOSOL-363 stick propellant

$\frac{\partial P}{\partial t}$, MPa/s	$(\Delta P)_{rupture}$, MPa	Source
0.00	2.76	Robbins (1983)
0.05	3.45	Present
77,220.00	32.7	Present

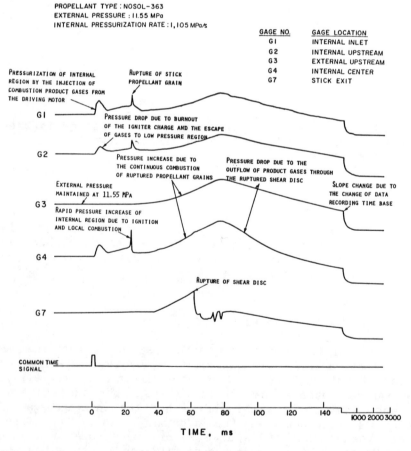

Fig. 7 Typical set of measured pressure-time traces of dynamic
pressurization of stick propellant by controlled rate of
injection of combustion product gases from igniter (test No.DADP-11).

Figures 4, 5 and 7 show three sets of typical P-t
traces. In Fig. 4, the internal and external regions of
the stick propellant were pressurized at the same rate to
prevent grain rupture between time = 0 s. and time = 6.2 s.
Then both internal and external pressures were maintained
at 8.0 MPa (1.16 x 10^3 psi) until 15.8 s., when the flying
pin was initiated and nitrogen gas injection was introduced. The internal pressurization rate introduced by the
nitrogen injection was 34.3 MPa/s (4.97 x 10^3 psi/s).
This action can be seen clearly in the abrupt pressure
increase at time = 15.8 s. After the rapid internal pressurization that created the pressure difference between
the internal and external region of the stick propellant,
the propellant ruptured at time = 16.1 s. (featuring a
rapid pressure drop following the spike). After propellant rupture, the entire test chamber was pressurized by a
continuous supply of nitrogen gas until the rupture disk
located at the aft end of the test chamber broke at time =
21.8 s. At time = 33.7 s., the supply of nitrogen gas was
stopped and the entire test event ended.

In Fig. 5, the external and internal pressures were
kept at 0.1 MPa (14.5 psi) until combustion product gases
generated by the driving motor (igniter) were injected
into the internal perforation of the stick propellant.
The injection features a rapid pressure increase beginning
at time = 0 s. At time = 2,65 ms, the pressure difference
across the propellant web reached 32.7 MPa, and the propellant grain ruptured. Following rupture of the propellant,
the internal pressure dropped drastically to 0.1 MPa (14.5
psi). The dependence of critical rupture pressure on the
pressurization rate of internal perforation of NOSOL-363
stick propellant is shown in Table 1. It is quite evident
that due to the inertia effect, $(\Delta P)_{rupture}$ increases with
pressurization rate. Similar to NOSOL-363 propellant,
JA-2 also exhibits the dependence of critical rupture pressure on internal pressurization rate. Figure 6 shows that
$(\Delta P)_{rupture}$ monotonically increases with respect to internal pressurization rate. The effect of initial temperature cannot be evaluated from the limited information
available in Fig. 6. To assess the temperature effect,
additional data are needed.

In order to study the interaction of flame spreading,
combustion, and fracture of stick propellants, the external pressures of some test runs were set at elevated
levels. The mass rate of injection of high-temperature
gases was also kept at low values. A set of P-t traces
for such a test is shown in Fig. 7. High-speed motion
pictures were taken during this test run, and some selec-

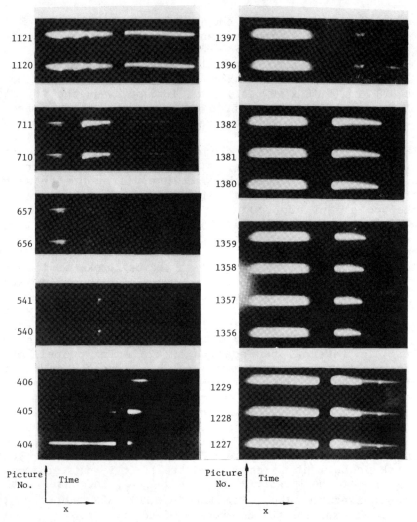

Fig. 8 High-speed motion pictures showing interesting events
(DADP-11).

ted portions of film showing interesting events are presented in Fig. 8. P-t traces and the physical interpretation of the phenomena that occurred during the transient
event are given in Fig. 7. Detailed observations of the
film record are summarized in Table II.

A number of polaroid pictures of recovered propellant
pieces were taken using a scanning electron microscope
(SEM). Surfaces of ruptured propellant pieces showed different patterns for different pressurization rates. Fig-

Table 2 Film interpretation of flame spreading,
combustion, and fracture of single-perforated stick
propellant (test firing No. DADP-11)

Interesting event No.	Picture	Time,ms	Observations and interpretations
1	0 ↓ 396	0.0 ↓ 22.50	Hot igniter gases enter internal perforation of stick propellant.
2	397 ↓ ↓ ↓ ↓ ↓ 412	22.56 ↓ ↓ ↓ ↓ ↓ 23.41	Combustion product gases become visible in first window for a period of 340 μs. Then propellant cracks near center part of first window, followed by depressurization and extinction.
	413 ↓ 515	23.47 ↓ 29.26	Propellant is momentarily extinguished, and film is very dark.
3	516 ↓ ↓ ↓ ↓ 624	29.32 ↓ ↓ ↓ ↓ 35.45	Propellant grain reignites near center support. Gasification processes continue, as indicated from P-t traces, and both external and internal pressures increase.
4	625 ↓ 703	35.51 ↓ 39.94	Bright flame reappears near upstream end of first window.
5	704 ↓ ↓ 1015	40.00 ↓ ↓ 57.66	A dark region in center portion of first window region is observed. This is due to rupture of propellant.
6	1016 ↓ ↓ ↓ ↓ ↓ ↓ ↓ ↓ ↓ 1219	57.72 ↓ ↓ ↓ ↓ ↓ ↓ ↓ ↓ ↓ 69.16	As combustion becomes more severe, hot product gases generated at upstream and downstream locations flow to center part of the first window. This causes the dark region to disappear. First window shows continuous illumination due to external and internal surface burning, while second window region shows only internal burning.
7	1220 ↓ 1335	69.32 ↓ 76.99	External region of second window starts to burn.

(Table continued on next page)

Table 2 (cont.) Film interpretation of flame spreading,
combustion, and fracture of single-perforated stick
propellant (test firing No. DADP-11)

Interesting event No.	Picture	Time, ms	Observations and Interpretations
8	1356 ↓ ↓ 1375	77.05 ↓ ↓ 78.125	Propellant grain located at downstream of second window shows momentary extinction due to rupture of shear disk.
9	1376 ↓ 1395	78.18 ↓ 79.26	Severe burning occurs throughout chamber.
10	1396 ↓ ↓ ↓ ↓ ↓ ↓ 2516	79.32 ↓ ↓ ↓ ↓ ↓ ↓ 142.95	Propellant located at second window burns with a lower luminosity due to outflow of gases; propellant located at first window still exhibits severe burning. Phenomena continues until the end of entire event.

ure 9a shows the surface condition of a long slit opened
in a NOSOL-363 stick propellant under a very low pressurization rate. Figure 9b shows the surface condition of a
shattered piece from the same propellant which underwent a
highly dynamic fracture process. It is evident from the
figure that the surface of the shattered piece contains
many microcracks that are the result of very rapid stress
loading (similar to impact loading).

Theoretical Results

   Theoretical simulation was performed for a typical
dynamic pressurization test with no combustion. The coupled computer code was used to evaluate stresses and displacements. The web was divided into two rows of elements
along the radial direction and 30 elements along the axial
direction. This configuration gives three rows of nodal
points along radial directions for which displacements
were calculated. Figure 10 shows the radial displacements
of different rows of nodal points at different times. The
deformations are quite small as compared to the original
web thickness. The displacements for the outer row are
restricted at three holder locations; therefore, the displacement distributions have two peak regions. The theo-

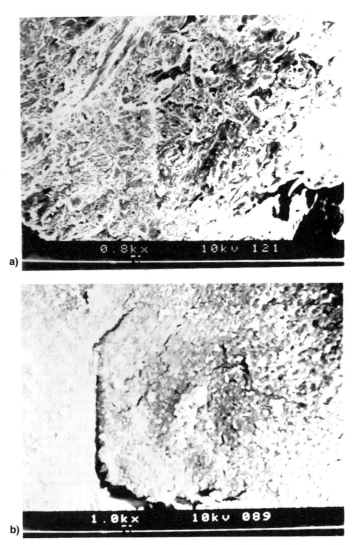

Fig. 9 Micrographs of surfaces of ruptured NOSOL-363 propellant grains. a) Propellant surface of a long slit opened under low pressurization rate at 0.055 Mpa/s (DADP-1). b) Propellant surface of a shattered piece recovered under high pressurization rate at 77,215 Mpa/s (DADP-9).

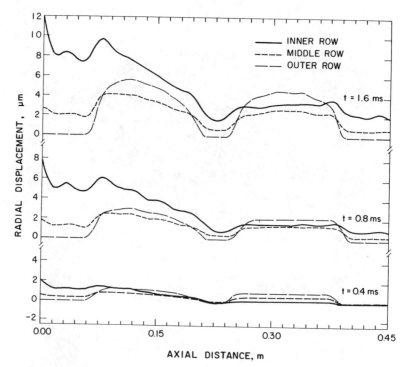

Fig. 10 Calculated radial displacement of nodal points at
different times (DADP-9).

retical simulation results for cases with combustion will
be reported in future papers.

Conclusions

Some of the major results obtained are as follows:
1) The critical pressure differential increases monotonically with pressurization rate, and the pressure differential can be substantially higher at rapid pressurization rates than at steady operating conditions.
2) Recovered test samples show that at lower pressurization rates, the grain fractures with one or more longitudinal slits; at very rapid pressurization, the propellant
shatters into several pieces. This fact is important,
since fragments of the shattered propellant generate a
significantly higher total burning surface area, which, in
turn, leads to enhanced burning of the propellant.
3) Fracture surfaces of shattered propellant pieces
show the existence of microcracks; such microcracks are

absent in the case of propellant ruptured under low pressurization rates.

4) A theoretical model has been proposed to analyze
the interaction between combustion processes and grain
deformation during combustion inside the perforation of
single-perforated stick propellants. Further work is in
progress for verification of the model.

Acknowledgments

This research represents a part of the results obtained under Contract DAAG29-83-K-0081, sponsored by the
Engineering Science Division, Army Research Office, Research Triangle Park, N.C., under the management of David
M. Mann. The authors would also like to acknowledge the
encouragement and support of David Downs of ARDC-Dover and
F. Robbins of Ballistic Research Laboratory. The propellants were supplied by J. Kennedy of Honeywell, Inc., and
F. Robbins of Ballistic Research Laboratory.

References

Chang, T. Y., Chang, J. P., Kumar, M., and Kuo, K. K. (1980),
Structural interaction in a viscoelastic material. Research in
Nonlinear Structural and Solid Mechanics, pp. 67-90. NASA
Conference Publication 2147.

Gough, P.S. (1983), Continuum modeling of stick charge
combustion. Proc. 20th JANNAF Combustion Meeting, Vol. I, pp.
555-567. CPIA Publication 383.

Horst, A. W., Robbins, F. W., and Gough, P. S. (1983), Multidimensional, multi-phase flow analysis of flamespreading in a
stick propellant charge. Proc. 20th JANNAF Combustion Meeting,
Vol. I, pp. 365-383. CPIA Publication 383.

Key, S. W., Beisinger, Z. E., and Kreig, R. D. (1978), HONDO II,
a finite element computer program for the large deformation
dynamics of axisymmetric solids. SAND 778-0422, Sandia National
Laboratories, Livermore, Calif.

Kuo, K. K., Hsieh, K. C., and Athavale, M. M. (1984), Modeling
of combustion processes of stick propellants via combined
Eulerian-Lagrangian approach. Proc. 8th International Symposium
on Ballistics, Orlando, Fla.,pp. 155-168.

Minor, T. C. (1982), Ignition phenomena in combustion-cased
stick propellant charges. Proc. 19th JANNAF Combustion Meeting,
Vol. I, pp. 555-567. CPIA Publication 366.

Peretz, A., Kuo, K. K., Caveny, L. H., and Summerfield, M.
(1972), The starting transient of solid-propellant rocket
motors with high internal gas velocities. AIAA Journal, Vol.
11, No. 12, pp. 1719-1727.

Robbins, F. W. and Horst, A. W. (1983),   Slotted stick
propellant study.   Proc. 20th JANNAF Combustion Meeting, Vol. I,
pp. 377-386.   CPIA Publication 383.

Robbins, F. W. and Horst, A. W. (1981),   A simple theoretical
analysis and experimental investigation of burning processes for
stick propellant.   Proc. 18th JANNAF Combustion Meeting, Vol.
II, pp. 25-34.   CPIA Publication 347.

Robbins, F. W., Kudzal, J. A., McWilliams, J. A., and Gough, P.
S. (1980), Experimental determination of stick charge flow
resistance.   Proc. 17th JANNAF Combustion Meeting, Vol. II, pp.
97-118.   CPIA Publication 329.

Robbins, F. W. (1982), Continued study of stick propellant
combustion processes.   Proc. 19th JANNAF Combustion Meeting,
Vol. I, pp. 443-459.   CPIA Publication 366.

Robbins, F. W. and Einstein, S. I. (1984), Workshop report:
Stick propellant combustion processes.   Proc. 21st JANNAF
Combustion Meeting, Vol. II, p. 215, CPIA Publication 412.

Wu, X., Kumar, M., and Kuo, K. K. (1983), A comprehensive
erosive-burning model for double-base propellants in strong
turbulent shear flow.   Combust. Flame 53 (1-3), 49-60.

Chapter III. Combustion Modeling and Kinetics

Interactions Between a Laminar Flame and End Gas Autoignition

William J. Pitz* and Charles K. Westbrook*
Lawrence Livermore National Laboratory, Livermore, California

Abstract

A numerical model combining one-dimensional fluid mechanics and detailed chemical kinetics is used to examine the interactions between laminar flame propagation and end gas autoignition at high temperatures and pressures of approximately 30 atm. The flame is found to have very little influence on the computed rates of fuel-air autoignition in the end gas, but the subsequent high rate of heat release during autoignition is shown to produce strong acoustic waves in the burned gases.

Introduction

Increases in the costs of petroleum fuels in recent years have created a need to improve fuel economy of internal combustion engines. This can be achieved by increasing the compression ratio of the engine, but such increases are severely limited by the onset of engine knock. Environmental concerns have necessitated the virtual elimination of the most effective metal-based antiknock additives such as tetraethyl lead (TEL), resulting in a great deal of interest in finding effective alternative methods of controlling engine knock. However, a more thorough understanding of the physics and chemistry of the autoignition process is needed before significant progress in this complex area can be achieved.

Presented at the 10th ICDERS, Berkeley, California, August 4-9, 1985. Copyright © 1986 by the American Institute of Aeronautics and Astronautics, Inc. All rights reserved.
*Physicist, Computational Physics Division.

One technique that recently has been applied to the autoignition problem is that of detailed numerical modeling of the chemical kinetics of hydrocarbon fuel ignition (Smith et al. 1985; Pitz and Westbrook 1986; Leppard 1985; Dimpelfeld and Foster 1985). These models use experimentally observed temperatures, pressures, and fuel-air mixture ratios from typical automobile engines to predict the time required for the unburned fuel-air mixture (the so-called "end gas") to ignite spontaneously. If the time required for self-ignition is relatively long, then there is usually sufficient time for this end gas to be consumed by the conventional flame front in the engine prior to autoignition, and knocking is avoided. However, if the autoignition occurs rapidly, then there is the possibility that some fraction of the end gas will ignite before it is consumed by the flame front, resulting in knocking conditions. This overall approach has demonstrated some success in describing part of the engine knock problem, including the ability to reproduce the time of knock occurrence and variations in tendency to knock with temperature, pressure, residence time, and fuel-air equivalence ratio. Furthermore, this approach has also helped to identify the most important parts in the mechanism by which metal-based antiknock additives such as TEL may slow the autoignition process and reduce the tendency of engines to knock (Pitz and Westbrook 1986). However, many questions still remain pertaining to both the physical and chemical processes occurring in knocking engines.

When knock occurs, its most immediate manifestation is the onset of a distinct sound signal, a "ringing" from the engine. This response is a result of interactions between fluid mechanical motions within the engine chamber and the metal components of the engine. There has been a great deal of discussion as to the exact nature of these fluid motions in the engine chamber, including how they are generated by the autoignition of the end gas and whether or not shock waves are present. The present paper represents an attempt to address some of these questions under highly idealized conditions.

Previous modeling efforts (Smith et al. 1985; Pitz and Westbrook 1986) using the HCT code (Lund, 1978) have focused on the autoignition of the end gas, dealing only with a homogeneous medium that effectively neglected fluid mechanical motions within the reacting gas mixture. The compression and heating of the end gas resulting both from piston motion and from flame propagation in the combustion chamber were included in the model as externally applied

boundary conditions. Aside from this formulation of the
boundary conditions, there was no way to examine the fluid
mechanical coupling between the burned gas and the end
gas. However, there are two major mechanisms by which this
coupling can take place, one dealing with the influence of
the burned gases on the rate of autoignition, and the
second dealing with the generation of pressure
inhomogeneities within the combustion chamber by the
autoignition of the end gas. Within the limitations of the
present model, both types of fluid mechanical coupling are
considered in the present paper.

Flame propagation and heat transfer in actual engines
take place under highly turbulent conditions.
Unfortunately, current abilities to simulate detailed
chemical kinetics and flame propagation in turbulent
environments are still very limited. In contrast, these
processes can be simulated quite accurately and completely
under laminar conditions. The problem selected for study
therefore was that of an end gas igniting in the presence
of a flame front and burned combustion products, all taking
place under laminar conditions. The possible relevance of
the computed results to conditions in a turbulent engine is
then discussed.

Numerical Model and Chemical Kinetic Mechanism

Numerical calculations were carried out using the HCT
computer code (Lund 1978), which solves the coupled
conservation equations of mass, momentum, energy and each
chemical species. The thermal diffusivity and molecular
diffusivities are based on a simplified treatment of the
general multicomponent diffusion problem and are discussed
in a previous paper (Westbrook and Dryer 1980).

The chemical kinetic mechanism for propane and
n-butane oxidation was employed in this study. This
reaction mechanism has been shown (Westbrook and Pitz 1984;
Pitz et al. 1985) to reproduce combustion rates, ignition
delays, and intermediate species concentrations in shock
tubes, laminar flames, and the turbulent flow reactor. The
temperature regime covered by these experiments is
approximately 900-2000 K. The reaction mechanism has been
described in detail in the above references and, due to
space limitations, the rate parameters will not be included
here.

During the compression stroke of the engine, the end
gas is subjected to temperatures in the range of 400-900 K,
where additional reactions not included in the above
mechanism may be important. The most important feature of

these reactions is the addition of molecular oxygen to various alkyl radicals. An additional set of reactions needed to describe low temperature behavior has been added to the above chemical kinetic mechanism, and the combined mechanism has been used to successfully describe the negative temperature coefficient regime of acetaldehyde (Kaiser et al. 1984).

Refinements are being made to the mechanism in order to reproduce the low temperature (550-900 K) oxidation of propene and propane (Wilk et al. 1985). The effects of these low temperature reactions on autoignition times in the end gas have been considered in a previous paper (Pitz and Westbrook 1986). Inclusion of a set of low temperature reactions had little effect on autoignition times for n-butane/air mixtures under temperatures and pressures simulating engine conditions. In the calculations presented here, these additional reactions have not been included, since this previous work indicates that their inclusion would insignificantly affect calculated autoignition times under engine conditions.

Homogeneous gas phase chemical kinetics calculations, in which the fluid motions were neglected, have been reported previously (Smith et al. 1985; Pitz and Westbrook 1986). In the current model computations, a one-dimensional domain is considered in which the majority of the combustible mixture has already been burned. In most of the calculations, the linear extent of the

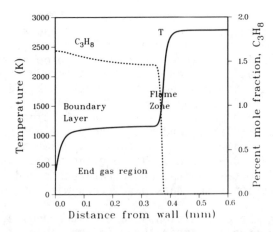

Fig. 1   Schematic diagram of flame-end gas configuration, showing spatial variations of temperature and fuel concentration. The thermal boundary layer is also indicated.

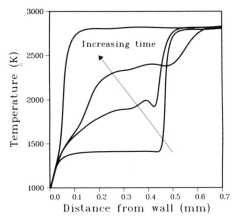

Fig. 2   History of the spatial temperature profile with a wall
temperature of 1000 K.   The end gas undergoes autoignition prior
to flame arrival.

combustion chamber is 7 mm. A laminar flame is propagating
toward the combustion chamber boundary, with an unreacted
"end gas" between the flame and the wall. The fuel
fraction and gas temperature in a typical case of
propane/air are shown in Fig. 1. At this particular time,
the flame is located at a distance of approximately 0.4 mm
from the wall of the combustion chamber. The end gas is at
a temperature of about 1100 K, and there is a thermal
boundary layer caused by heat transfer to the wall which is
maintained at 400 K.
     The simultaneous processes of end gas autoignition and
laminar flame propagation were then calculated, while
several of the most important parameters were varied in
order to examine their relative importance. In some of the
calculations, heat transfer from the reacting end gas to
the colder wall is included, in order to assess the
relative importance of heat losses in the autoignition
rate. In another series of calculations, the relative
timing of the flame arrival and time of autoignition were
changed, ranging from a condition in which no autoignition
occurred to one in which as much as 20% of the total
fuel-air mixture was consumed by the autoignition.
     For all of the model calculations, an initial
combustion chamber pressure of 30 atm was assumed. End gas
initial temperatures were 1000 K, and wall temperatures of
400 and 1000 K were used. All of these parameters were
chosen to approximate thermodynamic conditions in end gases
of reciprocating engines.

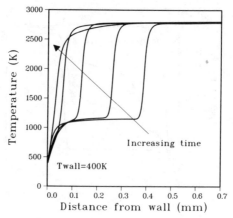

Fig. 3   History of the spatial temperature profile with a wall
temperature of 400 K.   No end gas autoignition occurs.

Results

     The most immediate conclusion reached in these
calculations was that the presence of the flame in the
combustion chamber has relatively little direct influence
on the autoigniti⌐n of the end gas. Of course, the
continuing heat release and pressure increase produced by
the flame continue to compress and heat the end gas, and
this does increase the rate of autoignition. However,
transport of heat and radical species from the flame were
found to have essentially no effect on the autoignition.
These transport processes, which are the central features
of flame propagation, really have an effect only over a
spatial range which is usually referred to as the flame
thickness or the diffusion length. Beyond that distance,
the end gas does not "see" the flame itself, although it
"feels" the flame through the compressional work done by
the flame. In the computed model results for cases in
which autoignition occurred, the autoignition time for a
given sample of end gas was found to be the same,
regardless of whether or not a flame was present. A
sequence of temperature profiles in this regime is shown in
Fig. 2. The ignition of the end gas is quite uniform,
although heat transfer to the wall and from the flame
contribute to some degree of nonuniformity. In this case
the autoignition of the end gas is promoted by a reduction
in the rate of heat transfer to the cold wall (the wall
temperature was maintained at 1000 K, rather than 400 K as
in the case shown in Fig. 1).

In those cases in which the time for consumption by the flame was substantially shorter than the time for autoignition, the presence of the flame did not accelerate the autoignition process. This case is illustrated by the temperature profiles in Fig. 3. Here the autoignition is retarded by the enhanced heat transfer to the colder wall at 400 K. The two extremes represented by Figs. 2 and 3 are clearly very distinguishable. Furthermore, the time scales for the two cases are quite different, with the flame propagation process being slower for the case shown in Fig. 3 than the more abrupt autoignition event shown in Fig. 2.

In the transition regime in which the two time scales were comparable, the combustion occurred as a composite event which actually looked very much like an accelerated flame (Maly and Ziegler 1982). A series of spatial plots of temperature in the end gas region are presented in Fig. 4 for this intermediate regime. If an equivalent "flame position" is defined as the point at which the gas temperature is 2000 K, then it is clear from Fig. 4 that the "flame" is accelerating as the end gas ignites. However, this is not actually a flame, since the advancing temperature front is not the product of transport processes as in a conventional flame, but rather an artifact of the simultaneous processes of ignition and flame propagation.

These results have two significant implications. First, it appears that the same conclusion should apply to turbulent combustion in an automotive engine chamber. Under turbulent conditions, the analogous flame thickness is of course significantly larger, since the turbulent

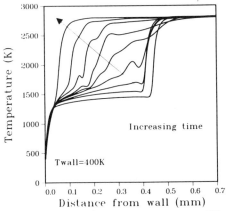

Fig. 4   History of the spatial temperature profile in the intermediate situation, with autoignition occurring on the same time scale as flame arrival.

transport coefficients are larger than the laminar values. However, beyond the turbulent flame thickness, the turbulent flame should have little or no influence on the autoignition rate of the end gas. The second implication of the above result, which should also apply to the turbulent case, is that since the flame and end gas regions are effectively decoupled, modeling of end gas ignition should easily be incorporated into most engine combustion models as a simple "satellite" submodel. This supports the approach taken in a number of existing submodels of engine knock (Natarajan and Bracco 1984; Hirst and Kirsch 1980).

      In those cases in which autoignition of the end gas occurs, the calculations indicate that significant pressure gradients can be produced within the combustion chamber. This results from the fact that the rate of heat release in the igniting end gas is larger than the rate of dissipation of the pressure gradients by convective heat and mass transfer. These pressure gradients then generate large gas velocities within the combustion chamber. In some of the cases examined, these velocities were as high as 150 m/s, 15% of the sound speed in the burned gases. An example of this pattern is presented in Fig. 5, with both the pressure and gas velocity profiles shown. These pressure waves travel back and forth across the combustion chamber at the local sound speed, eventually decaying. However, these waves have both the frequency and magnitude to be identified with the ringing sound that is associated with engine knock. That is, the relevant parameters in this system are the relative rates of heat release in the end gas as compared with the rate of propagation of sound waves, both of which should be approximately the same in laminar and in turbulent regimes. Neither the rate of flame propagation nor the flame thickness appear to play any significant role, at least in those cases in which end gas autoignition does take place. Therefore, these results show that the autoignition of the end gas does have a significant effect on the post-ignition motions of the burned gases in the combustion chamber.

      In the past, there have been speculations that engine knock might be the result of actual detonation in the end gas, reactive shock waves initiated by the autoignition of a significant fraction of the fuel-air charge. Although the current computations indicate that autoignition can generate strong acoustic waves, there is no indication that anything resembling actual detonation waves would be produced, consistent with recent experimental observations (Smith et al. 1985).

      As shown in Figs. 2-4, the temperature profile in the end gas is essentially a monotonic function of position, so

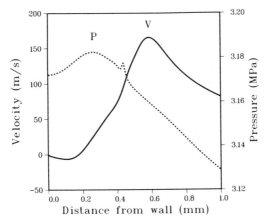

Fig. 5   Pressure and velocity profiles shortly after
autoignition has occurred.

long as the only processes included are flame propagation
and a thermal boundary layer. However, autoignition in
actual combustion chambers is often identified as occurring
at specific sites which vary from cycle to cycle. Under
the turbulent conditions in which this takes place, another
factor that may have some influence is the existence of
turbulent fluctuations in the end gas. Since these
fluctuations are stochastic in nature, they will occur at
different locations in the engine chamber on different
cycles. If it could be shown that such fluctuations can
lead to the existence of preferred sites for autoignition,
then this could explain the experimental observations of
seemingly random locations at which engine knock is
initiated.
     The reacting end gas region can be artificially
perturbed by introducing temperature fluctuations of
varying magnitude and spatial extent. An increase in local
temperature will produce an enhanced rate of autoignition,
but the competing processes of thermal conduction and
thermal diffusion will act to dissipate the fluctuation.
Further extensions of the model are planned to identify
critical values for the properties of the fluctuations,
above which the fluctuation provided a preferred site for
autoignition and below which the entire end gas region
ignited as a whole.

Summary
     These computations represent an attempt to couple a
sophisticated chemical kinetics treatment of end gas

autoignition with a reasonably detailed treatment of gas phase fluid mechanics. The model has examined several problems in which this coupling may be important. It was found that although the flame in the combustion chamber does not have a significant direct influence on the rate of end gas autoignition, the autoignition can generate pressure waves in the burned gases which are quite large. Generalizations of these results, in which only laminar flames were considered, to turbulent conditions were discussed. Since the important parameters in these results are usually relative rates of different processes, the conclusions reached here should be applicable to engine conditions, as long as the turbulent flame parameters and time scales are used for comparisons. These conclusions tend to support the logic used in the construction of knock submodels in engine combustion models in current use. It also suggests that it would be relatively simple to incorporate autoignition rates, computed using very detailed chemical kinetic mechanisms, into existing engine combustion models.

Acknowledgment

This work was performed under the auspices of the U.S. Department of Energy by the Lawrence Livermore National Laboratory under contract No. W-7405-ENG-48.

References

Dimpelfeld, P.M., and Foster, D.E. (1985) The prediction of autoignition in a spark ignited engine. In press.

Hirst, S.L., and Kirsch, L.J. (1980) The application of a hydrocarbon autoignition model in simulating knock and other engine combustion phenomena. Combustion Modeling in Reciprocating Engines (edited by J.N. Mattavi and C.A. Amann), pp. 193-229. Plenum Press, New York.

Kaiser, E.W., Westbrook, C.K., and Pitz, W.J. (1986) Acetaldehyde oxidation in the negative temperature coefficient regime: Experimental and modeling results. Paper presented at the Western States Section of the Combustion Institute, Stanford, California. Submitted for publication to Int. J. Chem. Kinet.

Leppard, W.R. (1985) A detailed chemical kinetics simulation of engine knock. Combust. Sci. Technol. 43, 1-20.

Lund, C.M. (1978) HCT-a general computer program for calculating time-dependent phenomena involving one-dimensional hydrodynamics, transport, and detailed chemical kinetics." Lawrence Livermore National Laboratory report UCRL-52504.

Maly, R. and Ziegler, G. (1982) Thermal combustion modeling--theoretical and experimental investigation of the knocking process. Society of Automotive Engineers paper 820759.

Natarajan, B. and Bracco, F.V. (1984) On multidimensional modeling of autoignition in spark-ignition engines. Combust. Flame 57, pp. 179-198.

Pitz, W.J., and Westbrook, C.K. (1986) Chemical kinetics of the high pressure oxidation of n-butane and its relation to engine knock. Combust. Flame, in press.

Pitz, W.J., Westbrook, C.K., Proscia, W.M. and Dryer, F.L. (1985) A comprehensive chemical kinetic reaction mechanism for the oxidation of n-butane, Twentieth Symposium (International) on Combustion, pp. 831-844. The Combustion Institute, Pittsburgh, PA.

Smith, J.R., Green, R.M., Westbrook, C.K., and Pitz, W.J. (1985) An experimental and modeling study of engine knock. Twentieth Symposium (International) on Combustion, pp. 91-100. The Combustion Institute, Pittsburgh, PA.

Westbrook, C.K. and Dryer, F.L. (1980) Prediction of laminar flame properties of methanol-air flames. Combust. Flame 37, 171-192.

Westbrook, C.K. and Pitz, W.J. (1984) A comprehensive chemical kinetic reaction mechanism for oxidation and pyrolysis of propane and propene. Combust. Sci. Tech. 37, 117-152.

Wilk, R.D., Cernansky, N.P., Pitz, W.J., and Westbrook, C.K. (1986) Submitted for publication.

On the Role of the Radical Pool in Combustion

J.R. Creighton*
Lawrence Livermore National Laboratory, Livermore, California
and
A.K. Oppenheim†
University of California, Berkeley, California

Abstract

We present numerical solutions of the detailed chemical
kinetics mechanisms for the reactions of hydrogen with
flourine and oxygen on a phase plane where temperature is
plotted vs the concentration of one of the radicals, F or
H. We observe that for given initial reactant concentrations, all solutions follow a common path, after an initial
transient, regardless of initial temperature or initial
radical concentration. We then analyze the reaction mechanism and find that the rate equations for radical species
rapidly reach a stable steady state that couples all of the
radicals into a pool. The pool concentration then grows
until onset of a reaction that consumes two radicals,
resulting in a steady state of the entire pool. Further
growth of the pool is due entirely to the temperature
dependence of the steady state, which has the form of an
equilibrium constant. We show that the rate of temperture
increase is proportional to the radical pool concentration,
particularly its steady-state value.

Introduction

In an earlier paper (Guirguis et al. 1981), we discussed the role of initial radical concentrations on the

Presented at the 10th ICDERS, Berkeley, California, August 4-9,
1985. Copyright © 1986 by The Regents of the University of California. Published by the American Institute of Aeronautics and
Astronautics, Inc. with permission.

*Chemist, Chemistry and Materials Science Department.
†Professor, Department of Mechanical Engineering.

ignition of methane-oxygen mixtures in the presence of
heat loss. In doing so, we found it illuminating to show
the reaction in the form of a plot of temperature versus
the logarithm of a selected radical concentration,
following the theory of Yang and Gray (1965). We discussed
the implications of this for ignition, but did not discuss
another significant aspect of that plot, namely, that after
an initial transient, all solutions seem to follow a common
path. Figure 1 shows a similar plot for hydrogen oxidation, based on numerical solutions of a complete kinetics
mechanism. Each curve in that figure represents a solution
starting at some initial temperature and initial radical
concentration. Because of a heat loss term, some initial
conditions resulted in ignition and others did not, but the
common path is essentially continuous for both cases.

The goal of this paper is to show that the common path
results from the behavior of the radical pool. Using
textbook reaction mechanisms, we will show how the radical
pool is formed, then grows until halted by some reaction
step that consumes two species in the pool, resulting in a
stable steady state of the entire radical pool. The common
path is simply this steady state, and its shape is determined by the temperature dependence of the reactions determining the steady state.

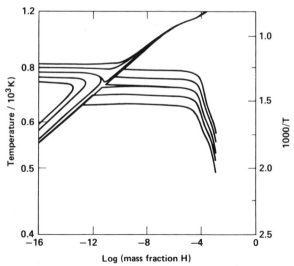

Fig. 1 Integral curves on a plane of temperature vs radical mass
fraction for a hydrogen-oxygen mixture at a pressure of 6 atm with
a thermal relaxation time of 1 s. Reciprocal temperature scale on
right is linear to create an Arrhenius plot.

We confine our attention to chemical reactions that are spatially uniform in order to concentrate on chemical effects without the complications introduced by temperature and concentration gradients. We build upon the theoretical foundation laid by Semenov (1935, 1943, 1944a, 1944b, 1958, 1959). His description of the chemical kinetics of spatially uniform reactions remains essentially valid. In the 1960's, Gray and Yang (1967) showed that the thermal and chain reaction theories of explosion limits could be unified if a simple heat loss term was considered along with the chemical kinetics. Guirguis et al. (1981) obtained numerical solutions using the Gray and Yang (1967) theory for the detailed chemical kinetics of methane oxidation and for a simplified kinetics model of methane oxidation equivalent to the radical pool developed in this paper. Oppenheim (1985) did a similar study using the detailed kinetics of hydrogen oxidation. The results of these recent works were presented on the phase plane and showed a common path independent of initial temperature or radical concentration.

These recent works concentrated on the conditions leading to ignition; in this paper, we will generally ignore the heat loss and concentrate on the kinetics. We will not generally present numerical results, but will use them as a guide in analyzing the reaction mechanism. Our goal is insight into the phenomena that control the behavior of the reaction.

The Radical Pool Concept

We have found that the idea of a radical pool is the key to understanding the behavior of combustion reaction mechanisms. A radical pool exists when the concentrations of two or more radicals or other highly reactive inter-mediate species track each other over a major portion of the reaction history. An example of this can be seen in Fig. 2, where the concentrations of H, O, and OH track each other over the entire history of both reactions, even though the ratio varies somewhat. A more precise math-ematical definition requires that there be algebraic equations coupling the concentrations of all species in the pool.

The concept of a radical pool first appeared in the literature several decades ago (Bulewicz et. al. 1956; Kaskan and Schott 1962), but the mathematical derivation was different from that which we will use. These earlier works derived it in terms of "partial equilibrium"; we will derive it in terms of steady-state approximations.

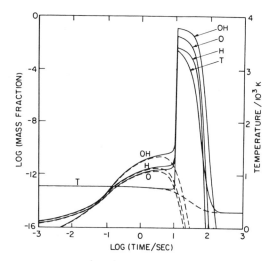

Fig. 2 Temperature and radical concentration profiles for a stoichiometric hydrogen-oxygen mixture initially at a pressure of 6 atm and a temperature of 819 K (————) and 818 K (- - - - -) with a thermal relaxation time of 1 s.

The "partial equilibrium" hypothesis starts from the principle of detailed balance, which asserts that the forward and reverse rates of each elementary reaction step must be equal at equilibrium. This gives algebraic equations coupling the species concentrations. The hypothesis asserts that selected reaction steps are balanced some time before equilibrium and the corresponding algebraic equations define the radical pool. This is a reasonably good approximation in the late stages of the hydrogen oxidation reaction, where the hypothesis yields insight.

The partial equilibrium hypothesis has proved less useful in interpreting the methane oxidation reaction. Although Peeters and Mahnen (1973) were able to explain much of the behavior of the post-burn region of the premixed methane flame using the partial equilibrium hypothesis, Biordi et al. (1976), who measured the radical concentrations throughout a low-pressure, laminar methane flame, could identify no significant reaction steps whose forward and reverse rates were in balance during the exothermic phase of the reaction. Numerical simulation of their experiment by Creighton and Lund (1979) confirmed this.   In any case, it is clear that the idea of partial equilibrium cannot be carried back to the beginning of the reaction.

Our derivation starts at the beginning of the reaction and is similar to the work of Brabbs and Brokaw (1974), Brokaw (1965), and Creighton (1977), which derived expressions for the growth of radical concentrations during the induction period but did not refer to the radical pool. The novel aspect of this paper is the extension of the radical pool concept to the remainder of the reaction history. A brief paper by Creighton (1980) discussed this extension primarily in terms of the simplified methane kinetics model used by Guirguis et al. (1981). This paper develops these concepts more fully and applies them to two textbook reaction mechanisms.

Numerical Calculations

The results presented in this paper are based on a large number of numerical solutions of detailed chemical kinetics equations done over a period of several year at Lawrence Livermore National Laboratory and at the University of California Berkeley. Several computer programs were used, all of them based on the method of Gear (1971) or equivalent fully implicit methods for handling "stiff" differential equations.

Most of the calculations were done using the reaction mechanism of Westbrook and Dryer (1979, 1984), which includes all of the reaction steps necessary to model the oxidation of methane, hydrogen, and methanol with extensions to cover the reaction of hydrogen with flourine. Some of the hydrogen oxidation calculations, taken from Oppenheim (1985), used the reaction mechanism of Warnatz (1984). The different reaction mechanisms gave results that were qualitatively the same, mainly because a few rate coefficients determine the major features of the reaction mechanism. We will identify critical rate coefficients and give their values where appropriate. Thermodynamic data were obtained from the JANAF Thermochemical Tables (1971).

A Chain Reaction: Hydrogen and Flourine

We will introduce our discussion of the radical pool with the $H_2 + F_2$ reaction system because it demonstrates all of the significant features with a minimum of mathematical detail.

Figure 3 shows several numerical solutions of this system, demonstrating the same general behavior as the solutions of the hydrogen oxidation system shown in Fig. 1. The principal difference is that solutions beginning with large initial radical concentrations move roughly verti-

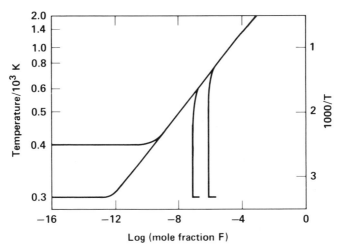

Fig. 3 Integral curves for a mixture consisting of 2.5 Torr H_2, 2.5 Torr F_2, and 10 Torr He, with no heat loss. Temperature scale is similar to Fig. 1.

cally to the common path rather than folding back, for reasons to be explained later.

The reaction mechanism consists of five bidirectional reactions:

$$F + H_2 \;=\; HF + H \qquad -31.9 \text{ kcal/mole} \qquad (1)$$

$$H + F_2 \;=\; HF + F \qquad -98.4 \text{ kcal/mole} \qquad (2)$$

$$F + F + M \;=\; F_2 + M \qquad -37.7 \text{ kcal/mole} \qquad (3)$$

$$H + H + M \;=\; H_2 + M \qquad -104.2 \text{ kcal/mole} \qquad (4)$$

$$H + F + M \;=\; HF + M \qquad -136.1 \text{ kcal/mole} \qquad (5)$$

where M represents any third body in a collision, and we ignore differences in chaperon efficiencies.

Pool Formation

In order to keep the algebra simple we will write rate equations for [F] and [H] using only the first three reactions. The rates of reactions (1) and (2) are orders of magnitude faster than the others, and the reverse of reaction (3) is the only significant source of radicals. It can be shown that the effects of reactions (4) and (5) are similar to reaction (3), and make only a small quantitative difference.   The rate equations are

$$\frac{d[H]}{dt} = k_1 [H_2][F] - k_{-1} [HF][H] - k_2 [F_2][H]$$

$$+ k_{-2} [HF][F] \qquad (6)$$

$$\frac{d[F]}{dt} = -k_1 [H_2][F] + k_{-1} [HF][H] + k_2 [F_2][H]$$

$$- k_{-2} [HF][F] - 2 k_3 [M][F][F] + 2 k_{-3} [M][F_2] \qquad (7)$$

We know from numerical solutions that the ratio [H]/[F] is maintained through most of the reaction so substituting [H] = r[F] in Eqs. (6) and (7) and setting d[H]/dt = r d[F]/dt, we obtain

$$\frac{d[F]}{dt} = \frac{1}{1 + r} \left(2 k_{-3}[M][F_2] - 2 k_3[M][F][F] \right) \qquad (8)$$

with

$$r = \frac{k_1 [H_2] + k_{-2} [HF]}{k_2 [F_2] + k_{-1} [HF]} \left[1 - \frac{2 Q_3}{(Q_1+Q_2)[F] + 2 Q_3} \right] \qquad (9)$$

where

$$Q_1 = k_2 [F_2] + k_{-1} [HF]$$

$$Q_2 = k_1 [H_2] + k_{-2} [HF]$$

$$Q_3 = k_{-3} [M][F_2] - k_3 [M][F][F]$$

Except for an initial transient, Q_3 is much smaller than the denominator of the last term, so the bracketed term approaches unity, reaching it when the forward and reverse rates of reaction (3) are balanced.   The value of r changes as the reaction proceeds, beginning at about 2.7 for a stoichiometric mixture at 300 K and approaching unity as the forward and backward rates of reactions (1) and (2) approach equality.

The only stable solution of these rate equations occurs when [H] = r[F], and recovery from a perturbation of the ratio takes only a microsecond or two. This transient can be seen in the solutions of Fig. 3 that begin with an initial F concentration. During the short horizontal portion some of the F is converted to H to achieve the proper ratio. Once the ratio of [H]/[F] is established, Eq. (9) is valid throughout the reaction history.

Because [H] is proportional to [F], Eq. (8) describes the growth of the entire radical pool, and henceforth we consider only the behavior of [F], which is representative of the entire pool. Equation (8) shows that growth and steady state of the radical pool depend on the ratio of the forward and reverse rates of reaction (3). When [F] is very small, the reverse direction of reaction (3) dominates, leading to growth of the radical pool. This is represented in Fig. 3 by the horizontal portions of the solutions with no initial [F]. As [F] grows the forward and reverse rates of reaction (3) approach equality and the growth of the radical pool stops.

Steady State

We will designate the value of [F] for which d[F]/dt equals zero as $[F]_s$, because it is a steady state for the entire radical pool. Setting d[F]/dt to zero gives the value

$$[F]_s = ([F_2][M] / K_3)^{1/2} \qquad (10)$$

with a corresponding $[H]_s$.

This is a new steady state for the entire radical pool and is to be distinguished from the steady state that couples the two radical species. The new steady state is also stable, though it requires about 10 microseconds to recover from a perturbation. The existence of the new steady state is due to the nonlinearity of Eq. (8), which has a term quadratic in [F]. In every reaction mechanism we have analyzed this new steady state is due to a nonlinear term in the rate equation.

Although Eq. (8) shows that d[F]/dt equals zero once [F] reaches $[F]_s$, Fig. 3 shows [F] continuing to increase. This is explained by reference to Eq. (10) which shows that $[F]_s$ has a temperature dependence through K_3. The common path for all solution curves is a result of Eq. (10) being independent of the initial temperature or initial radical concentration.

Agreement of Eq. (10) with the numerical solutions can be shown by substituting values for K_3 and taking the logarithm, yielding

$$\log [F]_s = 7.35 + 1/2 \log([F_2][M]) - 9425 / T \qquad (11)$$

The last term matches the slope of the common portion of the solution curves in Fig. 3.

Once d[F]/dt approaches zero it remains there to the end of the reaction, though the value of $[F]_s$ may vary due to consumption of F_2 and variation of the temperature. A zero value of d[F]/dt is a necessary condition for equilibrium, so our derivation of the radical pool by means of steady-state relationships is valid from the moment these relationships are established until final equilibrium. This differs from the "partial equilibrium" hypothesis, which is valid only near the end of the reaction. Both derivations give algebraic equations that couple the radical concentrations, but they only agree near equilibrium.

Exothermic Effects

Numerical solutions all lie near the common path regardless of the initial values of temperature or radical concentration, although the position of the common path does change if the ratio or amount of initial reactants is changed, as shown by Guirguis et al. (1981). A simple derivation shows why this is so.

The rate of temperature change is obtained from the energy equation, assuming no heat loss

$$C_p \frac{dT}{dt} = - H_f^0(HF) \frac{d[HF]}{dt} \qquad (12)$$

where $H_f^0(HF)$ is the standard enthalpy of formation of HF, and is negative, by convention. Terms for H_2 and F_2 do not appear because, by convention, their enthalpies of formation are zero. Terms for H and F are omitted because at steady state their derivatives are very small.

Equation (12) is easily integrated, giving

$$T - T_0 = - \frac{H_f^0(HF)}{C_p} [HF]$$

while conservation of atoms gives

$$[F_2] = [F_2]_0 - 1/2 [HF] - 1/2 [F]$$

When these are substituted in Eq. (10), neglecting 1/2 [F],

$$[F]_s = \left[\frac{[M]}{K_3}\right]^{1/2} \left[[F_2]_0 - \frac{C_p}{2 H_f} (T - T_0) \right]^{1/2} \qquad (13)$$

where H_f is an abbreviation for $H_f^0(HF)$. All variable terms
in this equation are functions only of temperature. The
dependence on $[F_2]_0$, the initial concentration, is clear.
The dependence on T_0, the initial temperature, only shows
up after a significant portion of reactants are consumed.

We can go a step further and show that dT/dt is a
function only of temperature with a weak dependence on T_0
and initial reactant concentrations. The rate equation for
[HF] is

$$\frac{d[HF]}{dt} = k_1 [H_2][F] + k_2 [F_2][H] = 2 k_1 [H_2][F]$$

where reverse rates and the minor contribution from reaction (5) have been ignored. The second equality is a
result of the steady state that forms the radical pool.
Substituting this in Eq. (12) gives

$$\frac{dT}{dt} = - 2 \frac{H_f}{C_p} k_1 [H_2][F] \tag{14}$$

demonstrating that the rate of temperature increase is
proportional to [F], which equals $[F]_s$ at steady-state.
Substituting $[F]_s$ from Eq. (13) and invoking atom conservation yields

$$\frac{dT}{dt} = -2 \frac{H_f}{C_p} k_1 \left[[H_2]_0 - 2 \frac{C_p}{H_f} (T - T_0) \right]$$

$$x \left[\frac{[M]}{K_3}\right]^{1/2} \left[[F_2]_0 - 2 \frac{C_p}{H_f} (T - T_0) \right]^{1/2} \tag{15}$$

This has the same dependence on temperature and initial
values as Eq. 13.

This is a major simplification because we can find the
temperature as a function of time by integrating Eq. (15).
This could be done in closed form if the temperature dependence of k_1 and K_3 were not so complex. However, numerical
integration of Eq. (15) is much easier than solving the set
of coupled rate equations. The complete solution requires
use of Eq. (8) to find the time when [F] becomes equal to
$[F]_s$.

At the beginning of this section we noted that
solutions with large initial radical concentrations
approached the common path in a nearly vertical direction,

as seen in Fig. 3. This is easily explained. The slope, $dT/d[F]$, is the ratio of Eq. (14) to Eq. (8). The terms in Eq. (8) are orders of magnitude smaller than those in Eq. (14) so the slope is essentially vertical. In the hydrogen oxidation reaction we will see that the corresponding numerator and denominator are of nearly equal magnitude and the radical concentration is reduced as the solution approaches the common curve, as seen in Fig. 1.

A Branching Chain Reaction: Hydrogen Oxidation

We will now examine a branching chain reaction and look for the same behavior found in the $H_2 + F_2$ chain reaction, namely,
1) early formation of a radical pool,
2) growth of the radical pool until it reaches a second steady state,
3) a steady-state value that is a function only of temperature and initial reactant concentrations, yielding reaction trajectories similar to those in Fig. 3, and
4) a rate of temperature increase proportional to the steady-state radical pool concentration, yielding a simplified description of the reaction.

Our prototype branching chain reaction will be the high-temperature hydrogen oxidation mechanism given in textbooks (Lewis and von Elbe 1961; Strehlow 1984).

$$H + O_2 \quad = \quad O + OH \qquad\qquad (16)$$

$$O + H_2 \quad = \quad H + OH \qquad\qquad (17)$$

$$OH + H_2 \quad = \quad H_2O + H \qquad\qquad (18)$$

With initial radicals provided by

$$H_2 + O_2 \quad = \quad HO_2 + H \qquad\qquad (19)$$

(We have followed Baulch et al. (1973) in rejecting the unlikely four center reaction which gives 2 OH as the product.) Numerical calculations show that this mechanism is valid at temperatures above 1200 K. Later we will comment briefly on a different set of reactions that dominate at lower temperatures.

The radical pool involves three species, O, OH, and H. Numerical solutions show that reaction (16) is the rate limiting step and that the growth of the radical pool can be described in terms of [H]. Forward rates are sufficient, except for reaction (16) whose reverse rate will be seen to be crucial. Numerical solutions of the complete mechanism show that the reverse rates of the other reaction steps are an order of magnituede smaller, so omitting them does not alter the qualitative results.

We can write the rate equations for the radicals and substitute $[O] = r_1[H]$ and $[OH] = r_2[H]$. The leading terms for these ratios are obtained by setting $d[O]/dt$ and $d[OH]/dt$ to zero.

$$r_1 = k_{16}[O_2] \; / \; (k_{17}[H_2] - r_2 k_{-16}[H]^2) \qquad (20)$$

$$r_2 = k_{16}[O_2] / (k_{18}[H_2]) \qquad (21)$$

(The second term in the denominator of Eq. (20) can be neglected because it is orders of magnitude smaller than the first.) These equations define the radical pool.

To examine the growth of the radical pool we eliminate $[O]$ and $[OH]$ in the rate equation for $[H]$, yielding

$$\frac{d[H]}{dt} = k_{19}[O_2][H_2] + 2\, k_{16}[O_2][H] - 2\, r_1 r_2 k_{-16}[H]^2 \qquad (22)$$

The three terms represent generation of radicals by the dissociation reaction, growth of radicals due to branching, and recombination of radicals, in that order. In the early stages of the reaction we can neglect the last term and integrate, yielding

$$[H] = [\, k_{19} \, / \, (2\, k_{16}) \,][\, \exp(\, 2\, k_{16}[O_2]t) - 1 \,]$$

This exponential growth law is typical of branching chain reactions.

Examination of Eq. (22) shows that growth continues until the last term cancels the other two. By that time the first term is negligible so the steady state value of $[H]$ is

$$[H]_s = K_{16}[O_2] / r_1 r_2 \qquad (23)$$

where K_{16} is the equilibrium constant for reaction (16). The principal temperature dependence is that of K_{16}, though $[O_2]$ will decrease as the reaction proceeds with increasing temperature. A more accurate derivation would include the reverse rates of reactions (17) and (18), but would not change the temperature dependence significantly.

The rate of temperature increase can be approximated by

$$C_p \frac{dT}{dt} = -H_f^o(H_2O) \frac{d[H_2O]}{dt} \qquad (24)$$

This is not as good an approximation as Eq. (12), for the $H_2 + F_2$ reaction, because [H] can rise to as high as 10 percent of the initial $[H_2]$.

Reaction 18 is the only one creating H_2O so

$$\frac{d[H_2O]}{dt} = k_{18} [H_2][OH] = r_2 k_{18} [H_2][H]$$

Substituting this into eq. 24 gives

$$\frac{dT}{dt} = - \frac{H_f^o(H_2O)}{C_p} (r_2 k_{18} [H_2] [H]) \qquad (25)$$

showing that the rate of temperature increase is proportional to the radical pool concentration. A derivation similar to that for Eq. (15) would show that dT/dt is a function only of T and the initial reactant concentrations, with a weaker dependence on T_0, once the radical pool concentration reaches $[H]_s$.

From our derivation of the steady-state for hydrogen oxidation we see that it is due to balancing of the forward and reverse rates of reaction (16). (A more precise derivation would include contributions from the reverse rates of reactions (17) and (18)). In the $H_2 + F_2$ system steady-state was reached when the forward and reverse rates of reaction (3) achieved balance. This appears similar to the "partial equilibrium" hypothesis which derives the radical pool by assuming that the forward and reverse rates of certain steps are in balance. However, the steady-state does not generally result from balance of a single reaction step. In a later paper we will show that the steady state in methane oxidation is achieved by a set of reaction steps all of which are proceeding almost exclusively in one direction. Preliminary investigation of other oxidation reactions indicates that this is more typical. Partial equilibrium is useful in analysing the hydrogen oxidation reaction because it happens that that reaction (16), and to a lesser extent, reactions (17) and (18), are approximately balanced in the latter half of the reaction.

The Reaction Mechanism Below 1000 K

We have not directly compared the results derived above with the numerical values displayed in Fig. 1 because the initial temperatures in Fig. 1 were below 1000 K. At those temperatures H reacts with O_2 by

$$H + O_2 + M = HO_2 + M \qquad (26)$$

instead of reaction (16). For conditions appropriate to Fig. 1 the following additional reactions control the behavior of the radical pool

$$HO_2 + H_2 = H_2O_2 + H \qquad (27)$$

$$HO_2 + HO_2 = H_2O_2 + O_2 \qquad (28)$$

$$H_2O_2 + M = OH + OH + M \qquad (29)$$

with reaction (18) completing the chain. There is no source of O atoms to give reaction (17) an appreciable rate. Hydrogen peroxide is not a radical because it has no unpaired electron spin, nor does it react with H_2 or O_2. Thus it is not part of the radical pool, but simply an intermediate product whose concentration rises to a peak and then falls. However, it does not break the chain as is often assumed. In spite of the relative stability of H_2O_2, the rate of reaction (29), with an activation temperature of 24,000 K, is only a few percent less than that of reaction (28) because the concentration of H_2O_2 is orders of magnitude larger than that of the radicals. In a later paper we will show that the slope of the common portion of the solutions in Fig. 1 is determined mainly by reaction (29).

We have shown that the branching chain reaction mechanism for high temperature oxidation of hydrogen behaves like the $H_2 + F_2$ reaction mechanism in that a radical pool is formed, grows until it reaches a steady state with a temperature dependence that has the form of an equilibrium constant and depends only on the temperature and the initial reactant concentrations, and that the rate of temperature increase is proportional to the radical pool concentration.

Conclusion

This paper was motivated by the desire to explain a phenomenon observed while doing numerical calculations of the chemical reactions leading to ignition. This phenomenon is shown in the phase plane plots of Figs. 1 and 3, where after a transient, all reaction paths lie on or near a common curve. To explain this phenomenon, we have analyzed the reaction mechanisms for a simple chain reaction, hydrogen with flourine, and a simple branching chain reaction, hydrogen oxidation.

In both cases, we found that the behavior was due to the formation and subsequent steady state of a radical pool. The radical pool was described in terms of steady-state approximations to the rate equations for the radical species, rather than being derived from the "partial equilibrium" hypothesis. The first steady state occurs within microseconds and couples the radical concentrations, which subsequently rise and fall together, constituting a radical pool. The pool grows until its growth is halted by a reaction step consuming two radical species.

This is a second steady state for the entire pool and is stable because two or more opposing reaction rates are in balance. This steady-state radical pool concentration has a temperature dependence that resembles an equilibrium constant because it depends on a balance between reaction rates. It is essentially independent of initial temperature or initial radical concentration, so after a transient growth period, all solutions follow a common path in the phase plane as shown in Figs. 1 and 3.

We also showed that the rate of temperature rise was proportional to the radical pool concentration. Because the pool concentration depends only on temperature and initial reactant concentrations once it reaches steady state, a differential equation for the temperature rise can be written that does not depend on any species concentrations except for a weak dependence on the initial reactant concentrations. Initial temperature and initial radical concentrations affect only the time required for the radical pool to grow to its steady-state concentration.

Acknowledgments

The portions of this work performed at Lawrence Livermore National Laboratory were supported by the U.S. Department of Energy under contract W-7405-Eng-48. The portions performed at the University of California, Berkeley were supported in part by the U.S. Department of Energy and by the National Science Foundation and by NASA Lewis Research Center.

References

Baulch, D.L., Drysdale, D.D., Horne, D.G., and LLoyd, A.C. (1973) Evaluated Kinetic Data for High Temperature, Vol. I, Butterworths, London.

Biordi, J.C., Lazzara, C.P., and Papp, J.F. (1976) An examination of the partial equilibration hypothesis and radical recombination in 1/20 atm methane flames. 16th Symposium (International) on Combustion, pp. 1097-1109. The Combustion Institute, Pittsburgh, Pa.

Brabbs T.A. and Brokaw, R.S. (1974) Shock tube measurements of specific reaction rates in the CH_4-CO-O_2 system. 15th Symposium (International) on Combustion, pp. 893-900. The Combustion Institute, Pittsburgh, Pa.

Brokaw, R.S. (1965) Analytic solutions to the ignition kinetics of the hydrogen-oxygen reaction. Tenth Symposium (International) on Combustion, pp. 269-278. The Combustion Institute, Pittsburgh, Pa.

Bulewicz, E.M., James, C.G. and Sugden, T.M. (1956) Photometric investigations of alkali metals in hydrogen flame gas: II. The study of excess concentrations of hydrogen atoms in burnt gas mixtures. Proc. R. Soc. London, Ser. A 235, 89-106.

Creighton, J. R. (1977) Some general principles obtained from numerical studies of methane combustion. J. Phys. Chem. 81(25), pp. 2520-2526.

Creighton, J.R. and Lund, C.M. (1979) Modeling study of flame structure in low-pressure, laminar, pre-mixed methane flames. Tenth Materials Research Symposium on Characterization of High Temperature Vapors and Gases, NBS Special Publication 561, Vol. 2, pp. 1223-1244. U.S. Government Printing Office, Washington, D.C.

Creighton, J.R. (1980) Rate of methane oxidation controlled by free radicals. Laser Probes for Combustion Chemistry (edited by D.R. Crosley), ACS Symposium Series No. 134, pp. 357-363. American Chemical Society, Washington, D.C.

Gear, C.W. (1971) Numerical Initial Value Problems in Differential Equations, Prentice/Hall, New York.

Gray, B.F. and Yang, C.H. (1967) The determination of explosion limits from a unified thermal chain theory. 11th Symposium (International) on Combustion, pp. 1099-1106. The Combustion Institute, Pittsburgh, Pa.

Guirguis R.H., Oppenheim A.K., Karasalo, I. and Creighton, J. R. (1981) Thermochemistry of methane ignition. Combustion in Reactive Systems: AIAA Progress in Astronautics and Aeronautics (edited by J.R. Bowen) Vol. 76, pp. 134-153. AIAA, New York.

JANAF Thermochemical Tables (1971) (edited by D.R. Stull). U.S. Government Printing Office, Washington, D.C.

Lewis, B. and von Elbe, G. (1961) Combustion, Flames and Explosions of Gases. Academic Press, New York.

Kaskan, W. E. and Schott, G. L. (1962) Requirements imposed by stoichiometry in dissociation-recombination reactions. Combust. Flame 6(1), pp. 73-75.

Oppenheim, A.K. (1985) Dynamic features of combustion. Philos. Trans. R. Soc. London, Ser. A, 315, 471-508.

Peeters, J. and Mahnen, G. (1973) Reaction Mechanisms and rate constants of elementary steps in methane-oxygen flames. 14th Symposium (International) on Combustion, pp. 134-146, The Combustion Institute, Pittsburgh, Pa.

Semenov, N.N. (1935) Chemical Kinetics and Chain Reactions. Clarendon Press, Oxford, England.

Semenov, N.N. (1943) On types of kinetic curves in chain reactions. I. Laws of the autocatalytic type. C. R. Dokl. Acad. Sci. URSS 42(8), 342-348.

Semenov, N.N. (1944a) On types of kinetic curves in chain reactions. II. Consideration of the interaction of active particles. C. R. Dokl. Acad. Sci. URSS 44(2), 62-66.

Semenov, N.N. (1944b) On types of kinetic curves in chain reactions. Allowance for chain rupture on walls of reaction vessels in the case of oxidation of hydrogen. C. R. Dokl. Acad. Sci. URSS 44(6), 241-245.

Semenov, N.N. (1958) Some Problems in Chemical Kinetics and Reactivity (translated by M. Boudart) Vol. 1, Princeton University Press, Princeton, N.J.

Semenov, N.N. (1959) Some Problems in Chemical Kinetics and Reactivity (translated by M. Boudart) Vol. 2, Princeton University Press, Princeton, N.J.

Strehlow, R.A. (1984) Combustion Fundamentals. McGraw Hill, New York.

Warnatz, J. (1984) Survey of rate coefficients in the C/H/O system. Chemistry of Combustion (edited by W.C. Gardiner, Jr.) pp. 197-360. Springer Verlag, New York.

Westbrook, C.K. and Dryer, F.L. (1979) A comprehensive mechanism for methanol oxidation. Combust. Sci. and Technol., 20(3), pp. 125-140.

Westbrook, C.K. and Dryer, F.L. (1984) Chemical kinetics modeling in hydrocarbon combustion. Prog. Energy Combust. Sci. 10(1) pp. 1-57.

Yang, C.H. and Gray, B.F. (1965) On the unification of the thermal and chain theories of explosion limits. J. Phys. Chem. 69(8), 2747-2750.

A Numerical Analysis of the Ignition of Premixed Gases by Heat Addition

Peter S. Tromans* and Ronald M. Furzeland†

Shell Research Ltd., Chester, England

Abstract

In order to illuminate the roles of physical processes such as curvature and flame stretch in spark ignition, the ignition of a combustible gas subjected to a local heat addition has been analyzed numerically. The conservation equations for reactant and sensible enthalpy have been solved and the gas allowed to undergo a one-step chemical reaction. The numerical technique involves a finite difference scheme with an adaptive grid. Three geometries have been studied. In the first and second, heat is added for a specified period at a planar boundary or a cylindrical one of small diameter: the various stages in the ignition process observed have been described in Kapila's asymptotic analysis. No minimum ignition energy can be identified. In the third geometry, heat is added at a spherical surface of small diameter: the first stages are similar to the other cases. However, as the flame propagates away from the source, it passes through a critical phase in which it may be extinguished. For this reason, the spherical model exhibits many of the phenomena of spark ignition: a minimum ignition energy, a critical radius, an optimum source duration, and a sensitivity to stretching.

Nomenclature

a	= radius of the cylindrical and spherical sources
B	= pre-exponential term in the reaction rate
C_p	= specific heat

Presented at the 10th ICDERS, Berkeley, California, August 4-9, 1985. Copyright © 1986 by Shell Internationale Research Mij. BV, The Hague. Published by the American Institute of Aeronautics and Astronautics, Inc. with permission.

*Senior Scientist; currently with Shell Research BV, Rijswijk, Holland.

†Senior Scientist; currently with Shell Research BV, Amsterdam, Holland.

e	=	dimensional strain rate
E	=	activation energy
E_A, E_L	=	ignition energy per unit area of planar source, per unit length of cylindrical source
E_S	=	ignition energy for spherical source
F, G, H	=	functions giving time dependence of the heat sources
J	=	monitor function
k	=	thermal conductivity
L	=	Lewis number
M	=	nondimensional reaction rate constant
N	=	number of mesh points in space
Q	=	heat release by chemical reaction
r	=	nondimensional space coordinate
R	=	temperature rise across an adiabatic flame
t	=	nondimensional time
T	=	temperature
u_l	=	adiabatic burning velocity
v	=	velocity
W	=	chemical reaction rate
x	=	dimensional space coordinate
Y	=	reactant concentration
β	=	nondimensional activation energy
δ	=	adiabatic flame thickness
ε	=	nondimensional strain rate

Introduction

To explain their observations of spark ignition, Lewis and von Elbe (1961) made intuitive arguments in terms of "excess enthalphy" and "flame stretch." Their description provides useful scales for the minimum ignition energy and predicts its relationship with pressure and burning velocity. More recently, Kapila (1981) has used activation energy asymptotics to analyze the ignition of a combustible occupying a semi-infinite space. He revealed several stages in the process, including inert heating, induction, local explosion, and flame propagation. None of the stages or transitions between stages provide an explanation for the minimum energy required for spark ignition. Further asymptotic analysis has been performed by Deshaies and Joulin (1984). They examined the stability of the steady solution that exists for small spherical flames. They found a minimum radius for unsupported propagation of a spherical flame and so demonstrated the importance of the physical and geometrical factors in controlling spark ignition.

Several numerical studies have been made of ignition initiated by small hot pockets; all involved complex

chemistry. Dixon-Lewis and Shepherd (1974, 1978) examined
the ignition of hydrogen/air mixtures by spherical and
cylindrical hot spots. They investigated the effect of
pocket size and radical addition on minimum ignition energy.
Overley et al. (1978) studied the influence of hot pocket
size and shape on minimum ignition energy. Kailasanath et
al. (1982) included the effect of adding heat to the hot
pocket over some prescribed period rather than instantaneously.

In the present analysis, ignition by heat addition at a
boundary is investigated. The boundary is either planar to
a semi-finite system or one formed by the outer surface of a
very small diameter cylinder or sphere. The objective is to
investigate the roles of flame curvature and stretch, so
only a one-step chemical reaction is considered. The
conservation equations are solved numerically using an
adaptive mesh technique.

Conservation Equations and Method of Solution

The purpose is to analyze the ignition of premixed gases
by planar, cylindrical and spherical sources of heat. A
one-step reaction controlled by the concentration of one
limiting reactant is studied. The conservation equations
for this reactant and enthalphy are

$$\frac{\partial \widetilde{Y}}{\partial \tau} + v \frac{\partial \widetilde{Y}}{\partial x} = \frac{1}{\widetilde{\rho} L C_p x^n} \frac{\partial}{\partial x} \left(k x^n \frac{\partial \widetilde{Y}}{\partial x} \right) - \frac{W}{\widetilde{\rho}} \tag{1}$$

and

$$\frac{\partial \widetilde{T}}{\partial \tau} + v \frac{\partial \widetilde{T}}{\partial x} = \frac{1}{\widetilde{\rho} C_p x^n} \frac{\partial}{\partial x} \left(k x^n \frac{\partial \widetilde{T}}{\partial x} \right) + \frac{QW}{\widetilde{\rho} C_p} \tag{2}$$

where

$$W = \widetilde{\rho} B \widetilde{Y} \exp \left(-E/R\widetilde{T} \right) \tag{3}$$

The exponent n takes the value zero for a planar system, 1 for
a cylindrical system and 2 for a spherical system. The
continuity equation is

$$\frac{\partial \widetilde{\rho}}{\partial \tau} + \widetilde{\rho} e + \frac{1}{x^n} \frac{\partial}{\partial x} \left(x^n \widetilde{\rho} v \right) = 0 \tag{4}$$

where e is the velocity gradient in a direction parallel to
the flame surface. It provides a flow generated flame

stretch in a problem with only one space dimension appearing
as an independent variable.

The heat addition takes place at the surface x = 0, the
cylindrical surface with its axis on x = 0, or a spherical
surface centred on X = 0. It is assumed that

$$\tilde{\rho}^{-1} \propto k \propto \tilde{T}. \tag{5}$$

Equations (1-4) may be nondimensionalized as follows

$$T = \tilde{T}/\tilde{T}* \qquad\qquad T_0 = \tilde{T}_0/\tilde{T}* \qquad Y = \tilde{Y}/\tilde{Y}_0$$

$$\delta = k_0/\tilde{\rho}_0 \, u_1 \, C_p \qquad r = x/\delta \qquad t = \tau \, u_1/\delta$$

$$\rho = \tilde{\rho}/\tilde{\rho}_0 \qquad\qquad \beta = E/R\tilde{T}* \qquad M = k_0 B/e^{\beta}\tilde{\rho}_0 \, u_1^2 \, C_p \, \beta^2$$

$$\tag{6}$$

where u_1 is adiabatic burning velocity and δ the adiabatic
flame thickness; the tilde indicates a dimensional variable,
subscript zero unburned gas conditions, and the superscript
asterisk conditions in the burned gas behind a planar,
adiabatic, laminar flame. The nondimensional conservation
equations are:

$$\frac{\partial Y}{\partial T} - \varepsilon\phi \, \frac{\partial Y}{\partial \phi} = \frac{1}{L} \frac{\partial}{\partial \phi} \, (r^{2n} \, \frac{\partial Y}{\partial \phi}) - \beta^2 \, M \, Y \, \exp \beta \, (1 - \frac{1}{T}) \tag{7}$$

and

$$\frac{\partial T}{\partial t} - \varepsilon\phi \, \frac{\partial T}{\partial \phi} = \frac{\partial}{\partial \phi} \, (r^{2n} \, \frac{\partial T}{\partial \phi}) + R\beta^2 \, M \, Y \, \exp \beta \, (1 - \frac{1}{T}) \tag{8}$$

The continuity equation has been eliminated by introducing
the mass weighted coordinate ϕ, such that

$$\frac{\partial \phi}{\partial r} = r^n \rho = r^n \, \frac{T_0}{T} \tag{9}$$

with r = 0 at ϕ = 0.

The heat addition takes place at a surface bounding the
gas at ϕ = 0 for the planar problem and a surface defined by
ϕ = a for the cylindrical and spherical problems. A small
value a = 0.05 was chosen so that the heat addition
reasonably models a line or point source. Thus, the boundary

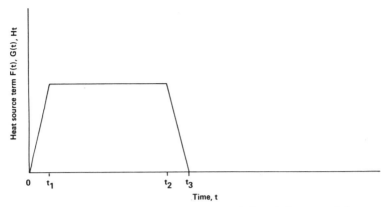

Fig. 1   Form of heat source terms F(t), G(t), and H(t).

conditions for the planar problem are

$$\frac{\partial T}{\partial \psi} = F(t) \; , \; \frac{\partial Y}{\partial \psi} = 0 \text{ at } \psi = 0$$

$$T = T_0, \quad Y = 1 \text{ at large } \psi \qquad (10)$$

The boundary conditions for the cylindrical problem are

$$\frac{\partial T}{\partial \psi} = \frac{T_0}{T} G(t) \; , \; \frac{\partial Y}{\partial \psi} = 0 \text{ at } \psi = a$$

$$T = T_0, \quad Y = 1 \text{ at large } \psi \qquad (11)$$

while for the spherical problem, they are

$$\frac{\partial T}{\partial \psi} = (\frac{T_0}{T})^{4/3} H(t)$$

$$T = T_0, \quad Y = 1 \text{ at large } \psi \qquad (12)$$

In both cases, the intital conditions are

$$T = T_0 \; , \quad Y = 1 \text{ for all } \psi$$

In general, T_0 was given the physically reasonable value of 0.14. F(t), G(t) and H(t) are proportional to the heat inputs of the respective sources. They were generally given the form shown in Fig. 1, although other distributions were investigated. It is straightforward to show that the energy added per unit area in the planar system is

$$E_A = \int_0^\infty F(t)\ dt \qquad (13)$$

and the energy added per unit length of the cylinder in the cylindrical system is

$$E_L = 4\pi a \int_0^\infty G(t)\ dt \qquad (14)$$

The total energy added in the spherical system is:

$$E_s = 4\pi\ (3a)^{4/3} \int_0^\infty H(t)\ dt \qquad (15)$$

All calculations reported in this paper were made with Lewis number L set equal to 1, the activation energy β set equal to 8, and the initial temperature T_0 set equal to 0.14, which is appropriate for atmospheric initial conditions.

The governing equations are discretized in space, using an adaptive mesh and corresponding finite difference approximations, to give a system of ordinary differential equations in time for T and Y (of order 2N, where N is the number of space points). The nonuniform space mesh is generated by the coordinate transformation

$$X(\psi,\tau) = \int_0^\psi J\ (\psi,t)\ d\ \psi\ /\int_0^{\psi N} J\ (\psi,t)\ d\psi$$

where the monitor function $J(\psi,\tau)$ is some measure of the space derivatives of T and Y. For a given uniform X mesh, $X_i = (i-1)\ \Delta X$ (where $i = 1,\ \ldots,\ N$, with $(N-1)\ \Delta X \equiv 1$, constant for all time) and the Eq. (15) adapts the spacing of the corresponding ψ_i points proportional to the gradients of T and Y. For the planar problem, the monitor function used herein is

$$J\ (\psi,t) = \left| \frac{\partial Y}{\partial \psi} \right| + 10^{-3} \left| \frac{\partial^2 Y}{\partial \psi^2} \right| \qquad (16)$$

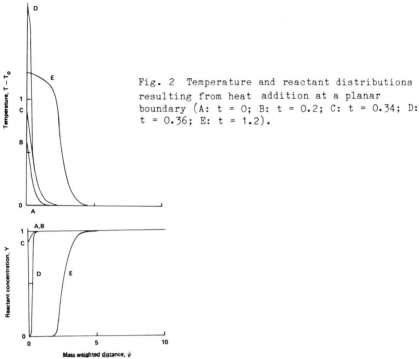

Fig. 2 Temperature and reactant distributions resulting from heat addition at a planar boundary (A: t = 0; B: t = 0.2; C: t = 0.34; D: t = 0.36; E: t = 1.2).

while for the cylindrical and spherical problems, it is

$$J\,(\psi,t) = \left|\frac{\partial Y}{\partial \psi}\right| + 0.5 \times 10^{-3} \left|\frac{\partial^2 Y}{\partial \psi^2}\right| \Big/ r \qquad (17)$$

The mesh allows accuracy and fine resolution of the flame to be combined with computational efficiency. The mesh spacing is automatically controlled to satisfy user-chosen criteria on the maximum and minimum mesh sizes $\Delta\psi_i$. The finite difference discretizations, space remeshing, and time integration are performed automatically by the SPRINT package (software for problems in time) developed jointly by Leeds University and Shell Research Ltd. (Berzins, et al., 1984).

Results

Planar Source

The results of the calculations for the planar heat source will be discussed in this section. The spatial distributions are shown for various times in Fig. 2. As

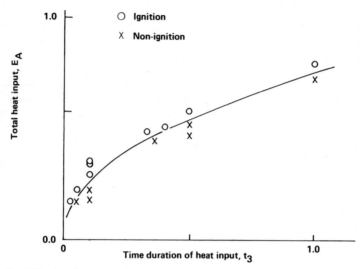

Fig. 3 Effect of source duration on minimum energy for ignition by the planar source.

indicated by Kapila (1981), ignition develops in stages. During the heating stage, the temperature in the region of the heat source increases, but there is no chemical reaction, distributions A to B in Fig. 2. Some chemical reaction takes place in the subsequent induction stage, distributions B to C. When the peak temperature is high enough, there is a local explosion: the temperature jumps quite suddenly to a high value. Simultaneously, the concentration locally drops to zero, as shown in distribution D. Next, the reaction zone separates from the hot boundary and a flame propagates away from the heat source, distribution E. The flame quite quickly relaxes to its adiabatic state. The same pattern of stages was observed for all successful ignitions whatever strength was set for the heat source. The mixture failed to ignite only if the heat supply was cut off before the explosion stage was reached. The nonuniformity in temperature and concentration would then diffuse away. The minimum energy for ignition was merely that required to initiate the explosion.

A series of calculations was performed in which the duration of the heat addition t_3 and the total energy added E_A were varied, while keeping $t_1 \simeq t_3/10$ and $t_2 \simeq 9t_3/10$. These reveal the relationship between minimum ignition energy and source duration shown in Fig. 3. Intense, short duration sources are the most effective. This is to be

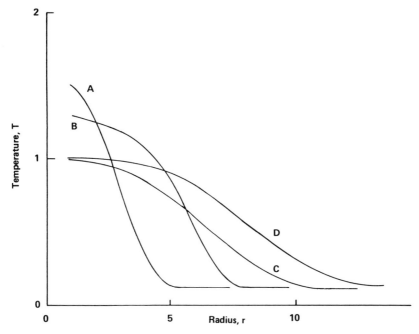

Fig. 4 Temperature distributions resulting from the spherical source (A: t = 0.4;   B: t = 1.1;   C: t = 2.6   D: t = 4.0).

expected, since these sources produce the highest temperatures at the heated boundary and the occurrence of the explosion essentially depends on that temperature.

The effect of changing the strain rate ε on the minimum energy for ignition when the source duration is held constant was investigated and found to be insignificant.

Cylindrical Source

Generally, results for this case are qualitatively similar to the planar one.

Spherical Source

For the system with a very small diameter spherical source of heat, the same stages are observed in the ignition process as for the planar and cylindrical sources. A series of temperature distributions is shown in Fig. 4. However, there is a crucial difference: ignition failure can occur as a result of flame extinction early in the propagation stage. Such a case is shown in Fig. 5. Position diagrams for flames initiated by heat sources of different strengths are

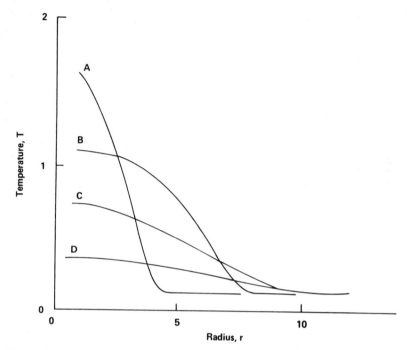

Fig. 5 Temperature distributions resulting from the spherical
source (A: t = 0.4;   B: t = 1.1;   C: t = 2.6;   D: t = 4.0).

plotted in Fig. 6. The flame position is defined as that
point at which the steepest tangent to the concentration
profile that can be drawn within the flame front cuts the
Y = 0 axis. These plots show that the flame propagates
quickly while it is driven by the heat source. After
cessation of the source, the flame speed is reduced to much
less than the planar, adiabatic value: it will not approach
this value until the flame has a radius of more than 50
flame thicknesses. There appears to be a critical radius of
approximately four flame thicknesses that the flame must
reach, aided by the heat source, before it can continue
unsupported. Such critical radii and reduced burning
velocities have been observed in the experiments of Lintin
and Wooding (1959), de Soete (1984), and Akindele et al.
(1982).

The relationships between the minimum energy for
ignition and the duration of the heat source obtained from
these calculations are shown in Fig. 7. There is a an
optimum duration of the heat source for ignition; this
duration is roughly equal to or less than the time scale of
an adiabatic flame. It might be inferred from Fig. 7 that

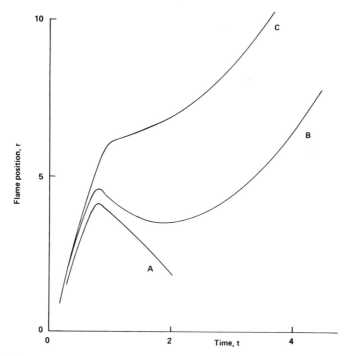

Fig. 6 Position diagrams for flames initiated by spherical sources (A,B for sources just below, above minimum for ignition, C for source well above minimum for ignition).

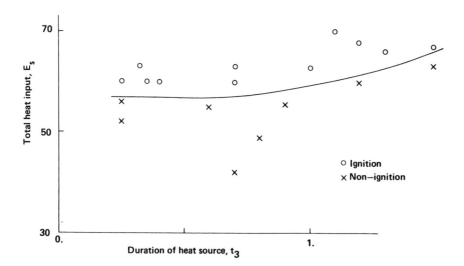

Fig. 7 Effect of source duration on minimum energy for ignition by spherical sources.

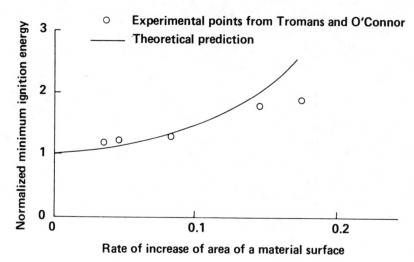

Fig. 8 Effect of strain rate of minimum ignition energy for
spherical sources.

the sources of this duration are the most effective in
driving the flame beyond its critical radius. This optimum
source duration is in reasonable agreement with the
experimental observations of spark ignition by Kono et al.
(1976), Ballal and Lefebvre (1974), de Soete (1983), and
Tromans and O'Connor (1983).
     The sensitivity of ignition to the strain rate ε was
examined. Increasing the strain rate increases the energy
required for ignition. This sensitivity to strain rate
reflects the relationship between the minimum energy for
spark ignition and the root mean square, turbulent strain
rate found experimentally by Tromans' and O'Connor (1983).
Fig. 8 shows a theoretical curve and Tromans and O'Connor's
experimental results for methane/air mixtures. The
independent variable used is the rate of increase of area of
a material surface which equals ε/3 in the numerical
analysis and approximates one-half of the root mean square,
turbulent strain rate in the experiments. The agreement is
encouraging.

                          Conclusions

     The main conclusions of this work are:

1) Detailed models for ignition by planar and slender
cylindrical heat sources have been solved, but they do not
predict the phenomena observed in spark ignition.

2) If the source is a small, spherical one, then flame extinction may occur below a critical radius. There is a minimum ignition energy, an optimum source duration, and a sensitivity to strain as found in experimental studies of spark ignition.

References

Akindele, O.O., Bradley, D., Mak, P.W., and McMahon, M. (1982) Spark ignition of turbulent gases. Combust. Flame 47, 129-155.

Ballal, D.R. and Lefebvre, A.H. (1974) The influence of flow parameters on minimum ignition energy and quenching distance. Fifteenth Symposium (International) on Combustion, pp.1473-1481. The Combustion Institute, Pittsburgh, PA.

Berzins, M., Dew, P.M., and Furzeland, R.M. (1984) Software for time-dependent problems, PDE Software Modules, Interfaces and Systems (edited by B. Engquist and T. Smedsaas), pp.309-324. North Holland Publishing Co., Amsterdam.

Deshaies, B., and Joulin, G. (1984) On the initiation of a spherical flame kernal. Combust. Sci. and Techno., 37, 99-116.

de Soete, G.G. (1983) Propagation behaviour of spark ignited flames in early stages. International Conference on Combustion in Engines, Vol.I, pp.93-100. Institution of Mechanical Engineers, London.

de Soete, G.G. (1984) Effects of geometrical and aerodynamical induced flame stretch on the propagation of spark fired premixed flames in early stages. Twentieth Symposium (International) on Combustion, pp.161-168. The Combustion Institute, Pittsburgh, PA.

Dixon-Lewis, G. (1978) Effect of core size on ignition energy by localized sources. Combust. Flame 33, 319-321.

Dixon-Lewis, G. and Shepherd, I.G. (1974) Some aspects of ignition by localized sources, and of cylindrical and spherical flames. Fifteenth Symposium (International) on Combustion, pp.1483-1491. The Combustion Institute, Pittsburgh, PA.

Kailasanath, K., Oran, E., and Boris, J. (1982) A theoretical study of the ignition of premixed gases. Combust. Flame 47, 173-190.

Kapila, A.K. (1981) Evolution of deflagration in a cold combustible subjected to a uniform energy flux. Int. J. Eng. Sci. 19, 495-509.

Kono, M., Kumagai, S., and Sakai, T. (1976) The optimum condition for ignition of gases by composite sparks. Sixteenth Symposium (International) on Combustion, pp.757-766. The Combustion Institute, Pittsburgh, PA.

Lewis, B. and von Elbe, G. (1961) Combustion, Flame and Explosion of Gases. Academic Press, New York and London.

Lintin, D.R. and Wooding, E.R. (1959) Investigation of the ignition of a gas by an electric spark. Br. J. Appl. Phys. 10, 159-166.

Overley, J.R., Overholser, K.A., and Reddien, G.W. (1978) Calculation of minimum ignition energy and time dependent laminar flame profiles. Combust. Flame 31, 69-83.

Tromans, P.S. and O'Connor, S.J. (1983) The influence of turbulent motion on spark ignition. AIAA Progress in Astronautics and Aeronautics: Dynamics of Flame and Reactive Systems, Vol.95, (edited by J.R. Bowen, N. Manson, A.K. Oppenheim, and R.I. Soloukhin), pp.421-432. AIAA, New York.

Thermal Ignition and Minimum Ignition Energy in Oxygen-Ozone Mixtures

B. Raffel,* J. Warnatz,† H. Wolff,‡ and J. Wolfrum§
*University of Heidelberg and Sonderforschungsbereich 123,
Heidelberg, West Germany*
and
R.J. Keeπ
Sandia National Laboratories, Livermore, California

Abstract

To study the influence of chemistry on the ignition process, thermal ignition in a one-dimensional configuration is studied experimentally and by computer simulation for O_2-O_3 mixtures. For this purpose, the decomposition of O_2-O_3 mixtures is studied experimentally by laser-induced ignition in the axis of a cylindrical combustion chamber. To interpret these measurements, the corresponding time-dependent, one-dimensional conservation equations (assuming uniform pressure) are integrated by various differential/algebraic solvers. Experimental results on thermal minimum ignition energies in O_2-O_3 mixtures can be simulated by the calculations within the limits of experimental error. On the basis of this "calibration," further simulations are carried out to study the dependence of minimum ignition energy on ignition source diameter and ignition time to get more information on the nature of the ignition process.

I. Introduction

Igniton processes, e.g., by sparks, have often been discussed in the literature in the last decades (Lewis and von Elbe, 1961). Newer approaches to this problem combining

---

Presented at the 10th ICDERS, Berkeley, California, August 4-9, 1985. Copyright © 1986 by J. Warnatz. Published by the American Institute of Aeronautics and Astronautics, Inc. with permission.
*Dipl.-Physiker, Institute of Physical Chemistry.
†Priv.-Dozent, Institute of Physical Chemistry.
‡Dipl.-Ingenieur, Institute of Physical Chemistry.
§Professor, Institute of Physical Chemistry.
πSupervisor, Computational Mechanics Division.

experimental results with theoretical interpretations using numerical simulations are described by Overley (1976), but failed because of an inadequate physical description of the ignition process. Furthermore, in recent years, realistic descriptions of the homogeneous ignition process, including detailed chemistry for the H_2-O_2 system, have become available (Kailasanath et al. 1982; Wiriyawit and Dabora 1984). However, there was a complete lack of experimental data comparable with these computer simulations.

To compensate for this lack of data and for the complexity of this process, the problem is limited here to thermal ignition in a one-dimensional configuration. This procedure then allows computer simulation and direct comparison with the experiment. Thus, an experimental setup for laser-induced thermal ignition in the axis of a cylindrical combustion chamber has been developed that permits the observation of the temporal development of the ignition process in a one-dimensional geometry. Using O_2-O_3 mixtures, these measurements can be used for comparison with calculated results to "calibrate" the models used in the computations and to validate further simulations, varying parameters like ignition time and source diameter systematically.

II. Experimental Method

First, experiments on laser-induced thermal ignition are carried out in the O_2/O_3 system because of its simple reaction mechanism and its easy experimental accessibility. Ozone has stong absorption bands in the range of the $00°1 - 02°0$ emission of a CO_2 laser. Thus, there is no need for additional absorbers in the gas mixture, and the gas may be heated by absorption of slightly focused CO_2 laser pulses at the axis of a cylindrical vessel filled with a given O_2/O_3 mixture. Provided that sufficient energy is deposited and the heating is homogeneous, O_3 decomposition starts into the radial direction uniformly over the whole axis. Because of its strong ultraviolet absorption, O_3 is detectable by absorption of the light of an Hg-Xe high-pressure lamp.

The experimental setup is shown in Fig. 1. O_3 decomposition is initiated by pulses of a Lumonics TEA 801-A CO_2 laser operating at the 9P20 line at $\lambda = 9.552$ μm in multimode oscillation. The laser light is focused with a concave mirror and shows a beam waist of 1.5 ± 0.2 mm diameter along the cell axis. With pulse energies up to 0.7 J, mean fluences of about 40 J/cm^2 are achieved in the cell. Due to multimode operation, the fluence is fairly constant over the whole profile, neglecting statistical variations. A

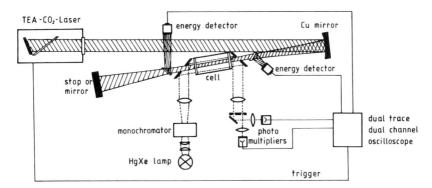

Fig. 1 Experimental setup.

part of the laser radiation is coupled out in front of and behind the cell and measured by two pyroelectric energy detectors. With regard to the reflectivity of the beam splitters and cell windows and the polarization of the laser light, the absorbed energy is calculated.

The radial propagation of the flame front is observed by uv absorption of O_3 at λ = 312.6 nm. To that end, a grating monochromator separates light of this wavelength from the radiation of an Hg-Xe lamp. Its exit slit is imaged to the center of the cell perpendicular to the cell axis. A double pinhole determines two beams parallel to the axis with a radial resolution of 1.5 $\pm$ 0.5 mm. The light intensity of the two beams is detected by photomultipliers (1P28A) and recorded by a dual trace oscilloscope (Tektronix 7623A). If the two beams are chosen at different radii, the radial propagation of the O_3 decomposition can be observed.

The ignition limit of an O_2/O_3 mixture, i.e., the minimum absorbed energy density leading to ignition, is investigated by approaching the limit from the low-energy side. For this purpose, the fluence of the laser pulse is raised by removing partial reflectors from the beam or raising the laser high voltage.

III. Calculation Method

After transformation to Lagrangian coordinates $(r,t) \longrightarrow (\Psi,t)$,

$$\left(\frac{\partial \Psi}{\partial r}\right)_t = \varrho\, r^i \qquad\qquad \left(\frac{\partial \Psi}{\partial t}\right)_r = -\varrho\, v\, r^i$$

The system of equations under consideration is given by

$$\varrho \, c_p \frac{\partial T}{\partial t} - \frac{\partial P}{\partial t} = \varrho \, \frac{\partial}{\partial \psi} \left(r^{2i} \varrho \lambda \, \frac{\partial T}{\partial \psi} \right)$$

$$- \varrho \, r^i \frac{\partial T}{\partial \psi} \, \sum_k j_k \, c_{p,k} - \sum_k R_k \, h_k$$

$$\frac{\partial Y_k}{\partial t} = - \varrho \, \frac{\partial}{\partial \psi} \left(r^i \, j_k \right) + R_k$$

Here, c_p is the specific heat capacity; h the specific enthalpy (Burcat 1984); and i = 0,1,and 2 for linear, cylindrical, and sperical geometry, respectively. P is the pressure, R the (molar scale) chemical rate of formation, r the Cartesian coordinate, T the temperature, v the flow velocity, Y the mass fraction, λ the mixture heat conductivity, and ϱ the density.

The diffusion fluxes and the heat conductivity are given by

$$j_k = - D_{M,k} \varrho \, \frac{\partial Y_k}{\partial r} - D_{T,k} \frac{\partial \ln T}{\partial r}$$

$$D_{M,k} = \frac{1 - Y_k}{\displaystyle\sum_{l \neq k} \frac{\mathscr{D}_{kl}}{x_l}}$$

$$\lambda = \frac{1}{2} \left(\sum_k x_k \lambda_k + \frac{1}{\displaystyle\sum_k x_k \lambda_k} \right)$$

where $\mathscr{D}_{kl}$ is the binary diffusion coefficient, $D_{T,k}$ the thermal diffusion coefficient, x the mole fraction, and λ_k the pure species heat conductivity. These transport properties are calculated according to the data given by Kee et al. (1983).

Together with the uniform pressure assumption

$$\frac{\partial P}{\partial \psi} = 0$$

and the boundary conditions (R = vessel radius)

$$\psi = 0 \quad : \quad r = 0 \, , \quad \frac{\partial T}{\partial \psi} = 0 \, , \quad \frac{\partial Y_k}{\partial \psi} = 0$$

$$\Psi = \Psi_R \quad : \quad r = R \ , \quad \frac{\partial T}{\partial \Psi} = 0 \ , \quad \frac{\partial Y_k}{\partial \Psi} = 0$$

the problem can be reduced to a differential/algebraic system by discretization using simple finite-difference methods (Warnatz 1978).

This system is then solved by Sandia National Laboratories' DASSL program (Petzold 1982) or Heidelberg University's LIMEX program (Deuflhard et al. 1986).

Because of the large ratio of vessel diameter and flame front thickness, in the simulations of the experimental runs (see Sec. IV), adaptive gridding must be used to describe flame propagation following the ignition process. An example is given in Fig. 3, using static regridding with the grid point density proportional to the temperature gradient, and local montone piecewise cubic interpolation (Fritsch and Butland 1984). However, during the ignition process itself, there is no flame propagation, and regridding is not essentially necessary during this early period. Therefore, for most of the calculations, a nonadaptive grid point system is used, concentrating about 50% of the grid points equidistantly in the range of large gradients enclosing the ignition pocket (see Fig. 3 of Sec. IV).

Calculations are performed using the reaction mechanism shown in Table 1. Typical calculation times amount to about 5 min for the O_2-O_3 system (IBM 3081 D).

IV. Comparison of Experiments and Calculations

Figure 2a shows an example of a set of results for the ignition of O_2/O_3 mixtures containing 23 % O_3 at a total pressure of 0.34 bar. The laser fluence varies between 3

Table 1   Mechanism of O_3 decomposition [a]

Reaction		A, cm, mole,s	E, kJ/mole
$O_3 + M^b$	$\longrightarrow O_2 + O + M$	3.8×10^{14}	95.0
$O + O_2 + M$	$\longrightarrow O_3 + M$	1.5×10^{13}	-1.9
$O + O_3$	$\longrightarrow O_2 + O_2$	5.2×10^{12}	17.4
$O_2 + O_3$	$\longrightarrow O + O + O_2$	4.7×10^{12}	416.1
$O_2 + M$	$\longrightarrow O + O + M$	1.4×10^{14}	451.0
$O + O + M$	$\longrightarrow O_2 + M$	6.2×10^{12}	-44.6

[a] Kailasanath et al. (1982); Baulch et al. (1976); Warnatz (1984).
[b] $M = 2.8\ [O] + 1.0\ [O_2] + 2.3\ [O_3]$.

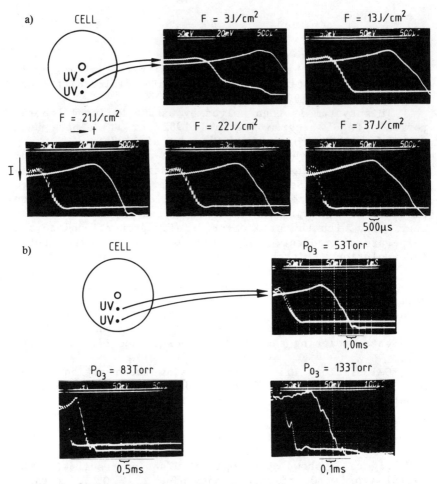

Fig. 2 Time dependence of transmitted uv intensity at r = 6.2 mm
and 12.4 mm following CO_2 laser pulse, O_2/O_3 mixture at P = 0.34
bar. a) At different fluences, 23 % ozone. b) At different ozone
partial pressures, fluence F = 38 J/cm^2.

and 37 J/cm^2. Given is the transmitted light intensity
following the laser pulse and recorded simultaneously at
two radial positions r = 6.2 mm and 12.4 mm. This intensity
is determined by the local O_3 concentration and by the
temperature, since the molar absorptivity of O_3 at λ = 312.6
nm strongly rises with temperature. The slight reduction of
transmitted intensity after the laser pulse reflects the
compression of the gas and the temperature rise at r_1 and
r_2, respectively. The increase to the maximum intensity is
given by the O decomposition at these radii. From the

signals, it may[3] be deduced that the velocity of the decomposition flame outside the irradiated volume is independent of the energy of the ignition source, whereas the arrival of the decomposition flame at r_1 occurs faster for higher energies, reflecting an induction period.

Figure 2b shows similar traces, this time for different partial pressures of ozone. The oscillations of the signals account for acoustic modes of the cell excited by the dissipation of the absorbed laser energy.

A typical simulation of ignition in oxygen-ozone mixtures is given in Fig. 3. After a relatively long induction period (confirming the corresponding experimental observation), there is a thermal explosion in the ignited

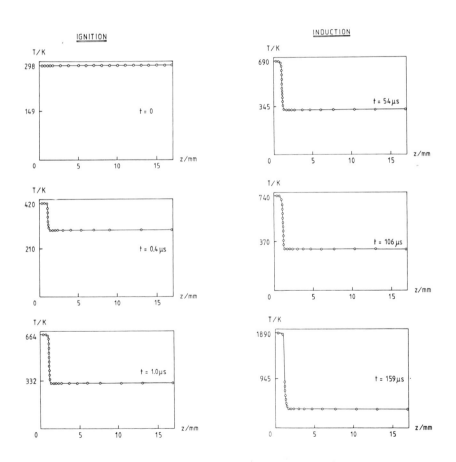

(Figure 3 continued on next page)

gas pocket. This explosion initiates flame propagation across the vessel, with subsequent equilibration by transport processes.

Calculated minimum thermal ignition energies are compared with the experimental ones in Fig. 4. The simulations clearly show a pressure dependence of the minimum ignition energy that cannot be resolved experimentally due to missing sensitivity of the energy detectors used in these measurements. Nevertheless, there is good agreement of calculations and measurements, confirming reasonable operation of the experiment. On the other hand, the experiment yields a verification of the calculation method, thus enabling reliable extrapolation to larger reaction systems (if a detailed reaction mechanism is available).

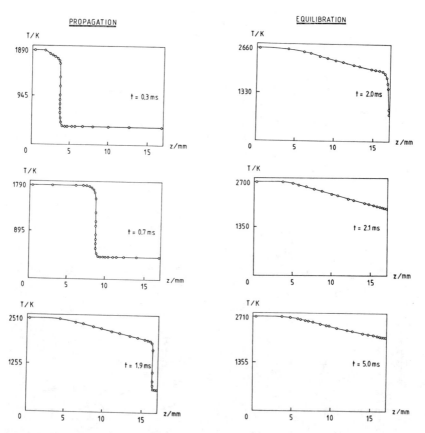

Fig. 3 Example of the simulation of the progress of the ignition process in equimolar O_2/O_3 mixtures, represented by calculated temperature profiles; $P = 0.34$ bar, $T = 298$ K, $t_{ign} = 1$ µs.

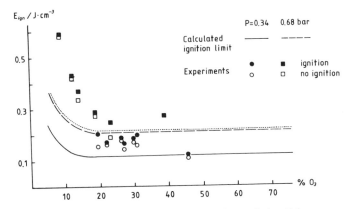

Fig. 4 Measured and calculated minimum thermal ignition energy in O_2/O_3 mixtures at P = 0.34 and 0.68 bar, T = 298 K, t_{ign} = 1μs. The dotted line represents calculations at 0.68 bar using a fully adaptive grid point system.

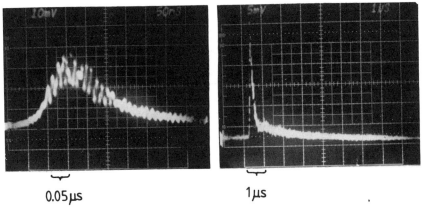

Fig. 5 Experimental laser fluence as function of time.

V. Dependence of the Minimum Ignition Energy on Ignition Time and Source Diameter

Some additional simulations have been carried out taking into consideration the dependence of the minimum ignition energy on ignition time and source diameter. Knowledge of the dependence of the minimum ignition energy on ignition time is important to confirm appropriate performance of the experiments. As shown in Fig. 5, the laser fluence is a complicated function of time, with a pulse length amounting to about 5 μs. Relaxation times for

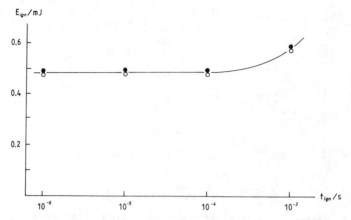

Fig. 6 Minimum ignition energy (spherical geometry) as function of
the ignition of the ignition source time, O_2/O_3 mixture with 50%
ozone, T = 298 K, P = 1 bar, source radius 0.75 mm, constant volume
V = 20.6 cm^3.

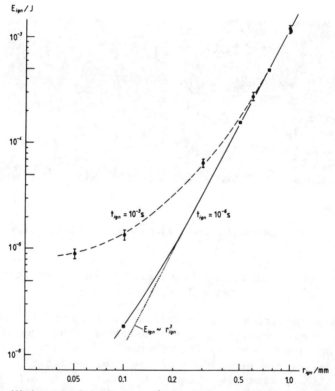

Fig. 7 Minimum ignition energy (spherical geometry) as function of
the ignition source radius. O_2/O_3 mixtures with 50% O_3, T = 298 K,
P = 1 bar, constant volume V = 20.6 cm^3; parameter is the ignition
source time t_{ign}.

the transition of this energy into thermal energy are in
the order of 5 µs, as well. However, simulations with
systematically varied ignition times (see Fig. 6) show that
this parameter does not influence minimum ignition energy,
as far as times shorter than the characteristic diffusion
time $\Delta t = \Delta x^2/2D$ (Δx = beam diameter, D = mean diffusion
coefficient) are considered, and this is the case in the
experiments.

A systematic variation of the ignition source diameter
(see Fig. 7, spherical symmetry) shows the influence of
transport on the ignition process. For short ignition times
(small if compared with the characteristic diffusion time,
e.g. 10^{-6} s), the energy density necessary for ignition is
practically independent of the ignition source diameter.
For large ignition times (comparable with characteristic
diffusion times, e.g. 10^{-3} s), there is loss of energy out
of the source diameter by transport processes, leading to
ignition energies several orders of magnitude larger than
for the case $E_{ign} \sim r_{ign}^3$, if the source is small enough. The
present calculations do not lead to a quenching distance as
predicted in simulations of the H_2-O_2 system (Kailasanath
et al. 1982). It should be mentioned, that this effect
should not be confused with quenching e.g. by electrode
surfaces in experiments on spark ignition (Lewis and von
Elbe 1961).

Acknowledgments

The financial support by the Deutsche Forschungsge-
meinschaft, the Stiftung Volkswagenwerk, the Fonds der
Chemischen Industrie, and (within the frame of the TECFLAM
project) the BMFT is cordially acknowledged. The authors
thank P. Deuflhard and U. Nowak for helpful discussions and
for providing the computer code LIMEX and U. Maas for help
in carrying out the calculations.

References

Baulch, D.L., Drysdale, D.D., Horne, D.G., and Lloyd, A.C. (1976),
    Evaluated Kinetic Data for High Temperature Reactions; Homo-
    geneous Reactions in the O-O System, the CO-CO-H System, and
    of Sulfur-Containing Species, Vol. 3. Butterworths, London.

Burcat, A. (1984), Thermodchemical data for combustion calcula-
    tions, Combustion Chemistry (edited by W.C. Gardiner),
    Springer, New York, pp. 455 – 473.

Deuflhard, P., Hairer, E., and Zugck, J. (1986), Linear-implicit
    differential equation systems by extrapolation methods (to be
    published).

Fritsch, F.N., and Butland, J. (1984), A method for constructing local monotone piecewise cubic interpolants, SIAM J. Sci. Stat. Comput., Vol. 5, pp. 300 - 304.

Kailasanath, K., Oran, E.S., Boris, J.P., and Young, T.R. (1982), Time-dependent simulation of flames in hydrogen-oxygen-nitrogen mixtures, Numerical Methods in Laminar Flame Propagation (edited by N. Peters and J. Warnatz), Vieweg, Braunschweig, pp. 152 - 166.

Kee, R.J., Miller, J.A., and Warnatz, J. (1983), A Fortran computer code package for the evaluation of gas phase viscosities, conductivities, and diffusion coefficents. SAND 83 - 8209, Sandia National Laboratories, Livermore, Calif.

Lewis, B., and von Elbe, G. (1961), Combustion, Flames, and Explosions in Gases. Academic Press, New York.

Overley, J.R. (1976), Establishing criteria for ignition: An experimental and theoretical study of the effect of spark geometry on the minimum ignition energy of hydrazine. Ph.D.-thesis, Vanderbilt University, Nashville, Tenn.

Petzold, L. (1982), A description of DASSL: A differential/algebraic system solver. IMACS World Congress, Montreal, Canada.

Warnatz, J. (1978), Calculation of the structure of laminar flat flames I: Flame velocity of freely propagating ozone decomposition flames, Ber. Bunsenges. Phys. Chem., Vol. 82, pp. 193 - 200.

Warnatz, J. (1984), Survey of rate coefficients in the C/H/O system. Combustion Chemistry (edited by W.C. Gardiner), Chap. 5, Springer, New York.

Wiriyawit, S., and Dabora, E.K. (1984), Modelling the chemical effects of plasma ignition in one-dimensional chamber, 20th Symposium (International) on Combustion, pp. 179 - 186.

Chemical Effects on Reflected-Shock Region in Combustible Gas

Yasunari Takano*
Tottori University, Tottori, Japan
and
Teruaki Akamatsu†
Kyoto University, Kyoto, Japan

Abstract

In this paper, the effects of chemical reactions on reflected-shock flowfields in shock tubes are studied by comparison of the results of experiments, numerical simulations, and analysis. Pressure measurements have been made in the reflected-shock region in a mixture of hydrogen and oxygen diluted in argon with a diaphragmless shock tube equipped with a pneumatic valve in place of diaphragms. Numerical simulations have been made with a detailed modeling that combines the random choice method of gasdynamics with a chemical kinetics simulation code. Comparisons are excellent between the simulations and the experiments for highly diluted combustible gas. For moderately diluted cases, the simulations successfully predict averaged flowfields over which some disturbance waves are observed to develop. The results of these simulations satisfy the analytical relations for perturbation in reactive reflected-shock flowfields that were derived by the present authors by applying the method of linearized characteristics. At the end of shock tube the analytical results hold satisfactorily for highly diluted cases, but some discrepancies appear between the analysis and the simulations for moderately diluted cases. The discrepancies can be estimated by an analysis of the acceleration effect of the reaction front.

Presented at the 10th ICDERS, Berkeley, California, August 4-9, 1985. Copyright © 1985 by the American Institute of Aeronautics and Astronautics, Inc. All rights reserved.

*Associate Professor, Department of Applied Mathematics and Physics.

†Professor, Department of Mechanical Engineering.

347

Introduction

Because a uniform stationary gas is generated behind
a reflected-shock wave, a chemical process under a constant
physical condition can be observed in the shocked gas when
a test gas is added to the driven gas. In actual experiments, the constancy for the physical properties in the
reflected-shock region is rather limited because of gasdynamic disturbances caused by the boundary-layer effects and
the release of chemical energy.

The authors (1985) have analyzed the gasdynamic disturbances in the reflected-shock region due to chemical reactions of a highly diluted combustible gas in inert gas. The
method of linearized characteristics is applied to describe
perturbations of the properties in the flowfields. The
variations of the perturbed properties are expressed mainly
in terms of the chemical energy release function. The
analysis yields uniformly valid expressions for only very
small perturbations in combustible gas when the combustion
process has a very high activation energy.

The aim of the present paper is to examine whether
these analytical expressions are applicable to actual
disturbances due to combustion processes in the reflected-
shock region. For the $H_2/O_2/Ar$ mixture, the results of
shock tube experiments, numerical simulations, and analyses
are compared.

Many investigations utilizing reflected-shock waves
have been conducted to elucidate detonation-initiation
mechanisms in $H_2/O_2/Ar$ mixtures. For example, Strehlow and
Cohen (1962) and Gilbert and Strehlow (1966) classified
types of detonation initiation behind reflected-shock waves
and proposed models to explain the observed wave patterns.
Voevodsky and Soloukhin (1965) observed the explosion limits
of a hydrogen/oxygen mixture ignited by reflected-shock
waves and gave qualitative explanations of their observations on basis of chemical kinetics models. Meyer and
Oppenheim (1971) investigated in detail mild and strong
ignitions of a $H_2/O_2/Ar$ mixture behind reflected-shock
waves. They showed that the strong ignition caused one-
dimensional flowfields, while the mild ignition, starting
from distinct kernels, generated complicated multidimensional flowfields. Recently, Oran et al. (1982) simulated
with a detailed model the reflected-shock experiments for
strong ignition.

In the present paper, attention is focused on the one-
dimensional features of the reactive gasdynamics behind
reflected-shock flowfields in combustible gas. Strong ignition mode conditions and mixtures of hydrogen and oxygen

sufficiently diluted in argon minimize the effects of transverse detonation waves that destroy the one-dimensionality of the shock flowfield. The experiments performed in this work were limited to these conditions.

Theoretical Background

Most of theoretical investigations on the chemical effects of shock tube flowfields have been performed by using numerical procedures. Some of those on combustible gas are discussed in the section on simulations. As for analytical treatments, Spence (1961) applied the method of linearized characteristics to analyze unsteady flowfields in the relaxing gas behind a shock wave produced by an impulsively started piston. Also, Fickett (1984) studied shock initiation of detonation in an explosive for the limiting case of small heat release by a linear perturbation analysis. The authors (1985) also employed the method of linearized characteristics to investigate reflected-shock flowfields in combustible gas and to describe perturbations of the properties due to chemical processes of a highly diluted combustible gas in inert gas. That work is summarized here.

Let x and t be the distance from the end wall of shock tubes and the elapsed time after shock reflection, respectively. When the amplitudes of gasdynamic disturbances remain small, the physical properties in the reflected-shock region can be expressed as

$$\rho(x,t) = \rho_f[1 + \rho'(x',t')] \qquad (1)$$

$$u(x,t) = a_f u'(x',t') \qquad (2)$$

$$p(x,t) = \rho_f a_f^2[1/\gamma_f + p'(x',t')] \qquad (3)$$

$$t = t_{ig}t', \qquad x = a_f t_{ig}x' \qquad (4)$$

where ρ, u, and p are the density, velocity, and pressure, respectively, subscript f the frozen physical properties in the reflected-shock region immediately after the shock reflection, a_f the acoustic speed, γ_f the specific heat ratio, and t_{ig} the induction time that normalizes the coordinates. Perturbation terms to the first approximation satisfy the linearized equations

$$\frac{\partial \rho'}{\partial t'} + \frac{\partial u'}{\partial x'} = 0 \qquad (5)$$

$$\frac{\partial u'}{\partial t'} + \frac{\partial p'}{\partial x'} = 0 \tag{6}$$

$$\frac{\partial p'}{\partial t'} + \frac{\partial u'}{\partial x'} = q \tag{7}$$

where q is the release rate of the chemical energy in the flowfield.

If the combustible gas is highly diluted in the inert gas, the gasdynamic disturbances do not appreciably affect the chemical process and the chemical reactions proceed coherently in the mixture entering the reflected-shock front. In this case, the release rate of the chemical energy can be expressed as function of the time elapsed after heating of the reflected-shock wave,

$$q(x',t') = q_o(t'-x'/u_R') \tag{8}$$

where u_R' ($= u_R/a_f$) is the normalized velocity of the reflected-shock wave. Integration of p'+ u', p'- u', and p'- ρ' along their linearized characteristics yields the following expressions for the perturbed properties:

$$p'(x',t') = \frac{u_R'}{1-u_R'^2}Q(t'-x') - \frac{u_R'^2}{1-u_R'^2}Q(t'-\frac{x'}{u_R'}) + \ldots \tag{9}$$

$$u'(x',t') = \frac{u_R'}{1-u_R'^2}Q(t'-x') - \frac{u_R'}{1-u_R'^2}Q(t'-\frac{x'}{u_R'}) + \ldots \tag{10}$$

$$\rho'(x',t') = \frac{u_R'}{1-u_R'^2}Q(t'-x') - \frac{1}{1-u_R'^2}Q(t'-\frac{x'}{u_R'}) + \ldots \tag{11}$$

where Q is a heat release function representing the amount of released chemical energy in the flowfield, defined as

$$Q(t') = \int_o^{t'} q_o(t)dt \tag{12}$$

In addition, the variation of the temperature is written as

$$T'(x',t') = \gamma_f p'(x',t') - \rho'(x',t') + \ldots \tag{13}$$

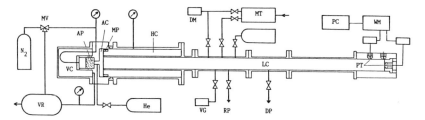

Fig. 1 Schematic diagram of shock-tube apparatus.

The variations of the properties at the shock tube end wall are obtained by putting $x' = 0$ into Eqs. (9), (11), and (13),

$$p'_w(t') = \frac{u'_R}{1+u'_R} Q(t') + \dots \tag{14}$$

$$\rho'_w(t') = -\frac{1}{1+u'_R} Q(t') + \dots \tag{15}$$

$$T'_w(t') = \frac{\gamma_f u'_R + 1}{1+u'_R} Q(t') + \dots \tag{16}$$

Experiments

Experiments of reflected-shock waves were performed in a 3 m long cylindrical driven section with a 40 mm inner diameter of a diaphragmless shock tube. The present apparatus used a pneumatic valve to replace the diaphragms and consequently had several advantages as a chemical shock tube. In conventional shock tubes, fragments of ruptured diaphragms are one of the sources for contaminations. Also, the exposure of the inside of the apparatus to the air while the used diaphragms are replaced allows impurities to enter the test gas. Such sources of contamination are not generated in the present apparatus. Figure 1 is a schematic diagram of the present apparatus in an initial setup for the experiments. The following is a brief description of the apparatus mechanism.

A main piston separates a driven section from a driver section. The driver section is connected to an actuating section through a leak between the outer diameter of the main piston and the inner diameter of the driven section. The driver gas is introduced after an auxiliary piston is pushed by a compressed gas, the pressure of which is set to be slightly higher than the maximum pressure of the driver

Table 1 Conditions for shock tube experiments and simulations

Case	Ar	p_1, kPa H$_2$	O$_2$	T_1, K	u_S, km/s Exp	Cal
I	12	0.27	0.13	298	0.692	0.740
II	12	0.53	0.27	301	0.701	0.735
III	12	0.80	0.40	296	0.711	0.725
IV	12	1.07	0.53	290	0.711	0.730
V	12	1.33	0.67	296	0.711	0.735

p_1, T_1 = pressure and temperature at test gas condition
u_S = incident shock speed

gas because the auxiliary piston is movable in a piston cylinder. When the gas in the piston cylinder is exhausted to a vacuum reservoir through a magnetic valve, the auxiliary piston moves back and the actuating chamber is linked to a vacuum chamber. As soon as the gas is released from the actuating chamber to the vacuum chamber, the main piston moves rapidly back toward the actuating chamber and opens the driven section to the driver gas. Hence, a shock wave is generated. The structure of the apparatus and its advantages have been reported in detail by the authors (1984).

Experiments on reflected-shock waves in a $H_2/O_2/Ar$ mixture were carried out for the conditions listed in Table 1. Piezoelectric transducers (PCB 111A24) are flush mounted on the end wall and on the side wall 54 mm from the end wall. Pressure profiles are recorded with a transient digital recorder (NF WM842) at a sampling period of 1 μs, and the data stored on floppy disks.

Figure 2 shows the pressure histories for the experiments and the results of the numerical simulations. In the end wall pressure records (lower traces), a pressure increase is generated by the heat release in an induction time after an abrupt pressure jump in the shock reflection. In the side wall pressure records, a pressure pulse due to the reaction front follows the pressure jumps in the incident- and reflected-shock waves. Comparisons between the experiments and the simulations will be discussed in the following section.

Simulations

Considerable effort has been devoted to simulating shock waves in combustible gas. Gilbert and Strehlow (1966) used the method of characteristics for one-dimensional flow

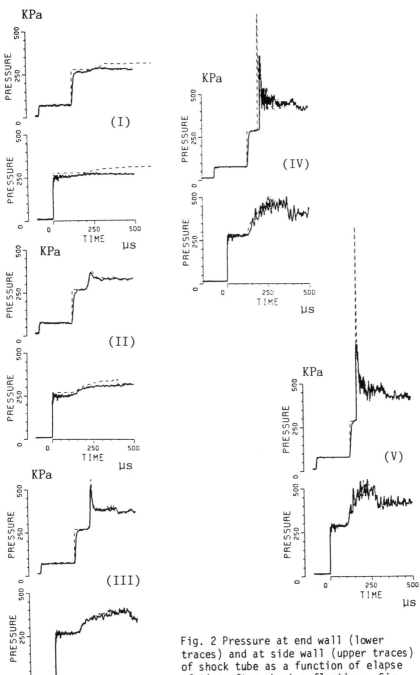

Fig. 2 Pressure at end wall (lower traces) and at side wall (upper traces) of shock tube as a function of elapse of time after shock reflection. Simulations (dotted lines) and experiments (solid lines).

Table 2 List of chemical reactions and rate coefficient data

$$k_r = A_r T^{n_r} \exp(- T_r/T)$$

r	Reactions				A_r [a]	n_r	T_r, K
1	H	+ O_2	= OH	+ O	0.145E15[b]	0	8250
2	O	+ H_2	= OH	+ H	0.180E11	1	4480
3	H_2	+ OH	= H_2O	+ H	0.378E 9	1.3	1825
4	O	+ H_2O	= OH	+ OH	0.590E14	0	9120
5	H	+ HO_2	= OH	+ OH	0.240E15	0	950
6	OH	+ HO_2	= H_2O	+ O_2	0.300E14	0	0
7	O	+ HO_2	= OH	+ O_2	0.480E14	0	500
8	H	+ HO_2	= H_2	+ O_2	0.240E14	0	350
9	H_2	+ HO_2	= H_2O	+ OH	0.600E12	0	9400
10	H	+ O_2 + Ar	= HO_2	+ Ar	0.160E16	0	−500
11	H	+ H + Ar	= H_2	+ Ar	0.110E19	−1	0
12	O	+ O + Ar	= O_2	+ Ar	0.725E17	−1	−200
13	H	+ OH + Ar	= H_2O	+ Ar	0.360E23	−2	0

a Units are $cm^3 mol^{-1} s^{-1}$ for r = 1,...,9, and $cm^6 mol^{-2} s^{-1}$ for r = 10,...,13.   b Read 0.145E15 as 0.145×10^{15}

with heat addition. Cohen et al. (1975) reported a computational technique that numerically integrates rate equations for chemical kinetics and conservation equations for gasdynamics expressed in a Lagrange form. Oran et al. (1978) developed a detailed model that combines a gasdynamic scheme with a chemical kinetics simulation code by use of the splitting technique. They employed the flux-corrected-transport (FCT) scheme for gasdynamics modeling (Book et al., 1975) and the CHEMEQ algorithm for chemical kinetics (Young and Boris, 1977).

In the present simulations, the time splitting technique is used to combine the random choice scheme (Chorin, 1976) with the CHEMEQ algorithm. The random choice method has several advantages (high resolutions and no artificial viscosity) that make it attractive for modeling reactive gasdynamics. The random choice scheme, a flow-chart of which is given by Sod (1978), has been extended to multicomponent fluids by Takano et al. (1985).

In this simulation, the reactive mechanism has 13 elementary reactions with 8 chemical species: Ar, H_2, O_2, H_2O,

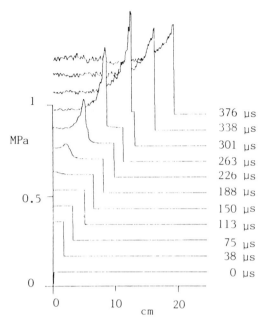

Fig. 3 Pressure profiles in the reflected-shock region (Case III) for specified elapsed time after shock reflection.

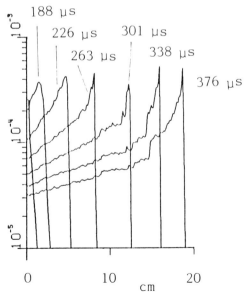

Fig. 4 Hydrogen atom mass fraction profiles in the reflected-shock region (Case III) for specified elapsed time after shock reflection.

Table 3 Physical properties in frozen reflected-shock regions

Case	ρ_f, kg/m^3	p_f, kPa	T, K	a_f, Km/s	γ_f	u_R'
I	1.02	277	1282	0.672	1.65	0.649
III	1.04	270	1172	0.651	1.63	0.651
V	1.09	287	1146	0.653	1.61	0.648

H, O, OH, and HO$_2$. The rate coefficient data used for the mechanism is shown in Table 2 (Jensen and Jones, 1978). Results are presented here for case III in Table 1.

Figure 3 shows pressure profiles plotted at every 400 steps of the computations. In the combustible mixture heated by reflected-shock wave, intermediate species such as H, O, and OH increase exponentially due to the induction reactions. Since these reactions release almost no chemical energy, no gasdynamic disturbance is generated in the reflected-shock flowfields for a short time after shock reflection. After the induction time is exceeded, ignition takes place and compression waves, generated by the heat release caused by exothermic reactions, propagate away from the end wall.

The hydrogen atom concentration peaks in Fig. 4 correspond to the ignition phase and also mark the reaction front where the formation of H$_2$O and the heat release occur rapidly. The hydrogen atom concentration decays behind the reaction front due to recombination reactions that release chemical heat.

As shown in Fig. 2, the agreement between experiment and simulation for the pressure histories on the end and side walls are excellent for cases II and III. On the end wall, some vibrations are observed to develop in the pressure profile after the induction time for cases IV and V. This is due to the occurrence of transverse waves, which characterize the detonation waves. The numerical simulations produce averaged pressure profiles above which the experimental profiles oscillate. These transverse waves are perturbing waves imposed on one-dimensional flows that are successfully simulated by the present model. As for case I, simulations must assume a higher shock speed than that of the experiment to match the predicted and observed induction times. The discrepancy in the induction time for highly lean combustible mixtures may be rectified if a more complete reaction model than that given in Table 2 is utilized.

Comparisons with Analytical Expressions

In the previous sections, the simulations are shown to predict adequately experimental results. In the following, a comparison is made between the simulations and the analytical expressions, and the limits for validity of the analysis are considered. Comparisons are made for the perturbed physical properties and the released chemical energy at the end wall of shock tubes ($x = 0$).

Perturbations in the properties are calculated from results of the simulations and are normalized according to Eqs. (1-4),

$$\rho_w'(t') = [\rho(0, t/t_{ig}) - \rho_f]/\rho_f \tag{17}$$

$$p_w'(t') = [p(0, t/t_{ig}) - p_f]/(\gamma_f p_f) \tag{18}$$

$$T_w'(t') = [T(0, t/t_{ig}) - T_f]/T_f \tag{19}$$

where the subscript w denotes the values at $x = 0$ and subscript f the properties in the frozen reflected-shock region immediately after the shock reflection shown in Table 3.

The heat release rate q_o can be determined from the computational data for the fluid properties and the concentrations for the chemical species at the end wall ($x = 0$). Integration of q_o with respect to the time yields the released chemical heat Q. Instead, in the present comparisons, the following relation was used:

$$Q(t') \cong \sum_j [c_j(0, t/t_{ig}) - c_j(0,0)]h_j^o/a_f^2 \tag{20}$$

to calculate released chemical energy per unit mass of mixture. Here, c_j and h_j are the mass fraction and the enthalpy of formation for the jth chemical species, respectively.

Figure 5 shows comparisons of perturbations of the physical properties with released chemical energy at the end wall between the simulations and the analysis. The perturbations satisfy Eqs. (14-16) for case I. As for case III, the variations of the properties obey the relations until they attain 20% of the frozen values ρ_f, p_f, and T_f; afterward, they begin to deviate from them according to an increase of Q.

Figure 6 shows relations among the end wall perturbed properties, where results of the simulations are plotted in

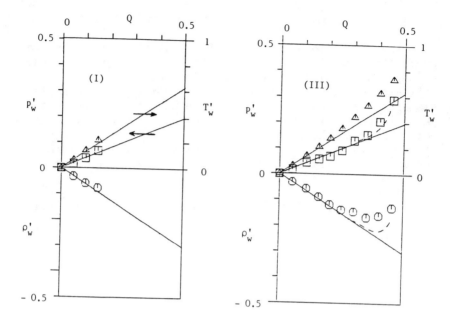

Fig. 5 End-wall density, temperature, and pressure pertubation dependence on relased chemical energy.
Simulations (symbols) and analysis (solid lines).

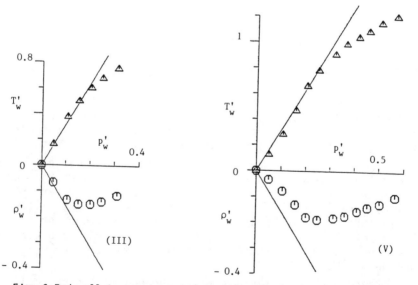

Fig. 6 End-wall temperature and density pertubation dependence on pressure pertubations. Simulation (symbols) and analysis (solid lines).

symbols. Solid lines refer to analytical relations such as

$$\rho_w'(t') = -1/u_R' \cdot p_w'(t') \tag{21}$$

$$T_w'(t') = (\gamma_f u_R' + 1)/u_R' \cdot p_w'(t') \tag{22}$$

which are obtained from Eqs. (14-16) by eliminating Q. Comparisons between the simulations and the analysis show that Eq. (22) holds fairly well until the pressure increases from the frozen value by 30% for case III and by 40% for case V.

With Eq. (14), the heat release rate of combustion process can be determined from the end wall measurement of the pressure in shock tube experiments. Also, the deviation of the temperature from the idealized value (the frozen temperature) due to chemical reactions can be estimated by measurement of the pressure at the shock tube end wall. There are certain limits beyond which the actual perturbations in the properties deviate from the analytical relations. In the next section, a possible reason is advanced for the breakdown of the analytical expressions.

Acceleration Effect of Reaction Front

In the framework of the analysis for coherent chemical processes in mixture heated by reflected-shock waves, the reaction fronts propagate at the same speeds as the reflected-shock waves. Figure 7 shows the x-t diagram of the reflected-shock region for case III, for which the reaction front is observed to accelerate considerably. To consider the effect of the acceleration, the quantity z, which represents reduction of the induction time and is assumed to be small, is introduced into the release rate of chemical energy as follows:

$$q(x',t') = \frac{\partial}{\partial t} Q[t' - \frac{x'}{u_R'} + z(x',t')] \tag{23a}$$

$$q(x',t') \cong q_o(t' - \frac{x'}{u_R'})(1+z_t) + z(x',t')\frac{d}{dt} q_o(t' - \frac{x'}{u_R'}) \tag{23b}$$

Substitution of Eq. (23b) instead of Eq. (8) into the expressions for the perturbations of pressure and density at x = 0, such as

$$p_w'(t') = \int_{t'/(1+u_R')}^{t'} q(t'-t,t)dt \tag{24}$$

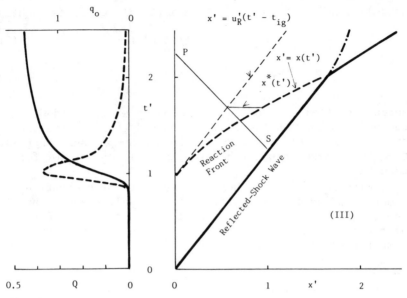

Fig. 7 Space time diagram for the reflected-shock region and temporal heat request, Q, and temporal release-rate of chemical energy, q_0 (Case III).

$$\rho'_w(t') = p'_w(t') - \int_0^{t'} q(0,t)dt \qquad (25)$$

which, respectively, refer to integrations of $p' - u'$ and $p' - \rho'$ along characteristic lines PS and PO in Fig. 7 (Takano and Akamatsu, 1985), yields

$$p'_w(t') = \frac{u'_R}{1+u'_R} Q(t') + I(t') \qquad (26)$$

$$\rho'_w(t') = - \frac{1}{1+u'_R} Q(t') + I(t') \qquad (27)$$

where:

$$I(t') = \frac{u'_R}{(1+u'_R)^2} \int_0^{t'} q_0(s)(\frac{\partial}{\partial t} + u'_R \frac{\partial}{\partial x}) z[\frac{u'_R}{1+u'_R}(t'-s), \frac{u'_R s+t'}{1+u'_R}]ds \qquad (28)$$

As shown in Fig. 7, q_0 has a sharp peak at $t' = t'_{ig} (= 1)$.

The approximate expression for $I(t')$ is

$$I(t') = \frac{u_R'}{(1+u_R')^2} Q(t')(\frac{\partial}{\partial t} + u_R'\frac{\partial}{R\partial x})z[\frac{u_R'}{1+u_R'}(t'-t_{ig}'), \frac{u_R't_{ig}'+t'}{1+u_R'}] \quad (29)$$

The values of $z(x',t')$ along $x'= u_R't' - t_{ig}'$, necessary to evaluate Eq. (29), are related to the trajectory of the reaction front. As shown in Eq. (23), the function $q(x',t')$ has peaks along a trajectory

$$t' - \frac{x'}{u_R'} + z(x',t') = t_{ig}' \quad (30)$$

The trajectory of the reaction front is denoted $x'= \bar{x}(t')$, which satisfies Eq. (30) and leads to

$$\frac{d\bar{x}}{dt} = u_R' \frac{1+z_t[\bar{x}(t'),t']}{1-u_R'z_x[\bar{x}(t'),t']} \quad (31)$$

For $|z| \ll 1$, it can be approximately expressed as

$$\frac{d\bar{x}}{dt} = u_R'(1 + z_t + u_R'z_x)|_{x=u_R'(t'-t_{ig}')} \quad (32)$$

where z_x and z_t denote partial derivatives of z with respect to x and t, respectively. The distance x^* by which the reaction front accelerates, as shown in Fig. 7, is defined as

$$x^*(t') = \bar{x}(t') - u_R'(t'- t_{ig}') \quad (33)$$

Then, Eq. (32) reduces to

$$\frac{dx^*}{dt} = u_R'(\frac{\partial}{\partial t} + u_R'\frac{\partial}{R\partial x})z[u_R'(t'- t_{ig}'),t'] \quad (34)$$

Consequently, Eq. (29) can be written as

$$I(t') = \frac{1}{(1+u_R')^2} Q(t')\frac{dx^*(t')}{dt} \quad (35)$$

resulting in the following expressions for Eqs. (26) and (27):

$$p_w'(t') = \frac{u_R'}{1+u_R'} Q(t')(1 + \frac{1/u_R'}{1+u_R'} \frac{dx^*}{dt}) \qquad (36)$$

$$\rho_w'(t') = - \frac{1}{1+u_R'} Q(t')(1 - \frac{1}{1+u_R'} \frac{dx^*}{dt}) \qquad (37)$$

The acceleration distance x^* is obtained from the x-t diagram in Fig. 7 numerically for case III and Eqs. (36) and (37) are shown as dotted lines in Fig. 5. Deviations of results of the simulations from the analytical relations can be explained by considering the acceleration effect of the reaction front.

Conclusions

Chemical effects on reflected-shock flowfields in combustible gas are considered by comparing results of shock tube experiments, detailed simulations and perturbation analyses for a combustible $H_2/O_2/Ar$ mixture. To investigate the one-dimensional features of the reactive gasdynamics, conditions for reflected-shock waves are chosen for the strong ignition mode and hydrogen and oxygen sufficiently diluted in argon gas is used as test gas.

Results of the simulations, performed by applying a detailed model that combines the random choice method with a chemical kinetics simulation code, are compared to results of pressure measurements for reflected-shock waves in a shock tube. The comparisons are excellent between the experiments and the simulations for a highly diluted combustible gas. Also, for moderately diluted gases, the simulations are shown to successfully predict averaged flowfields over which some disturbing waves are observed to develop.

The results of the simulations are compared to predictions of the analytical relations for perturbations in reactive reflected-shock flowfields (Takano and Akamatsu, 1985). Comparisons between the analysis and the simulations are made with respect to the end wall relations. The analytical relations are shown to hold satisfactorily for highly diluted combustible gas. For considerably diluted cases, the results of the simulations are observed to deviate from the analytical relations. Discrepancies between the analysis and the simulation are estimated by considering the acceleration of the reaction front.

References

Book, D. L., Boris, J. P., and Hain, K. (1975) Flux-corrected transport II: Generalizations of the method. J. Comput. Phys. 18, 248-283.

Chorin, A. J. (1976) Random choice solution of hyperbolic systems. J. Comput. Phys. 22, 517-533.

Cohen, L. M., Short, J. M., and Oppenheim, A. K. (1975) A computational technique for the evaluation of dynamic effects of exothermic reactions. Combust. Flame 24, 319-334.

Fickett, W. (1984) Shock initiation of detonation in a dilute explosive. Phys. Fluids 27, 94-105.

Gilbert, R. B. and Strehlow, R. A. (1966) Theory of detonation initiation behind reflected shock waves. AIAA J. 4, 1777-1783.

Jensen, D. E. and Jones, G. A. (1978) Reaction rate coefficients for flame calculations. Combust. Flame 32, 1-34.

Meyer, J. M. and Oppenheim, A. K. (1971) On the shock-induced ignition of explosive gases. Thirteenth Symposium (International) on Combustion, pp. 1153-1164. The Combustion Institute, Pittsburgh, PA.

Oran, E. S., Young, T. R., and Boris, J. P. (1978) Application of time-dependent numerical methods to the description of reactive shocks. Seventeenth Symposium (International) on Combustion, pp. 43-54. The Combustion Institute, Pittsburgh, PA.

Oran, E. S., Young, T. R., and Boris, J. P. (1982) Weak and strong ignition. I. Numerical simulations of shock tube experiments. Combust. Flame 48, 135-148.

Sod, G. A. (1978) A survey of several finite difference methods for systems of nonlinear hyperbolic law. J. Comput. Phys. 27, 1-13.

Spence, D. A. (1961) Unsteady shock propagation in a relaxing gas. Proc. R. Soc. London 262, 221-234.

Strehlow, R. A. and Cohen, A. (1962) Initiation of detonation. Phys. Fluids 5, 97-101.

Takano, Y. and Akamatsu, T. (1984) A diaphragmless shock tube. J. Phys. E 17, 644-646.

Takano, Y. and Akamatsu, T. (1985) Analysis of chemical effects on reflected-shock flowfields in combustible gas. J. Fluid Mech. 160, 29-45.

Takano, Y., Kittaka, S., and Murata, Y. (1985) Chemical effects on reflected-shock flowfields in combustible gas, 1st report: Numerical calculations for reactive gas by use of random choice method (in Japanese). Trans. JSME 51-469B, 2956-2963.

Voevodsky, V. V. and Soloukhin, R. I. (1965) On the mechanism and explosion limits of hydrogen-oxygen chain self-ignition in shock waves. Tenth Symposium (International) on Combustion, pp. 279-283. The Combustion Institute, Pittsburgh, PA.

Young, T. R. and Boris, J. P. (1977) A numerical technique for solving stiff ordinary differential equations associated with the chemical kinetics of reactive-flow problems. J. Phys. Chem. 81, 2424-2427.

Systematic Development of Reduced Reaction Mechanisms for Dynamic Modeling

M. Frenklach*
Pennsylvania State University, University Park, Pennsylvania
and
K. Kailasanath† and E.S. Orant‡
Naval Research Laboratory, Washington, D.C.

Abstract

A method for systematically developing a reduced chemical reaction mechanism for dynamic modeling of chemically reactive flows is presented. The method is based on the postulate that if a reduced reaction mechanism faithfully describes the time evolution of both thermal and chain reaction processes characteristic of a more complete mechanism, then the reduced mechanism will describe the chemical processes in a chemically reacting flow with approximately the same degree of accuracy. Here this postulate is tested by producing a series of mechanisms of reduced accuracy, which are derived from a full detailed mechanism for methane-oxygen combustion. These mechanisms were then tested in a series of reactive flow calculations in which a large-amplitude sinusoidal perturbation is applied to a system that is initially quiescent and whose temperature is high enough to start ignition processes. Comparison of the results for systems with and without convective flow show that this approach produces reduced mechanisms that are useful for calculations of explosions and detonations. Extensions and applicability to flames are discussed.

Introduction

Computer modeling is a powerful tool for studying combustion. Prediction of explosive properties, combustion-generated polution, and automotive engine knock are just a few examples where computer modeling contributes to the

---
Presented at the 10th ICDERS, Berkeley, California, August 4-9, 1985. This paper is a work of the U.S. Government and therefore is in the public domain.
*Associate Professor, Fuel Science Program, Department of Materials Science and Engineering.
†Research Scientist, Laboratory for Computational Physics.
‡Senior Scientist, Laboratory for Computational Physics.

solution of current problems of concern to our society.
The most successful predictive models for combustion usually require a fairly detailed description of the important physical and chemical processes. Significant progress
has been achieved in this direction as computers have become faster, more computer memory has become available,
and numerical algorithms have become progressively more
sophisticated.

Even with the major advances in hardware and software,
there are significant difficulties in producing the type
of detailed models, i.e., those which combine the detailed
chemical kinetics and multidimensional fluid dynamics,
that we would like to use. One of these difficulties,
and perhaps the one that is most limiting, is the numerical stiffness problem associated with the integration of
the ordinary differential equations needed to describe the
chemical kinetics.

This problem was first defined as such in studies of
chemical kinetics and since then has become an active field
of research. The stiffness property of the equations puts
rather severe restrictions on the methods we can use and
the timestep we can take to obtain answers of the required
accuracy. The practical result is that the computations
are slowed down, and therefore, the overall practicality
of using the model depends primarily on the size of the
chemical reaction mechanism. We wish to consider mechanisms for successively more complicated and larger molecules
with a larger number of elementary reactions[1-3] coupled
to multidimensional fluid dynamics[4,5]. This presents a
dilemma, from both practical and philosophical points of
view.

Thus, until we have the essentially unlimited computer
time and memory that would be required to solve a detailed
multidimensional reactive flow model, we can consider a
number of possible strategies. We could simply wait until
a computer with a sufficient speed and memory for a given
problem is developed; or until we have a computer dedi-
cated entirely to our problem. We could use the approach
of global chemical modeling, in which the major features
of the chemical reactions (induction times, rates of heat
release, final temperatures, etc.) are fit to some types
of simplified equations.

This approach, which emphasizes the fluid dynamics,
has been proposed by a number of authors[6-9]. It has been
applied to multidimensional simulations of detonations[6,10]

and produces reasonably quantitative results. However,
not including the details of the chemistry, and thus the
changing specific features of the chemical kinetic-fluid
dyamics interactions, precludes modeling many processes.
Another approach is to model the problem using a fluid
dynamics model with reduced dimensionality from the real
problem but still keeping the full chemical kinetics mechanism. To date, the best we can do on this score is a
one-dimensional model with almost any currently proposed
mechanism, or a two-dimensional model with a full hydrogen
combustion mechanism. These require the largest, fastest
computers available.

The approach we are testing in this paper is to develop a reduced chemical mechanism, that is, a set of minimum number of elementary chemical reactions that are sufficient to describe selected properties of the process
with a prescribed accuracy[11]. A method for developing and
testing such reduced mechanisms in a systematic manner is
presented in this paper.

Statement of the Problem

Given a large reaction mechanism, which presumably
contains all necessary elementary chemical reactions, we
wish to know what reaction subset is sufficient to describe the dynamic behavior of a reactive flow system with
a given accuracy. The method is based on the postulate
that a reaction mechanism faithfully describing the dynamics of both thermal and chain reaction processes when
there is no convective transport present will describe the
chemical processes to the same degree of accuracy in reactive flows such as flames and detonations. Furthermore,
the accuracy of the reactive flow modeling with this reduced mechanism will be established a priori, i.e., in the
model with no convective transport.

The approach is introduced and tested by producing a
series of reduced chemical mechanisms of accuracy determined by how well they reproduce certain specified features of the time-dependent properties of the chemical
system. The reduced mechanisms are derived by eliminating reactions from the full mechanism, not by changing any
chemical rates or thermophysical properties. Some of the
chemical species, however, are eliminated automatically
when they do not appear in any of the chemical reactions
in the reduced mechanism. These reduced mechanisms are
then tested in a series of reactive flow calculations in

which a large-amplitude sinusoidal perturbation is applied
to a system that is initially quiescent and whose temperature is high enough to start ignition processes.

The particular case used here to test the hypothesis
is ignition in a 9.5% methane - 19% oxygen - 71.5% argon
mixture. The computations were performed with a reaction
mechanism developed recently by Frenklach and Bornside[12].
Table 1 summarizes the chemical reactions in the mechanism. The rate coefficients and equilibrium constants
used can be found in Frenklach and Bornside.

Chemical Mechanism Reduction

The criteria imposed for reducing the number of chemical reactions are based on requiring accurate predictions
of 1) ignition delay times and 2) temperature profiles.
These criteria are based on the following considerations.
The dynamic development is determined primarily by the
rate of chain reactions during the preignition period and,
at later times, by the rate of energy release. The ignition delay time is a measure of the dynamics of chain
reactions, while the temperature profile is the result of
heat release. Meeting both of these criteria should assure the faithful description of the entire process. Furthermore, exactly these considerations test the main postulate of the paper: If a reaction mechanism is capable
of describing both kinetic and thermal aspects of the process, it should be able to perform with the same degree of
faithfulness under various gasdynamic conditions.

Following the first criterion, the reaction rates (or
pR values[11]) are screened against the rate of the rate-
determining step. The rate-determining reaction for the
preignition oxidation of methane is R8:

$$CH_3 + O_2 \rightarrow CH_3O + O$$

for the first part of the induction period, and R17:

$$CH_3 + HO_2 \rightarrow CH_3O + OH$$

for the remainder of the the induction period[12]. Thus,
the absolute values of the reaction rates are compared to
those of reactions R8 or R17, whichever is faster, multi-

Table 1   Methane reaction mechanism

Number	Reaction	Index[a]
	CH_x reactions	
1	$CH_4 + M = CH_3 + H + M$	10
2	$CH_4 + O_2 = CH_3 + HO_2$	10
3	$CH_4 + H = CH_3 + H_2$	10
4	$CH_4 + O = CH_3 + OH$	10
5	$CH_4 + OH = CH_3 + H_2O$	10
6	$CH_4 + HO_2 = CH_3 + H_2 + O_2$	10
7	$CH_3 + M = CH_2 + H + M$	
8	$CH_3 + O_2 = CH_3O + O$	10
9	$CH_3 + O_2 = CH_2O + OH$	10
10	$CH_3 + O_2 = CH_2 + HO_2$	
11	$CH_3 + H = CH_2 + H_2$	5
12	$CH_3 + O = CH_2O + H$	10
13	$CH_3 + O = CH_2 + OH$	5
14	$CH_3 + OH = CH_3O + H$	10
15	$CH_3 + OH = CH_2O + H_2$	10
16	$CH_3 + OH = CH_2 + H_2O$	10
17	$CH_3 + HO_2 = CH_3O + OH$	10
18	$CH_3 + CH_2O = CH_4 + CHO$	10
19	$CH_3 + CHO = CH_4 + CO$	1
20	$CH_2 + O_2 = CH_2O + O$	10
21	$CH_2 + O_2 = CO_2 + H + H$	10
22	$CH_2 + H = CH + H_2$	
23	$CH_2 + O = CH + OH$	
24	$CH_2 + OH = CH + H_2O$	
25	$CH + O_2 = CO + OH$	1
26	$CH + O_2 = CHO + O$	10
	CH_xO reactions	
27	$CH_3OH + M = CH_3 + OH + M$	5
28	$CH_3O + M = CH_2O + H + M$	10
29	$CH_3O + O_2 = CH_2O + HO_2$	
30	$CH_2O + M = CHO + H + M$	
31	$CH_2O + M = CO + H_2 + M$	1
32	$CH_2O + O_2 = CHO + HO_2$	
33	$CH_2O + H = CHO + H_2$	10
34	$CH_2O + O = CHO + OH$	10
35	$CH_2O + O = CO_2 + H + H$	10
36	$CH_2O + OH = CHO + H_2O$	10
37	$CH_2O + HO_2 = CHO + H_2O_2$	1
38	$CHO + M = CO + H + M$	10
39	$CHO + O_2 = CO + HO_2$	10
40	$CHO + H = CO + H_2$	1
41	$CHO + O = CO + OH$	1
42	$CHO + OH = CO + H_2O$	10
	CH_2H_x reactions	
43	$CH_4 + CH_3 = C_2H_6 + H$	
44	$CH_4 + CH_3 = C_2H_5 + H_2$	1
45	$CH_4 + CH_2 = CH_3 + CH_3$	10

(Table continued on next page)

Table 1 (cont.) Methane reaction mechanism

Number	Reaction	Index[a]
	CH_2H_x reactions	
46	$C_2H_6 = CH_3 + CH_3$	10
47	$CH_3 + CH_3 = C_2H_5 + H$	10
48	$CH_3 + CH_3 = C_2H_4 + H_2$	10
49	$CH_3 + CH_2 = C_2H_4 + H$	
50	$CH_2 + CH_2 = C_2H_2 + H_2$	
51	$C_2H_6 + CH_3 = C_2H_5 + CH_4$	10
52	$C_2H_6 + H = C_2H_5 + H_2$	10
53	$C_2H_6 + O = C_2H_5 + OH$	10
54	$C_2H_6 + OH = C_2H_5 + H_2O$	10
55	$C_2H_5 = C_2H_4 + H$	10
56	$C_2H_5 + C_2H_3 = C_2H_4 + C_2H_4$	
57	$C_2H_5 + O_2 = C_2H_4 + HO_2$	1
58	$C_2H_5 + H = C_2H_4 + H_2$	
59	$C_2H_4 + M = C_2H_3 + H + M$	
60	$C_2H_4 + M = C_2H_2 + H_2 + M$	1
61	$C_2H_4 + CH_3 = C_2H_3 + CH_4$	1
62	$C_2H_4 + H = C_2H_3 + H_2$	10
63	$C_2H_4 + O = CH_3 + CHO$	10
64	$C_2H_4 + O = CH_2 + CH_2O$	5
65	$C_2H_4 + OH = C_2H_3 + H_2O$	10
66	$C_2H_4 + OH = CH_3 + CH_2O$	10
67	$C_2H_3 + M = C_2H_2 + H + M$	10
68	$C_2H_3 + H = C_2H_2 + H_2$	1
69	$C_2H_2 + M = C_2H + H + M$	
70	$C_2H_2 + O_2 = CHO + CHO$	5
71	$C_2H + H_2 = C_2H_2 + H$	5
72	$C_2H_2 + O = CH_2 + CO$	10
73	$C_2H_2 + O = C_2H + OH$	
74	$C_2H_2 + O = C_2HO + H$	10
75	$C_2HO + H = CH_2 + CO$	5
76	$C_2HO + O = CHO + CO$	1
77	$C_2HO + OH = CHO + CHO$	1
78	$C_2H_2 + OH = CH_3 + CO$	1
79	$C_2H_2 + OH = C_2H + H_2O$	1
80	$C_2H_2 + C_2H_2 = C_4H_3 + H$	
81	$C_2H_2 + C_2H = C_4H_2 + H$	
82	$C_4H_3 + M = C_4H_2 + H + M$	
83	$C_4H_2 + M = C_4H + H + M$	
84	$C_2H + O_2 = CHO + CO$	10
85	$C_2H + O = CH + CO$	
	H_2-CO-O_2 reactions	
86	$CO + O_2 = CO_2 + O$	
87	$CO + O + M = CO_2 + M$	1
88	$CO + OH = CO_2 + H$	10
89	$CO + HO_2 = CO_2 + OH$	10
90	$O_2 + M = O + O + M$	
91	$O_2 + H = OH + O$	10

(Table continued on next page)

Table 1 (cont.)   Methane reaction mechanism

Number	Reaction	Index[a]
H_2-CO-O_2 reactions		
92	$O_2 + H + M = HO_2 + M$	10
93	$H_2 + M = H + H + M$	1
94	$H_2 + O_2 = OH + OH$	1
95	$H_2 + O = OH + H$	10
96	$H_2 + HO_2 = H_2O + OH$	
97	$H_2O + H = H_2 + OH$	10
98	$H_2O + O = OH + OH$	5
99	$H_2 + O_2 + M = OH + OH + M$	10
100	$H_2O_2 + O_2 = HO_2 + HO_2$	5
101	$H_2O_2 + H = H_2 + HO_2$	
102	$H_2O_2 + OH = H_2O + HO_2$	
103	$HO_2 + H = OH + OH$	10
104	$HO_2 + H = O_2 + H_2$	10
105	$HO_2 + O = O_2 + OH$	5
106	$HO_2 + OH = O_2 + H_2O$	10
107	$H + O + M = OH + M$	10
108	$H + OH + M = H_2O + M$	10

[a] The value of the index indicates the following: 1 = the reaction required for 1% reduced model; 5 = the reaction for 1% and 5% models; 10 = the reaction of 1%, 5% and 10% models; blank = the reaction is not required for any of the cases.

plied by a given factor, ϵ_R,

$$| R_i | < \epsilon_R \max(R_8, R_{17}) \qquad (1)$$

where R_i is the rate of reaction i.

Following the second criterion, the rates of energy accumulation or disappearance are screened according to the following condition:

$$| R_i Q_i | < \epsilon_Q \max(| R_j Q_j |, j = 1, 2, \ldots, 108) \qquad (2)$$

where Q_i is the heat of reaction i and ϵ_Q is a given factor.

To remove a reaction from the initial 108-reaction set, inequality (1) must be satisfied over the entire induction period, and inequality (2) must be satisfied before, during, and after ignition. The degree of accuracy of the reduced reaction mechanism is defined by the values assigned to ϵ_R and ϵ_Q. In the present work, we assumed $\epsilon_R = \epsilon_Q = \epsilon$ and considered three cases (i.e., three degrees of accuracy): ϵ = 0.01, 0.05, 0.10. For conven-

ience, these cases will be referred to in the following
text as the 1%, 5% and 10% reduced models, respectively.
The reactions required to be removed in order to obtain
each reduced model are identified in Table 1.

Numerical Model

The reactive flow calculations described below were
performed using a model that solves the time-dependent,
multispecies conservation equations for total density,
momentum, energy, and individual species densities. The
particular version used is one dimensional, with optional
Cartesian, cylindrical, spherical, or nozzle coordinates.
The solutions of the convective terms were done using a
variant of the flux-corrected transport algorithm (FCT), a
conservative monotonic, compressible algorithm with fourth-
order accuracy[13,14]. The ordinary differential equations
describing the chemical kinetics are solved using VSAIM, a
vectorized version of the CHEMEQ algorithm[15,16]. The convective and chemical calculations are coupled by timestep
splitting[4]. The individual algorithms, as well as the
results of coupling them, have been tested extensively
against both theoretical predictions and experimental observations[17-19].

Results

Consider a background, ambient mixture of 9.5% CH_4-19%
O_2-71.5% Ar at an initial temperature of 1500 K and total
pressure of 2.46 atm. Left alone, this mixture ignites
in approximately 530 μs and reaches a final temperature
of approximately 3200 K, as predicted by the full set of
chemical reactions given in Table 1.

First consider the 1%, 5%, and 10% reduced mechanisms
produced using the criteria described above. Results from
such constant volume calculations are compared to those
obtained using the full mechanism in the curves marked
"unperturbed" in Fig. 1. With decreased accuracy, the calculated ignition times are increased. The temperatures
reach the same final temperature.

Now consider such an initial ambient mixture in which
there is an imposed sinusoidal velocity perturbation with
a large amplitude, 100 m/s, and with a wavelength of 38 cm.
To study this system, the numerical model used a computational grid with 50 equally spaced cells covering 19 cm.
This is similar in spirit to the perturbation imposed in
the work on weak and strong ignition in hydrogen-oxygen

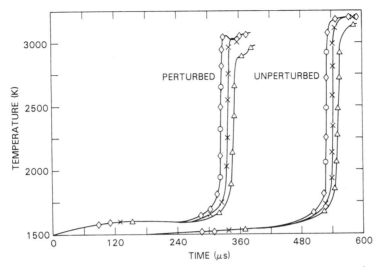

Fig. 1 Temperature vs time for the unperturbed and perturbed (sinusoidal velocity perturbations) calculations for four mechanisms: ○ full mechanism; reduced mechanisms: ◇ 1%, × 5%, and △ 10%.

mixtures[17,18]. For this set of parameters, the period of the oscillation is close in value to the unperturbed chemical induction time of the mixture. The curves marked "perturbed" are plotted at the first location of ignition in the mixture for the four sets of reaction mechanisms used in the "unperturbed" case. The presence of the perturbation has clearly reduced the ignition time, as is expected[18].

In this case, the ignition time is reduced substantially because the perturbation is large and close to resonance with the induction time. More important to our immediate concerns, however, is that the reduced mechanisms mimic the trends in the perturbed cases as in the unperturbed cases: The change in ignition delay times and in the rate of approach to the final temperature are approximately identical for the same reduced mechanisms. It is also interesting to note that the 1% reduced mechanism performs as well as the full 108-reaction mechanism. The deviations obtained with the 5% and 10% reduced mechanisms are of the order of the increase in the value of ϵ, i.e., of the order of the reduction in accuracy of the mechanism.

Our results support the conclusion that the accuracy
of a reduced model, developed by the described procedures,
can be set a priori. More testing, however, is necessary
to state this conclusion with confidence. Also, the reduction with ϵ_R not equal to ϵ_Q should be analyzed. The
separate assignment of ϵ_R and ϵ_Q values would provide a
more flexible means for developing reduced mechanisms ded-
icated to a specific problem. Additional criteria can
also be added, if necessary. For example, we might want
to apply inequalities similar to (1) and (2) for the concentration of certain species (e.g., hydrogen atom) in
flame modeling.

Summary and Conclusions

A method for systematically developing a reduced chemical reaction mechanism for dynamic modeling has been presented. The method is based on the postulate that a reaction mechanism faithfully describing the dynamics of both
thermal and chain reaction processes when there is no convective transport present will describe the chemical processes to the same degree of accuracy in reactive flows
such as flames and detonations. The technique and ideas
were tested here on the example of a large-amplitude sinusoidal perturbation on a reacting methane-oxygen-argon
system. This example is a fairly severe test of chemical
kinetics - fluid dynamic coupling. Thus we feel very con-
fident that this approach will be valuable for detonation
and explosion calculations.

The next test is to include diffusive transport effects and thus test how well the hypothesis works for modeling flames. Here we have to consider the addition of
thermal conductivity, molecular diffusion, and thermal
diffusion. In the reactive flow model, the sensitivity
of the results to these diffusion processes is determined
by the individual and mixture diffusion coefficients. In
general, the overall sensitivity of a mixture to these coefficients depends on the amounts and types of 1) species
initially present, 2) very light materials, e.g., H atom,
produced in the course of the reactions, and 3) the final
products. If, in addition to the dynamics of thermal and
chain reaction processes, the concentration of required
species are adequately predicted by the reduced mechanism,
then this reduced mechanism will produce accurate results
for flame modeling. Our preliminary results from modeling
a steady-state premixed stoichiometric methane-air flame

indicate that the criteria 1 and 2 may be sufficient for flame dynamics. This is not surprising, because the chemistry of flame and shock-tube ignition phenomena are both controlled by the same reactions: the branching reaction of hydrogen atom with molecular oxygen[2,12].

Acknowledgments

The work performed at Louisiana State University was sponsored by NASA-Lewis Research Center, Grant NAG 3-477, and the work performed at the Naval Research Laboratory was supported by the Office of Naval Research and the Naval Research Laboratory.

References

[1]Frenklach, M., Clary, D.W., Gardiner, W.C. Jr., and Stein, S.E. (1985) Detailed kinetic modeling of soot formation in shock-tube pyrolysis of acetylene. 20th Symposium (International) on Combustion. The Combustion Institute, Pittsburgh, Pa. (in press).

[2]Warnatz, J. (1984) Rate coefficients in the C/H/O system. Combustion Chemistry (edited by W.C. Gardiner Jr.), Chap. 5. Springer-Verlag, New York.

[3]Westbrook, C.K. and Dryer, F.L. (1984) Chemical kinetic modeling of hydrocarbon combustion. Prog. Energy Combust. Sci. 10, 1-57.

[4]Oran, E.S. and Boris, J.P. (1981) Detailed modelling of combustion processes. Prog. Energy Combust. Sci. 7, 1-71.

[5]Dixon-Lewis, G. (1984) Computer modeling of combustion reactions in flowing systems with transport. Combustion Chemistry (edited by W.C. Gardiner Jr.) Chap. 2. Springer-Verlag, New York.

[6]Oran, E.S., Boris, J.P., Young, T.R., Flanigan, M., Burks, T., and Picone, M. (1981) Numerical simulations of detonations in hydrogen-air and methane-air mixtures. 18th Symposium (International) on Combustion, pp. 1641-1649. The Combustion Institute, Pittsburgh, Pa.

[7]Coffee, T.P., Kotlar, A.J., and Miller, M.S. (1983) The overall reaction concept in premixed, laminar, steady-state flames. I. Stoichiometries. Combust. Flame 54, 155-169.

[8]Coffee, T.P., Kotlar, A.J., and Miller, M.S. (1984) The overall reaction concept in premixed, laminar, steady-state flames. II. Initial temperatures and pressures. Combust. Flame 58, 59-67.

[9]Hautman, D.J., Dryer, F.L., Schug, K.P., and Glassman, I. (1981) A multiple-step overall kinetic mechanism for the oxidation of hydrocarbons. Combust. Sci. Technol. 25, 219.

[10]Taki, S. and Fujiwara, T. (1981) Numerical simulation of triple shock behavior of gaseous detonation. 18th Symposium (International) on Combustion, pp. 1671-1681. The Combustion Institute, Pittsburgh, Pa.

[11]Frenklach, M. (1984) Modeling. Combustion Chemistry (edited by W.C. Gardiner Jr.), Chap. 7. Springer-Verlag, New York.

[12]Frenklach, M. and Bornside, D.E. (1984) Shock-initiated ignition in methane-propane mixtures. Combust. Flame 56, 1-27.

[13]Boris, J.P. and Book, D.L. (1976) Solution of continuity equations by the method of flux-corrected transport. Methods in Computational Physics, Vol. 16, pp. 85-129. Academic Press, New York.

[14]Boris, J.P. (1976) Flux-corrected transport modules for solving generalized continuity equations. NRL Memo. Report 3237, Naval Research Laboratory, Washington, D.C., 20375.

[15]Young, T.R. and Boris, J.P. (1977) A numerical technique for solving stiff ordinary differential equations associated with chemical kinetics of reactive-flows problems. J. Phys. Chem. 81, 2424-2427.

[16]Young, T.R. (1980) CHEMEQ--A Subroutine for solving stiff ordinary differential equations. NRL Memo. Report 4091, Naval Research Laboratory, Washington, D.C.

[17]Oran, E.S., Young, T.R., Boris, J.P., and Cohen, A. (1982) Weak and strong ignition. I. Numerical simulations of shock tube experiment. Combust. Flame 48, 135-148.

[18]Oran, E.S. and Boris, J.P. (1982) Weak and strong ignition. II. Sensitivity of the hydrogen-oxygen system. Combust. Flame 48, 149-161.

[19]Kailasanath, K. and Oran, E.S. (1983) Ignition of flamelets behind incident shock waves and the transition to detonation. Combust. Sci. Technol. 34, 345.

Kinetic Study of Ethylene Oxidation
in a Jet-Stirred Reactor

P. Dagaut,* M. Cathonnet,† F. Gaillard,† J.C. Boettner,† J.P. Rouan‡
and
H. James§
*Centre de Recherches sur la Chimie de la Combustion
et des Hautes Températures, CNRS, Orleans, France*

Abstract

A jet-stirred reactor designed for the study of oxidation of hydrocarbons in the intermediate temperature range up to 10 atm is described. The advantages of the technique are discussed and the results of an experimental study of ethylene oxidation are presented. The analysis of the reaction evolution from low to high fuel conversion has been performed by successive steady states, the mixing of reactants allowing uniformity of temperature and concentration in the whole volume. A detailed chemical kinetic reaction mechanism is used to model the experimental data. To illustrate the difference between results obtained with the jet-stirred reactor method and with the plug flow reactor method, two simulations are performed and compared for the same ethylene/oxygen mixture in both reactors under identical conditions. In the stirred reactor, the induction period is suppressed, but the concentration profiles are very similar to those observed in the plug flow reactor up to approximately 50% conversion. However, the evolution of the last stages of the reaction is slower than in the plug flow reactor.

Introduction

The use of detailed mechanisms to model the oxidation of several fuels, typically from hydrogen to butane, is

---

Presented at the 10th ICDERS, Berkeley, California, August 4-9, 1985. Copyright © 1985 by the American Institute of Aeronautics and Astronautics, Inc. All rights reserved.

*Boursier M.R.T.
†Chargés de Recherches.
‡Ingénieur.
§Directeur de Recherches.

377

now possible because of the availability of kinetic data
for the elementary reaction steps (Warnatz, 1984). The
validity of such mechanisms is established by comparison
of the model results with experimental data obtained for a
wide range of physical conditions in shock tubes, plug
flow reactors and flames. As the transport effects are
negligible for the first two experimental techniques, the
model equations are substantially simplified. The perfectly stirred reactor is another system in which the transport effects can be neglected. The rapid mixing achieves
spatial homogeneity inside the reactor when it is operated
at steady state. The stirring is mechanical or can be
obtained by the turbulence created by the gas jets flowing
through a number of inlets.

The mechanically stirred reactors used by Gray and
Felton (1974), Gray et al. (1981), and Caprio et al.
(1977) are operated at residence times of several seconds,
which is the characteristic time scale of the cool flame
induction period at low pressure. In the jet-stirred reactors used for the study of high temperature flames, the
order of magnitude of the residence time is about 1 ms
(Longwell and Weiss, [1955]; Hottel et al., [1965]; and
Duterque et al., [1981]). However, to stabilize a flame in
such a reactor and to prevent blowout, most of the heat
release must occur within the reactor. As a result, the
residence time must be comparable with the time scale of
the most exothermic stage of the reaction, i.e., the
oxidation of CO, which is considerably longer than the
time scale of the hydrocarbon oxidation. The consequence
is that the kinetics of the first stages of the reaction
(oxidation of the primary fuel and of the intermediate
hydrocarbons) cannot be determined in the Longwell reactor.

Matras and Villermaux (1973) and David and Matras
(1975) have designed a new jet-stirred reactor for mean
residence times of around 1 s. In this reactor, stirring
is achieved by means of four turbulent jets issuing from
the nozzles of a cross-shaped injector. The authors have
defined construction rules to obtain good macromixing in
their reactor (David and Matras, 1975). But it was shown
that preheating of the reactants and/or of a carrier gas
was necessary to operate without temperature gradient
(Come, 1983). Such reactors have been used for the study
of hydrocarbon pyrolysis (Come, 1983; Juste et al., 1981)
and low temperature oxidation (Ferrer et al., 1982).

This work reports experiments on ethylene oxidation
around 1000 K in a jet-stirred reactor similar to those
used by Come, Juste et al., and Ferrer et al., and is an

extension of previous experimental measurements on hydrocarbon oxidation in a laminar flow tubular reactor (Cathonnet et al., 1981, 1984). Because of the existence of a
radial temperature gradient at high conversion, these
experiments were restricted to a limited conversion range.
The jet-stirred reactor technique provides supplementary
data on hydrocarbon oxidation, particularly at higher
conversion in the intermediate temperature range where
most of the consumption of the fuel and intermediate
hydrocarbons takes place (Westbrook and Dryer, 1984).

Reactor design

The reactor is a fused silica sphere of 4 cm diameter
with four nozzles of 1 mm diameter for injection of the
reactants. Figure 1 shows the orientation of the injec-
tors. The four jets, issuing from the nozzles, achieve

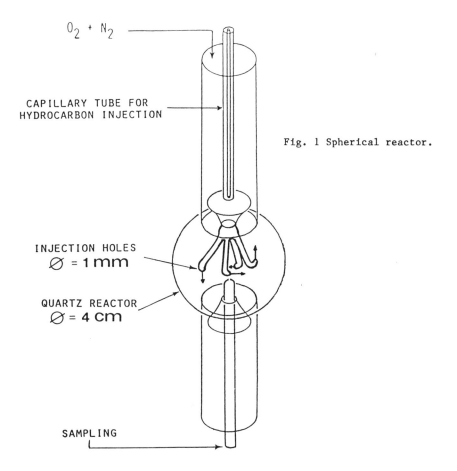

$O_2 + N_2$

CAPILLARY TUBE FOR
HYDROCARBON INJECTION

Fig. 1 Spherical reactor.

INJECTION HOLES
$\emptyset$ = 1 mm

QUARTZ REACTOR
$\emptyset$ = 4 cm

SAMPLING

turbulent mixing and recycling of the gases within the reactor. The gas flow is preheated by an electrical wire-wound resistor before entry into the reactor. The fuel, diluted in N_2, is introduced separately through a capillary and mixed with oxygen and the dilutant at the entrance of the injector (Fig. 1). As the residence time of the mixture in the injector is more than 300 times lower than the mean residence time in the reactor, pyrolysis or oxidation of the fuel inside the injector is minimal. The reactor is surrounded by an electrical heater and located in a steel vessel that can be pressurized, the pressure inside and outside the reactor being at all times equili-brated (Fig. 2).

A residence time distribution study was performed by pulsed injection of a tracer at the inlet of the reactor at room temperature. In the pressure range of this study (1-10 atm), the reactor is perfectly stirred for mean residence times varying from few milliseconds to several seconds. The reactor homogeneity during a reaction is

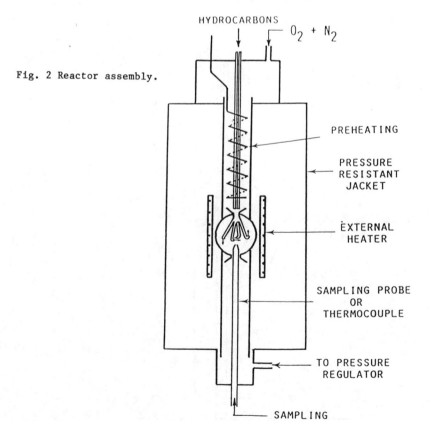

Fig. 2 Reactor assembly.

verified by measurement of the temperature of the gases
with a chromel-alumel thermocouple that can be moved along
a whole diameter. The chemical composition of the reacting
gases is determined from samples taken with a quartz probe
similar to those previously used in the tubular reactor
(Cathonnet et al., 1981,1984).

Experimental results

The oxidation of ethylene was studied in the temperature range of 900-1200 K at a pressure of 1-10 atm, the
mean residence time being varied between 0.02 and 3 s.

Before each experiment, after the nitrogen and oxygen
streams were flowing, the power inputs of the reactor
heater and of the preheating of the gases were adjusted to
obtain a constant temperature in the whole reactor (less
than 2 K deviation). These adjustments were made before
injection of hydrocarbons for each given mean residence
time. The nitrogen + oxygen flow rate represented more
than 99.5% of the total flow rate nitrogen + oxygen +
hydrocarbon. When the hydrocarbon entered the vessel
(0.15-0.50 % by volume), a temperature rise was observed.

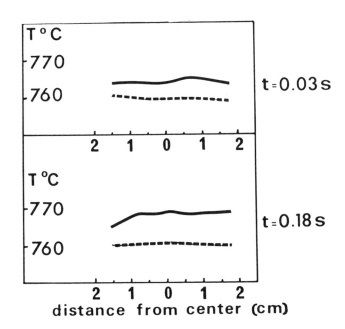

Fig. 3 Temperature profiles in the reactor at two mean residence
times (P = 1 atm): - - - - - without reaction:
——————— in the presence of 0.15% C_2H_4 (Φ = 0.4).

Because of the high dilution of the reactant, the temperature variation was quite small (< 20 K) and the temperature gradient in the reactor never exceeded 5 K (Fig. 3).

The evolution of a given ethylene/oxygen mixture was followed by variation of the mean residence time of the gases in the reactor at constant pressure and initial temperature. The composition of the gas mixture was determined from a chemical sample that was aspirated at low pressure through a small sonic quartz probe. The samples were analyzed by a multicolumn (Porapak R, Porapak Q, molecular sieve) multidetector (thermal conductivity, flame ionization) gas chromatograph (Hewlett Packard 5840) with column switching and temperature programming and

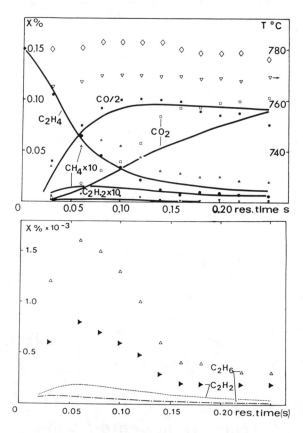

Fig. 4 Evolution of a $C_2H_4/O_2/N_2$ mixture (0.15% C_2H_4) as a function of the mean residence time: comparison between experimental data (points) and computed profiles (lines) (∇ T, $\Diamond$ $\Sigma C/2$). Φ = 0.4, 1atm.

helium as the carrier gas. For a better sensitivity, the
hydrogen content of the gases was determined on another
gas chromatograph equipped with molecular sieve column and
Carle microthermistances with nitrogen as carrier gas. The
main detected products were CO, CO_2, CH_4, and H_2 for rich
mixtures. Small quantities of C_2H_2 and C_2H_6 were also
observed, as well as trace quantities of CH_3CHO, C_3H_6,
C_3H_8, allene, propyne, C_4H_8, and C_4H_6. In Figs. 4-7, the
fuel and product concentrations are plotted vs the residence times for lean to rich ethylene/oxygen mixtures at 1
and 5 atm. The concentration time curves are very similar
to those obtained in a tubular reactor.

   Compared with the previously reported experimental
data from the laminar flow tubular reactor by Cathonnet et

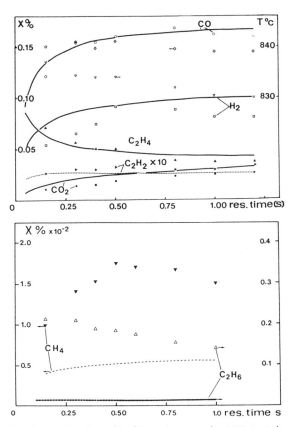

Fig. 5 Evolution of a $C_2H_4/O_2/N_2$ mixture (0.15% C_2H_4) as a func-
     tion of the mean residence time: comparison between experi-
     mental data (points) and computed profiles (lines)   (∇ T,
     $\diamond$ $\Sigma C/2$). Φ = 2.0, 5atm.

al. (1984), the obvious advantage of the present technique
is the greater conversion range available for experimental
investigation. Figures 4-7 show that the reaction can be
followed up to nearly complete fuel consumption (for lean
and stoichiometric mixtures) and to significant yields of
carbon dioxide.

The data obtained in this investigation are comparable to those reported by Westbrook et al. (1983) for a
turbulent flow reactor. An important difference is that
the jet-stirred reactor is not adiabatic, but because of
its good efficiency as a heat exchanger (David et al
1979), the temperature is uniform. As a consequence, the
temperature rise produced by the reaction in the jet-
stirred reactor is lower than that produced in an adiaba-

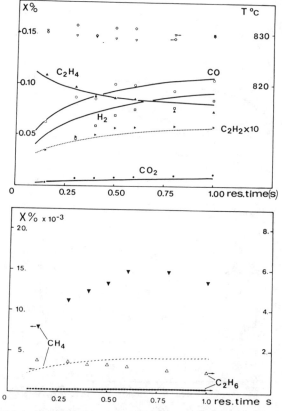

Fig. 6 Evolution of a $C_2H_4/O_2/N_2$ mixture (0.15% C_2H_4) as a function of the mean residence time: comparison between experimental data (points) and computed profiles (lines)   (∇ T,
$\Diamond$ $\Sigma C/2$). Φ = 4.0, 5atm.

tic reactor and the last step of the oxidation (conversion of CO to CO_2) is consequently more gradual.

Kinetic Modeling

At steady state, the gas concentrations within the reactor are computed from the balance between the net rate of production of each species by chemical reaction and the difference between the input and the output flow rate of the species,

$$\dot{m} (\sigma_i - \sigma_i^*)/v = R_{i+} - R_{i-}$$

where σ_i and σ_i^* are the concentrations (in $mol \cdot g^{-1}$) of the species i on output and on input, $\dot{m}$ the overall

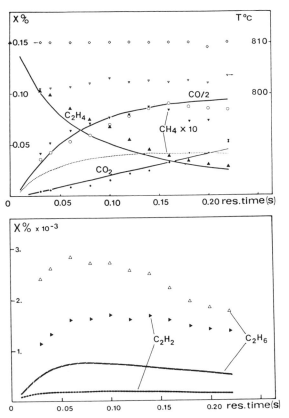

Fig. 7 Evolution of a $C_2H_4/O_2/N_2$ mixture (0.15% C_2H_4) as a function of the mean residence time: comparison between experimental data (points) and computed profiles (lines)   (∇ T, $\Diamond$ ΣC/2). Φ = 1.0, latm.

flow rate of gases (in $g \cdot s^{-1}$), and v the volume of the reactor (in ml). R_{i+} is the sum of the rates of elementary reactions forming i and R_{i-} the sum of the rates of elementary reactions consuming i. R_{i+} and R_{i-} are computed from the kinetic scheme and the rate constants of the elementary reactions at the reactor temperature. The set of equations for the i species forms a system of algebraic equations, instead of the differential ones formed in cases for the plug flow reactors or many other reactive systems (Westbrook and Dryer, 1984).

The kinetic scheme used in this study has been taken primarily from Westbrook et al. (1983, 1984). This mechanism was used previously to model experiments on ethylene oxidation in tubular flow reactors. To reduce a disagreement concerning the computed concentration level of C_2H_2, which was 20-100 times higher than the experimental one, a sensitivity study was performed and led to slight modification of the mechanism (Cathonnet et al., 1984). In this mechanism, the main reaction forming C_2H_2 was the oxidation of the vinyl radical

$$C_2H_3 + O_2 = C_2H_2 + HO_2 \qquad\qquad (a)$$

However, Baldwin and Walker (1981) consider this reaction path for formation of C_2H_2 to be of negligible importance, the main path being

$$C_2H_3 + O_2 = CH_2O + HCO \qquad\qquad (b)$$

Recently, Park et al.(1984) have confirmed the mechanism proposed by Baldwin and Walker.

In the modified mechanism in Table 1,reaction (a) has been replaced by reaction (b), with a rate constant k_b as determined by Slagle et al. (1984). Other modifications to the mechanism of Westbrook et al. (1983) are the addition of reaction 54 (Walker, 1975) and reactions 77-80 used by Miller et al. (1983) to model oxidation of C_2H_2. The rate constants of reactions 76 and 90-92 from the acetylene submechanism are also taken from Miller et al. (1983). The other rate constants modified from Westbrook et al. are those of reaction 22 (from Dixon-Lewis, 1981), reaction 34 (from Colket et al., 1977), and reaction 40 (from Walker, 1975).

The computer program evaluates the rate constants of the reverse of reactions 1-109 from the equilibrium constants computed from thermodynamical data (Bahn, 1973). The rate constants are computed at the experimentally measured temperature of the gases in the reactor. They are

Table 1 Etylene oxidation mechanism
(reaction rate constants in $cm^3.mol.s.kcal$)
$$k = A\ T^n \exp(-E/RT)$$

No	Reaction	A	n	E
1	H2 + M = H + H + M	+.2200E+15	+0.0	+96.00
2	H + HO2 = H2 + O2	+.2500E+14	+0.0	+0.70
3	H + HO2 = OH + OH	+.2500E+15	+0.0	+1.90
4	H + OH + M = H2O + M	+.1400E+24	-2.0	0.00
5	O + H2 = OH + H	+.1800E+11	+1.0	+8.90
6	HO2 + HO2 = H2O2 + O2	+.1000E+14	+0.0	+1.00
7	OH + OH + M = H2O2 + M	+.9100E+15	+0.0	-5.00
8	H + O2 + M = HO2 + M	+.1500E+16	+0.0	-1.00
9	H + O2 = OH + O	+.2200E+15	+0.0	+16.80
10	H2O2 + OH = H2O + HO2	+.1000E+14	+0.0	+1.80
11	H2O2 + H = H2 + HO2	+.1700E+13	+0.0	+3.75
12	H2 + OH = H2O + H	+.2200E+14	+0.0	+5.15
13	H2O + O = OH + OH	+.6800E+14	+0.0	+18.35
14	HO2 + OH = H2O + O2	+.5000E+14	+0.0	+1.00
15	HO2 + O = OH + O2	+.5000E+14	+0.0	+1.00
16	O2 + M = O + O + M	+.5100E+16	+0.0	+115.00
17	O + H + M = OH + M	+.1000E+17	+0.0	+0.00
18	CO2 + O = CO + O2	+.2700E+13	+0.0	+43.80
19	CO + HO2 = CO2 + OH	+.1500E+15	+0.0	+23.60
20	CO + O + M = CO2 + M	+.5900E+16	+0.0	+4.10
21	CO + OH = CO2 + H	+.1500E+08	+1.3	-0.77
22	HCO + O2 = CO + HO2	+.3000E+13	+0.0	+0.00
23	HCO + M = H + CO + M	+.1450E+15	+0.0	+19.00
24	HCO + OH = CO + H2O	+.1000E+15	+0.0	+0.00
25	HCO + H = CO + H2	+.2000E+15	+0.0	+0.00
26	HCO + O = CO + OH	+.1000E+15	+0.0	+0.00
27	CH3 + HCO = CH4 + CO	+.3000E+12	+0.5	+0.00
28	CH4 + M = CH3 + H + M	+.1400E+18	+0.0	+88.40
29	CH4 + OH = CH3 + H2O	+.3500E+04	+3.1	+2.00
30	CH4 + H = CH3 + H2	+.1300E+15	+0.0	+11.90
31	CH4 + O = CH3 + OH	+.1600E+14	+0.0	+9.20
32	CH4 + HO2 = CH3 + H2O2	+.2000E+14	+0.0	+18.00
33	CH3 + O2 = CH3O + O	+.4800E+14	+0.0	+29.00
34	CH3 + HO2 = CH3O + OH	+.2000E+14	+0.0	+0.00
35	CH3 + HO2 = CH4 + O2	+.1000E+13	+0.0	+0.40
36	CH3 + OH = CH2O + H2	+.4000E+13	+0.0	+0.00
37	CH3 + O = CH2O + H	+.1300E+15	+0.0	+2.00
38	CH3O + M = CH2O + H + M	+.5000E+14	+0.0	+21.00
39	CH3O + O2 = CH2O + HO2	+.1000E+13	+0.0	+6.00
40	CH2O + O2 = HCO + HO2	+.2040E+14	+0.0	+39.00
41	CH2O + M = HCO + H + M	+.3300E+17	+0.0	+81.00
42	CH2O + OH = HCO + H2O	+.7600E+13	+0.0	+0.17
43	CH2O + H = HCO + H2	+.3300E+15	+0.0	+10.50
44	CH2O + HO2 = HCO + H2O2	+.1000E+13	+0.0	+8.00
45	CH2O + O = HCO + OH	+.5000E+14	+0.0	+4.60
46	CH2O + CH3 = CH4 + HCO	+.1000E+11	+0.5	+6.00
47	C2H6 + OH = C2H5 + H2O	+.8700E+10	+1.1	+1.81
48	C2H6 + O = C2H5 + OH	+.2500E+14	+0.0	+6.40
49	C2H6 + H = C2H5 + H2	+.5300E+03	+3.5	+5.20
50	C2H6 + CH3 = C2H5 + CH4	+.5500E+00	+4.0	+8.30
51	C2H6 = CH3 + CH3	+.2200E+20	-1.0	+88.40
52	C2H5 + M = C2H4 + H + M	+.2000E+16	+0.0	+30.00
53	C2H5 + O2 = C2H4 + HO2	+.1000E+13	+0.0	+5.00
54	C2H4 + O2 = C2H3 + HO2	+.4000E+14	+0.0	+61.50
55	C2H4 + M = C2H2 + H2 + M	+.9300E+17	+0.0	+77.20
56	C2H4 + M = C2H3 + H + M	+.6300E+19	+0.0	+108.70
57	C2H4 + C2H4 = C2H5 + C2H3	+.5000E+15	+0.0	+64.70
58	C2H4 + OH = C2H3 + H2O	+.4780E+13	+0.0	+1.20
59	C2H4 + OH = CH3 + CH2O	+.2000E+13	+0.0	+0.96
60	C2H4 + O = CH2O + CH2	+.2500E+14	+0.0	+5.00
61	C2H4 + O = CH3 + HCO	+.3300E+13	+0.0	+1.13
62	C2H4 + H = C2H3 + H2	+.4000E+07	+2.0	+6.00
63	C2H3 + M = C2H2 + H + M	+.7900E+15	+0.0	+31.50
64	C2H3 + O2 = CH2O + HCO	+.4000E+13	+0.0	+0.25
65	C2H3 + H = C2H2 + H2	+.2000E+14	+0.0	+2.50
66	C2H2 + O2 = HCO + HCO	+.1000E+13	+0.0	+28.00
67	C2H2 + M = C2H + H + M	+.1000E+15	+0.0	+114.00
68	C2H2 + OH = C2H + H2O	+.6300E+13	+0.0	+7.00
69	C2H2 + OH = CH2CO + H	+.3000E+12	+0.0	+0.20
70	C2H2 + H = C2H + H2	+.2000E+15	+0.0	+19.00
71	C2H2 + O = CH2 + CO	+.2200E+11	+1.0	+2.60
72	C2H2 + O = C2H + OH	+.3000E+16	-0.6	+17.00

(Table 1 continued on next page.)

P. DAGAUT ET AL.

Table 1 (cont.) Etylene oxidation mechanism
(reaction rate constants in cm^3.mol.s.kcal)
$$k = A\, T^n \exp(-E/RT)$$

No	Reaction						A	n	E
73	C2H2 +	O	=	HCCO +	H		+.3550E+05	+2.7	+1.40
74	C2H +	O2	=	HCO +	CO		+.1000E+14	+0.0	+7.00
75	C2H +	O	=	CO +	CH		+.5000E+14	+0.0	+0.00
76	CH2 +	O2	=	HCO +	OH		+.4300E+11	+0.0	-0.50
77	CH2 +	O2	=	CO2 +	H2		+.6900E+12	+0.0	+0.50
78	CH2 +	O2	=	CO2 +	H	+ H	+.1600E+13	+0.0	+1.00
79	CH2 +	O2	=	CO +	H2O		+.1900E+11	+0.0	-1.00
80	CH2 +	O2	=	CO +	OH	+ H	+.8700E+11	+0.0	-0.50
81	CH2 +	O	=	CH +	OH		+.1900E+12	+0.7	+25.00
82	CH2 +	H	=	CH +	H2		+.2700E+12	+0.7	+25.70
83	CH2 +	OH	=	CH +	H2O		+.2700E+12	+0.7	+25.70
84	CH +	O2	=	CO +	OH		+.1300E+12	+0.7	+25.70
85	CH +	O2	=	HCO +	O		+.1000E+14	+0.0	+0.00
86	CH2CO +	H	=	CH3 +	CO		+.1100E+14	+0.0	+3.40
87	CH2CO +	O	=	HCO +	HCO		+.1000E+14	+0.0	+2.40
88	CH2CO +	OH	=	CH2O +	HCO		+.2800E+14	+0.0	+0.00
89	CH2CO +	M	=	CH2 +	CO	+ M	+.2000E+17	+0.0	+60.00
90	CH2CO +	O	=	HCCO +	OH		+.5000E+14	+0.0	+8.00
91	CH2CO +	OH	=	HCCO +	H2O		+.7600E+13	+0.0	+3.00
92	CH2CO +	H	=	HCCO +	H2		+.7600E+14	+0.0	+8.00
93	HCCO +	OH	=	HCO +	HCO		+.1000E+14	+0.0	+0.00
94	HCCO +	H	=	CH2 +	CO		+.5000E+14	+0.0	+0.00
95	HCCO +	O	=	HCO +	CO		+.3400E+14	+0.0	+2.00
96	CH3OH +	M	=	CH3 +	OH	+ M	+.3000E+19	+0.0	+80.00
97	CH3OH +	OH	=	CH2OH +	H2O		+.4000E+13	+0.0	+2.00
98	CH3OH +	O	=	CH2OH +	OH		+.1700E+13	+0.0	+2.30
99	CH3OH +	H	=	CH2OH +	H2		+.3000E+14	+0.0	+7.00
100	CH3OH +	H	=	CH3 +	H2O		+.5200E+13	+0.0	+5.34
101	CH3OH +	CH3	=	CH2OH +	CH4		+.1800E+12	+0.0	+9.80
102	CH3OH +	HO2	=	CH2OH +	H2O2		+.6300E+13	+0.0	+19.36
103	CH2OH +	M	=	CH2O +	H	+ M	+.2500E+14	+0.0	+29.00
104	CH2OH +	O2	=	CH2O +	HO2		+.1000E+13	+0.0	+6.00
105	C2H3 +	C2H4	=	C4H6 +	H		+.1000E+13	+0.0	+7.30
106	C2H2 +	C2H2	=	C4H3 +	H		+.1000E+14	+0.0	+45.00
107	C4H3 +	M	=	C4H2 +	H	+ M	+.1000E+17	+0.0	+60.00
108	C2H2 +	C2H	=	C4H2 +	H		+.4000E+14	+0.0	+0.00
109	C4H2 +	M	=	C4H +	H	+ M	+.3500E+18	+0.0	+80.00

incorporated in the system of algebraic equations describing the mass balance for each of the 30 species involved in the kinetic scheme.

Discussion

Figure 4 shows a typical comparison between model computations based on the mechanism of Table 1 and experimental results for lean mixtures. The model correctly predicts the concentration time profiles of C_2H_4, CO, and CO_2. As the computed concentrations of CH_4 and C_2H_2 are lower than the experimental ones, further refinement of the mechanism is still necessary.

For rich and stoichiometric mixtures, this mechanism predicts higher reactivities than experimentally observed and the kinetic scheme has to be supplemented with reactions of propane and propene (Westbrook et al., 1984). These supplementary reactions are required because minor products such as C_3H_6, C_3H_8, and C_4H_8 are present at higher concentrations in rich mixtures. As their reactions are no longer negligible, additional chain breaking processes are introduced. To improve further the agreement between predictions and experimental data for these mixtures, a sensitivity study was made to show the importance of reaction 62,

$$C_2H_4 + H = C_2H_3 + H_2$$

The pre-exponential factor of this rate constant was adjusted to reproduce the experimental H_2 concentration profiles. At 1000 K, the value of the rate constant used in this work is very near that reported by Skinner et al. (1971).

Figures 5-7 show the comparison between computed data obtained with the mechanism of Table 1 and Table 2 (reactions 1-168) and experimental results for the oxidation of stoichiometric and rich mixtures in the jet-stirred reactor at 1 and 5 atm. In Figs. 5 and 6 the computed concentrations of CO, CO_2, H_2, and C_2H_2 are very close to the experimental data, but the predicted CH_4 and C_2H_6 concentrations are smaller than the observed data. In Fig. 7 the C_2H_4, CO, and CO_2 concentration time profiles are better predicted than the C_2H_2, C_2H_6, and CH_4 ones. In the case of lean mixtures, the addition of reactions of Table 2 leads to the same predictions as the use of reactions 1-109 alone.

However, this mechanism does not include the reactions forming C_4H_6, which has been detected in trace quantities. Further extension of the mechanism will include these reactions, and sensitivity studies will be performed to improve the agreement between predictions and experimental concentrations of minor species.

It is interesting to compare the evolution of a reactive mixture in the stirred reactor with the evolution of the same mixture at the same initial temperature and pressure in another system such as the plug flow reactor. Figure 8 shows the computed concentrations vs time during the isothermal oxidation of ethylene at 765°C, 1 atm, and $\Phi = 0.4$.

The reaction begins sooner in the well-stirred reactor, because of the presence of radical species suppressing the initiation period. With time translation, C_2H_4 and CO concentration profiles are almost identical in both reactors up to 50% conversion, while CO_2 concentrations are slightly higher in the stirred reactor. However, at higher conversion, the concentration time curves are very different as follows. In the plug flow reactor, the C_2H_4 concentration falls almost to zero at the time of the CO maximum. This sharp maximum of CO concentration is accom-

Table 2 Supplementary reactions for the oxidation of propane (reaction rate constants in cm^3.mol.s.kcal)
$$k = A \, T^n \exp(-E/RT)$$

No	Reaction			A	n	E
110	C3H8	=	C2H5 + CH3	.1700E+17	+0.0	+84.84
111	C3H8 + O2	=	NC3H7 + HO2	.4000E+14	+0.0	+47.50
112	C3H8 + O2	=	IC3H7 + HO2	.4000E+14	+0.0	+47.50
113	C3H8 +C2H5	=	NC3H7 +C2H6	.1000E+12	+0.0	+10.40
114	C3H8 +C2H5	=	IC3H7 +C2H6	.1000E+12	+0.0	+10.40
115	C3H8 + CH3	=	NC3H7 + CH4	.1090E+16	+0.0	+25.14
116	C3H8 + CH3	=	IC3H7 + CH4	.1090E+16	+0.0	+25.14
117	C3H8+IC3H7	=	C3H8+NC3H7	.3000E+11	+0.0	+12.90
118	C3H8 +C2H3	=	NC3H7 +C2H4	.1000E+12	+0.0	+10.40
119	C3H8 +C2H3	=	IC3H7 +C2H4	.1000E+12	+0.0	+10.40
120	C3H8 +C3H5	=	NC3H7 +C3H6	.4000E+12	+0.0	+16.20
121	C3H8 +C3H5	=	IC3H7 +C3H6	.4000E+12	+0.0	+16.20
122	C3H8 + H	=	NC3H7 + H2	.5600E+08	+2.0	+7.70
123	C3H8 + H	=	IC3H7 + H2	.8700E+07	+2.0	+5.00
124	C3H8 + OH	=	NC3H7 + H2O	.5700E+09	+1.4	+0.85
125	C3H8 + OH	=	IC3H7 + H2O	.4800E+09	+1.4	+0.85
126	C3H8 + O	=	NC3H7 + OH	.5000E+07	+2.0	+3.00
127	C3H8 + O	=	IC3H7 + OH	.5000E+07	+2.0	+3.00
128	C3H8 + HO2	=	NC3H7 +H2O2	.5000E+13	+0.0	+18.00
129	C3H8 + HO2	=	IC3H7 +H2O2	.5000E+13	+0.0	+18.00
130	NC3H7	=	C2H4 + CH3	.9500E+14	+0.0	+31.00
131	IC3H7	=	C2H4 + CH3	.2000E+11	+0.0	+29.50
132	NC3H7	=	C3H6 + H	.1250E+15	+0.0	+37.00
133	IC3H7	=	C3H6 + H	.6300E+14	+0.0	+36.90
134	NC3H7 + O2	=	C3H6 + HO2	.1000E+13	+0.0	+5.00
135	IC3H7 + O2	=	C3H6 + HO2	.1000E+13	+0.0	+5.00
136	C3H6	=	C3H5 + H	.1000E+14	+0.0	+78.00
137	C3H6	=	C2H3 + CH3	.6300E+16	+0.0	+85.80
138	C3H6 +C2H5	=	C3H5 +C2H6	.1000E+12	+0.0	+9.20
139	C3H6 + CH3	=	C3H5 + CH4	.8900E+11	+0.0	+8.50
140	C3H6 + OH	=	CH3CHO + CH3	.3500E+12	+0.0	+0.00
141	C3H6 + OH	=	C3H5 + H2O	.4000E+13	+0.0	+0.00
142	C3H6 + OH	=	C2H5 +CH2O	.7900E+13	+0.0	+0.00
143	C3H6 + H	=	C3H5 + H2	.5000E+13	+0.0	+1.50
144	C3H6 + O	=	C2H5 + HCO	.3600E+13	+0.0	+0.00
145	C3H6 + O	=	C2H4 +CH2O	.5900E+14	+0.0	+5.00
146	C3H6 + O	=	CH3+CH3CO	.1200E+14	+0.0	+0.60
147	C3H5	=	C3H4 + H	.4000E+14	+0.0	+70.00
148	C3H5 + O2	=	C3H4 + HO2	.6000E+12	+0.0	+10.00
149	C3H5 + H	=	C3H4 + H2	.1000E+14	+0.0	+0.00
150	C3H5 + CH3	=	C3H4 + CH4	.1000E+13	+0.0	+0.00
151	C3H4 + OH	=	CH2O +C2H3	.1000E+13	+0.0	+0.00
152	C3H4 + OH	=	HCO +C2H4	.1000E+13	+0.0	+0.00
153	C3H4 + O	=	CH2O +C2H2	.1000E+13	+0.0	+0.00
154	C3H4 + O	=	HCO +C2H3	.1000E+13	+0.0	+0.00
155	CH3CHO	=	CH3 + HCO	.7080E+16	+0.0	+81.78
156	CH3CHO + O2	=	CH3CO + HO2	.2000E+14	+0.5	+42.20
157	CH3CHO + CH3	=	CH3CO + CH4	.1700E+13	+0.0	+8.43
158	CH3CHO + H	=	CH3CO + H2	.4000E+14	+0.0	+4.20
159	CH3CHO + O	=	CH3CO + OH	.5000E+13	+0.0	+1.80
160	CH3CHO + OH	=	CH3CO + H2O	.1000E+14	+0.0	+0.00
161	CH3CHO + HO2	=	CH3CO +H2O2	.1700E+13	+0.0	+10.70
162	CH3CO	=	CH3 + CO	.3000E+14	+0.0	+17.24
163	C4H8	=	C3H5 + CH3	.1500E+20	-1.0	+73.40
164	C4H8	=	C2H3 +C2H5	.100E+19	-1.0	+96.77
165	C4H8 + O	=	CH3CHO +C2H4	.1300E+14	+0.0	+0.85
166	C4H8 + O	=	CH3CO +C2H5	.1300E+14	+0.0	+0.85
167	C4H8 + OH	=	CH3CHO +C2H5	.1000E+14	+0.0	+0.00
168	C4H8 + OH	=	CH3CO +C2H6	.1000E+14	+0.0	+0.00

panied by a rapid rise in the CO_2 concentration. In the stirred reactor, the reaction rate of the last stages is much slower. The C_2H_4 concentration remains at a low, but not negligible value, the maximum of CO concentration is very smooth, and the increase of CO_2 concentration is gradual. This low rate of CO oxidation is a consequence of the competition between this product and the unburned hydrocarbon toward radicals (namely, OH, O, HO_2), the hydrocarbon acting as an inhibitor for CO oxidation.

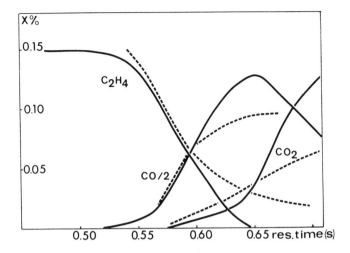

Fig. 8 Computed evolution of a $C_2H_4/O_2/N_2$ mixture (T = 765°C, P = 1 atm, Φ = 0.4) as a function of the residence time :
——————— in a plug flow reactor
— — — — in a jet-stirred reactor (after a translation)

Conclusion

The jet-stirred reactor presented in this paper is a valuable tool for the study of oxidation reactions in the intermediate temperature range (900-1300 K). It allows the analysis of the different stages of process evolution from low to high conversion range by successive steady states at uniform temperature and concentration. The main results of an ethylene oxidation study have been reproduced through a kinetic scheme previously used to model the oxidation of the same alkene in tubular reactors.

Acknowledgment

This work was supported by D.R.E.T. under contract n° 81/140.

References

Bahn, G.S. (1973) : Approximate thermochemical tables for some C-H and C-H-O species, NASA CR 2178.

Baldwin, R.R. and Walker, R.W. (1981) Elementary reactions in oxidation of alkenes. Eighteenth Symposium (International) on Combustion, pp. 819-829, The Combustion Institute, Pittsburgh, PA.

Caprio, V., Insola, A., and Lignola, P.G. (1977) Isobutane cool flames investigation in a continuous stirred tank reactor. Sixteenth Symposium (International) on Combustion, pp. 1155-1163, The Combustion Institute, Pittsburgh, PA.

Cathonnet, M., Boettner, J.C., and James, H. (1981) Experimental study and numerical modeling of high temperature oxidation of propane and n-butane. Eighteenth Symposium (International) on Combustion, pp. 903-913, The Combustion Institute, Pittsburgh, PA.

Cathonnet, M., Gaillard, F., Boettner, J.C., Cambray, P., Karmed, D., and Bellet, J.C. (1984) Experimental study and kinetic modelling of ethylene oxidation in tubular reactors. Twenth Symposium (International) on Combustion, pp. 819-829, The Combustion Institute, Pittsburgh, PA.

Colket, M.D., Naegeli, D.W., and Glassman, I. (1977) High Temperature oxidation of acetaldehyde. Sixteenth Symposium (International) on Combustion, pp. 1023-1039, The Combustion Institute, Pittsburgh, PA.

Come, G.M. (1983) Laboratory reactors for pyrolysis reactions. Pyrolysis: Theory and Industrial Practice (edited by L.F. Albright, B.L. Crynes, and W.H. Corcoran), pp. 255-275, Academic Press, New York.

David, R., and Matras, D. (1975) Règles de construction et d'extrapolation des réacteurs auto-agités par jets gazeux. Can. J. Chem. Eng. 53, 297-300.

David, R., Houzelot, J.L., and Villermaux, J. (1979) Gas mixing in jet-stirred reactors with short residence times. Third European Conference on Mixing, pp. 113-124.

Dixon-Lewis, G. (1981) Aspects of the kinetic modelling of methane oxidation in flames. First Specialists Meeting (International) of the Combustion Institute, pp. 284-289, The Combustion Institute, Pittsburgh, PA.

Duterque, J., Borghi, R., and Tichtinski, H. (1981) Study of quasi-global schemes for hydrocarbon combustion. Combust. Sci. Technol. 26, 1-15.

Ferrer, M., David, R., and Villermaux, J. (1982) Homogeneous oxidation of n-butane in a self stirred reactor. Oxidation Communications 4, 353-368.

Gray, B.F., and Felton, P.G. (1974) Low temperature oxidation in a stirred-flow reactor. I : Propane. Combust. Flame 23, 295-304.

Gray, P., Griffiths, J.F., Hasko, S.M., and Lignola, P.G. (1981) Oscillatory ignitions and cool flames accompanying the nonisothermal oxidation of acetaldehyde in a well stirred flow reactor. Proc. Roy. Soc. (London) A 374, 313-339.

Hottel, H.C., Williams, G.C., Nerheim, N.M., and Schneider, G.R. (1965) Kinetic studies in stirred reactors : combustion of carbon monoxide and propane. Tenth Symposium (International) on Combustion, pp. 111-121, The Combustion Institute, Pittsburgh, PA.

Juste, C., Scacchi, G., and Niclause, M. (1981) Minor products and initiation rate in the chain pyrolysis of propane. Int. J. Chem. Kine. 13, 855-864.

Longwell, J.P., and Weiss, M.A. (1955) High temperature reaction rates in hydrocarbons combustion. Ind. Eng. Chem. 47, 1634-1643.

Matras, D. and Villermaux, J. (1973) Un réacteur continu parfaitement agité par jets gazeux pour l'étude cinétique de réactions rapides. Chem. Eng. Sci. 28, 129-137.

Miller, J.A., Mitchell, R.E., Smooke, M.D., and Kee, R.J. (1983) Toward a comprehensive chemical kinetic mechanism for the oxidation of acetylene : comparison of model predictions with results from flames and shock tube experiments. Nineteenth Symposium (International) on Combustion, pp. 181-196, The Combustion Institute, Pittsburgh, PA.

Park, J.Y., Heaven, M.C., and Gutman, D. (1984) Kinetics and mechanism of the reaction of vinyl radicals with molecular oxygen. Chem. Phys. Let. 104, 469-474.

Skinner, G.B., Sweet, R.C., and Davis, S.K. (1971) Shock tube experiments on the pyrolysis of deuterium-substituted ethylenes. J. Phys. Chem. 75, 1-12.

Slagle, I.R., Park, J.Y., Heaven, M.C., and Gutman, D. (1984) Kinetics of polyatomic free radicals produced by laser photolysis 3. The reaction of vinyl radicals with molecular oxygen. J. Am. Chem. Soc. 106, 4356-4361.

Walker, R.W. (1975) Rate constants for gas-phase hydrocarbon oxidation. Reaction Kinetics, Vol. 1, pp.161-211, The Chemical Society, London.

Warnatz, J. (1984) Rate coefficients in the C/H/O system. Chemistry of Combustion Reactions, edited by W.C. Gardiner Jr., pp. 197-360, Springer-Verlag, New York.

Westbrook, C.K., Dreyer, F.L., and Schug, K. (1983) A comprehensive mechanism for the pyrolysis and oxidation of ethylene. Nineteenth Symposium (International) on Combustion, pp. 153-166, The Combustion Institute, Pittsburgh, PA.

Westbrook, C.K., and Dreyer, F.L. (1984) Chemical kinetic modeling of hydrocarbon combustion. Prog. Energy Comb. Sci. 10, 1-157.

Westbrook, C.K. and Pitz, W.J. (1984) A comprehensive chemical kinetic reaction mechanism for oxidation and pyrolysis of propane and propene. Comb. Sci. and Techno. 37, 117-152.

The Explosive Decomposition of Chlorine Dioxide Behind Shock Waves

C. Paillard,* S. Youssefi,† and G. Dupré‡
Université d'Orléans et Centre National de la Recherche Scientifique
Orléans, France

Abstract

Gaseous chlorine dioxide ClO_2 is an energetic and easily detonated system. When subjected to a shock wave, chlorine dioxide–argon mixtures can decompose in an explosive manner. The aim of the present study was to define the conditions of stability of such systems and to investigate the behavior of ClO_2 near the limits of a shock-induced detonation. Ignition delay times ahead of the explosive reaction were compared to the vibrational relaxation times of the reactant. The effects of reaction exothermicity and shock strength on shock velocities were studied to define the conditions of temperature and pressure for the onset of the detonation wave. At low temperatures, the explosive reaction was preceded by a long ignition delay time (several hundred μs), characteristic of chain reactions.

Introduction

Chlorine dioxide heated to temperatures higher than 50°C is readily decomposable in an explosive manner. In previous studies on the thermal decomposition of chlorine dioxide, the induction period has been measured as a

---

Presented at the 10th ICDERS, Berkeley, California, August 4-9, 1985. Copyright © 1985 by the American Institute of Aeronautics and Astronautics, Inc. All rights reserved.

*Maitre de Conferences, Chemical Kinetics Laboratory, University of Fundamental and Applied Sciences, and Researcher at the Research Center on the Chemistry of Combustion and of High Temperatures.

†Student, Research Center on the Chemistry of Combustion and of High Temperatures.

‡Chargée de Recherche, Research Center on the Chemistry of Combustion and of High Temperatures.

function of temperature, pressure, vessel size, irradiation, and added gases (Schumacher and Stieger 1930 ; McHale and von Elbe 1968). The explosive reaction is characterized by a long ignition delay time involving a mechanism of a degenerated chain-branching type. Semenov (1959) has shown explosions with induction periods exceeding a few seconds are due to the formation of a relatively stable intermediate. McHale and von Elbe proposed the intermediate to be Cl_2O_3; its presence seems to have been confirmed recently by Bethune et al. (1983). The mechanism deduced from the studies of McHale and von Elbe presumes that chain initiation and termination take place at the vessel walls. The temperature dependence on the induction period indicates that the activation energy of the explosive reaction would be small, of the order of 53 kJ/mole. This value does not differ too greatly from that of 39 kJ/mole estimated from measurements of the influence of tube diameter on the detonation velocities in ClO_2 (Ben Caïd 1966), although, in this case, the reactions at the wall cannot be taken into account.

The aim of the present work is to measure, independently of wall effect, ignition delay times of the explosive reaction in $|ClO_2-Ar|$ mixtures heated by a shock wave and to investigate the behavior of ClO_2 near the limits of the shock-induced detonation. Information on the induction period as a function of pressure and temperature is useful to predict the initiation, the stability, and the structure of detonation waves. From values of the induction period, Westbrook (1984) could verify some assumptions on reaction mechanism and calculate detonation parameters such as critical tube diameters for the transition to spherical detonation, cell sizes, and critical initiation energies for various mixtures (hydrocarbon-air). The coupling between the shock wave and the reaction zone required to obtain a detonation wave implies that the

Table 1   Characteristics of the shock tube

Shock tube diameter, mm	Driver section		Driven section		
	Nature	Length, m	Nature	Thickness, mm	Length, m
50	Stainless steel	2.00	Pyrex glass	4.5	6.44
22	Pyrex glass	1.15	Pyrex glass	2.0	2.15

induction period is very short. In this case, it can be compared to the vibrational relaxation time of the reactant. To explore this point, the vibrational relaxation time was measured for various pressures, temperatures, and dilutions in argon.

Experimental

Gaseous chlorine dioxide was generated from the reaction of concentrated sulfuric acid with potassium chlorate and stored at around 10°C in darkness. The gas-generating apparatus and storage vessel were made of Pyrex glass and Teflon to avoid corrosion and decomposition problems. The purity of ClO_2 was checked by means of mass spectrometry and absorption spectrophotometry after each preparation. Despite of the great care to keep ClO_2 as pure as possible, a noticeable partial decomposition could be observed after a few days' storage.

The characteristics of the two shock tubes used are summarized in Table 1. For each of them, the high and the low pressure sections were separated with a 0.018-mm-thick Terphane diaphragm. The initiation of the shock wave was produced by the sudden rupture of the diaphragm pierced by a four-bladed steel knife actuated by a pneumatic system.

The diagnostics included a series of tiny piezoelectric gages for shock velocity measurements and control of the time of arrival of the shock front along the observation windows. Smoked foils were used to observe the detonation structure in some experiments. The optical setup consisted of CaF_2 windows, a HgCdTe infrared detector in conjunction with a 8.89-μm filter (a major band of ClO_2), photomultipliers for recording overall visible and ultraviolet emissions, a monochromator for observation of the ultraviolet absorption at 360 nm, and a laser Schlieren system. This Schlieren system is based on the deflection of a He-Ne laser beam crossing the gaseous medium through which the shock wave passes. This technique, applied in the present study to measure vibrational relaxation times, has been thoroughly described and discussed by Kiefer (1981).

Results

Various systems $|x\ ClO_2 + (1-x)\ Ar|$ have been investigated to determine the conditions of the onset of detonation. Runs were made at ambient initial temperature T_1 over the initial pressure range P_1 = 5-50 Torr (6.67×10^2–6.67×10^3 Pa) and with initial pressure of helium P_4 in the driver tube between 0.2 and 2.3 bars (2×10^4–2.3×10^5 Pa).

The effects of reaction exothermicity and shock strength P_4/P_1 on shock velocities as well the influence of temperature and pressure on ignition delay times and vibrational relaxation times have been determined.

Ignition Delay Times

For an appreciable range of temperature and pressure behind the incident shock wave, an induction period τ_i (or ignition delay time) was observed before a rapid consumption of the reactant. An overall ultraviolet and visible emission signal was detected after the induction period, as shown in Fig. 1.

The thermal decomposition of ClO_2 was monitored at first by following the evolution of the infrared emission signal at 8.89 µm behind the incident shock wave. Because of the exothermicity of the reaction, the shock wave properties are changed, and the determination of shock parameters becomes inaccurate, especially for mixtures with ClO_2 concentration higher than 2%. Moreover, measurements of ignition delay times based on the i.r. emission at 8.89 µm are complex. As the signal is decreasing because

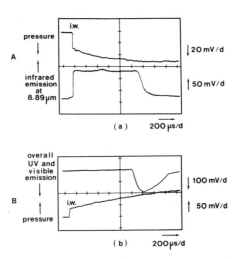

Fig. 1 Simultaneous recording of the pressure and the emission signals behind the incident wave (i.w.). a) i.r. emission at 8.89 µm ; b) overall visible and uv emission. A and B represent two observation stations 0.08 m apart. Mixture |0.10 ClO_2 + 0.90 Ar|. T_1 = 297 K, T_2 = 754 K, P_1 = 25 Torr, P_2 = 189 Torr, P_4 = 1.05 bars, P_2/P_1 = 2.980.

of ClO_2 decomposition, a second emission appears which is likely due to an intermediate species. As a consequence, the induction periods were determined from the variation of absorption at 360 nm of mixtures submitted to the reflected shock wave. A typical recording of such a signal is given in Fig. 2. Simultaneous recordings of the emission signal at 8.89 µm and of the absorption signal at 360 nm confirm the complexity of the i.r. signal. Experimental induction time data are shown in Fig. 3 for different mixtures (x_{ClO_2} = 0.01-0.02-0.10). Figure 3 gives the variation of the function $Y = \log \tau_i + 0.6 \log (P/T)$ with inverse shock temperature. More details of the data analysis have been recently given elsewhere (Paillard et al. 1985). In a range of temperature and pressure of respectively 1100-1450 K and 65-180 kPa, the relationship between the ignition delay time τ_i and pressure P, temperature T, and ClO_2 molar fraction x can be written as

$$\log \tau_i \text{ (s)} = \frac{4550}{T(K)} - 0.6 \log x \frac{P(Pa)}{T(K)} - 6.75$$

The small temperature coefficient leads to an activation energy of 87 kJ/mole.

Vibrational Relaxation Times

Vibrational relaxation times τ_v in $|ClO_2$-Ar$|$ mixtures containing 10, 25, and 50 mol.% ClO_2 were measured behind incident shock waves with the laser Schlieren technique in the temperature range 420-700 K. A representative

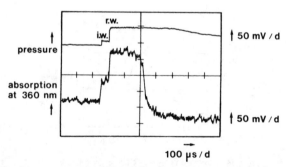

Fig. 2 Simultaneous recording of the pressure and the absorption signals at 360 nm, behind the incident (i.w.) and the reflected (r.w.) shock waves. The observation station is 0.025 m from the end of the shock tube. Mixture $|0.10 \ ClO_2 + 0.90 \ Ar|$. T_1 = 296 K, T_2 = 702 K, T_5 = 1193 K, P_1 = 25 Torr, P_2 = 169 Torr, P_5 = 649 Torr, P_4 = 1.4 bars, ρ_2/ρ_1 = 2.855, ρ_5/ρ_2 = 2.255.

Schlieren oscillogram showing axial density gradient and exponentially decaying relaxation signal is given in Fig.4 for a mixture containing 25 mol.% ClO_2 heated at T_2 = 537K. At temperatures higher than 600 K and for less dilute mixtures, a reaction can be initiated behind the shock. In this case, the shock front is curved and the relaxation time data become inaccurate. Figure 5 gives an example of a laser recording obtained when the shock wave and the reaction zone are coupled.

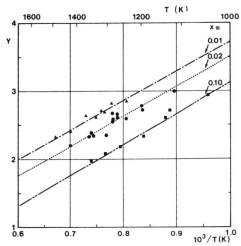

Fig. 3 Influence of shock temperature on the function $Y = \log \tau_i$ (µs) + 0.6 log P (Torr)/T(K) for different mixtures $|x \ ClO_2 + (1-x) \ Ar|$: x = 0.01, 0.02, 0.10. ▲ ● ■ : Experimental results. The lines are obtained by using the following expression : $Y = 4550/T(K) - 2.02 - 0.6 \log x$.

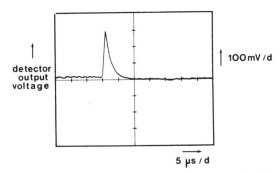

Fig. 4 Detector output voltage corresponding to the laser deviation vs laboratory time, in the case of a pure shock wave. Mixture $|0.25 \ ClO_2 + 0.75 \ Ar|$. T_1 = 293 K, T_2 = 537 K, P_1 = 25 Torr, P_2 = 125 Torr, P_4 = 0.4 bar, ρ_2/ρ_1 = 2.731.

To calculate T_v, a single relaxation time obeying the Bethe-Teller law was assumed. None of our recordings suggests multiple relaxation processes. When the shock is not disturbed by the exothermic reaction, the semi-log plot of the laser deviation amplitude V is a linear function of laboratory time (Fig. 6). From such a curve, the "laboratory" relaxation time T_1 can be deduced and, afterwards, the characteristic relaxation time T_v, as they are both related with the classical method of Blackman as follows :

$$T_v = T_1 \ (C_p/C_p^*) \ (\rho/\rho_o)$$

C_p is the total equilibrium heat capacity ; C_p^* is this one with vibrational frozen. The density ρ and the temperature T used for calculating C_p correspond to the mean of their initial and equilibrium values. The Landau-Teller plots are shown in Fig. 7 for three ClO_2 concentrations. For the mixture containing 25 mol.% ClO_2, a relation was found between the standardized relaxation time PT_v and $T^{-1/3}$ (P and T being the equilibrium pressure and temperature behind the incident shock front, respectively):

$$\log P \ T_v = -1.28 + 10 \ T^{-1/3}$$

With the mixtures containing 10 and 50 mol.% ClO_2, no accurate estimation of the relation $\log PT_v = f(T)$ could be obtained because of the small number of runs carried out.
The relaxation times in nondiluted ClO_2 ($T_{ClO_2-ClO_2}$) and in ClO_2 infinitely diluted in argon (T_{ClO_2-Ar}) can be

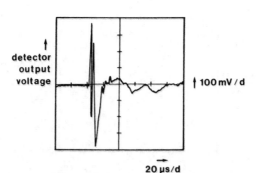

Fig. 5 Detector output voltage corresponding to the laser deviation vs laboratory time, for a shock wave tending to couple with the reaction zone. Mixture $|0.50 \ ClO_2 + 0.50 \ Ar|$. $T_1 = 294$ K, $T_2 = 645$ K, $P_1 = 25$ Torr, $P_2 = 223$ Torr, $P_4 = 0.4$ bar, $\rho_2/\rho_1 = 4.074$.

roughly estimated from the mixture rule

$$\frac{1}{\tau_M} = \frac{x}{\tau_{ClO_2 - ClO_2}} + \frac{1 - x}{\tau_{ClO_2 - Ar}}$$

over a small range of temperature. The corresponding values are reported in Fig. 7.

Effect of Reaction Exothermicity on Incident Shock Waves

Exothermic reactions can generate unstable incident shock waves. Fluctuations of the shock wave velocity are observed with ClO_2 mixtures when the time separating the shock front passage and the contact surface arrival at the observation point is greater than the ignition delay time.

At low shock strengths P_4/P_1 (that is, the initial pressure ratio across the diaphragm), there is no evidence of coupling between the shock wave and the reaction zone. As the shock strength is raised, the onset of coupling can be observed. The critical values of temperature and pressure above which coupling occurs, named "coupling temperature" T_c and "coupling pressure" P_c, have been determined from the effect of shock strength P_4/P_1 on shock velocity V_s.

Above T_c, three types of behavior are observed :
1) With mixtures containing 25 mol.% ClO_2 or less and at relatively low initial pressures ($P_1 \leqslant 25$ Torr), the shock velocity attains a higher value than that predicted in the absence of reaction but fairly below the theoretical value D_{CJ} for the one-dimensional Chapman-Jouguet detonation (Fig. 8).

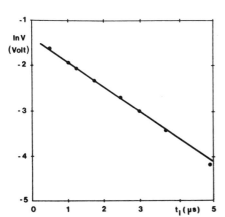

Fig. 6 Semilog plot of data from Fig. 4.

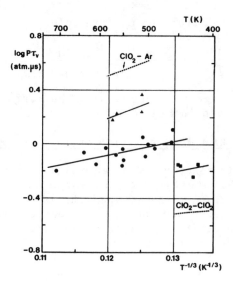

Fig. 7 Logarithmic Landau-Teller plots of the product of vibrational relaxation time into shock pressure vs shock temperature raised to the -1/3 power, for various $|x\ ClO_2 + (1-x)\ Ar|$ mixtures. ▲ $x = 0.10$; ● $x = 0.25$; ■ $x = 0.50$. The dashed lines represent values of vibrational relaxation times in nondiluted $ClO_2\ |\tau_{ClO_2-ClO_2}|$ and in ClO_2 infinitely diluted in Ar $|\tau_{ClO_2-Ar}|$.

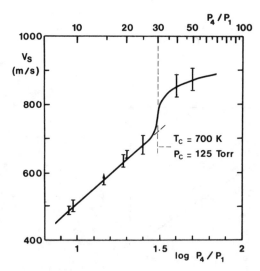

Fig. 8 Effect of shock strength on shock velocity. Mixture $|0.25\ ClO_2 + 0.75\ Ar|$. $P_1 = 15$ Torr, $T_1 = 293$ K. (T_c and P_c are the "coupling" temperature and pressure.)

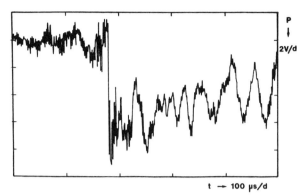

t → 100 µs/d

Fig. 9 Pressure recording ahead and behind the incident shock in the case of an unstable wave propagation. Mixture |0.50 ClO_2 + 0.50 Ar|. P_1 = 25 Torr, T_1 = 293 K.

Table 2 Experimental characteristics of the behavior of nondiluted ClO_2 submitted to an incident shock wave (T_1 = 295 K)

P_1, Torr	P_4/P_1	Smoked foil positions A	Smoked foil positions B	V_S, m/s 1-2	V_S, m/s 2-3	V_S, m/s 3-4	D_{CJ}, m/s
10	75	cells	spin	1023	1146	....	1220
25	15	no trace	no trace	504	494	494	1246
25	25.5	no trace	no trace	582	575	582	1246
25	30	cells	....	1389	1509	1386	1246
25	37.5	....	spin	1192	....	1318	1246
25	45	cells	spin	1186	1198	1180	1246
50	29	spin	spin	....	1231	....	1265
200	25	cells	cells	....	1280	....	1305

2) For weakly diluted mixtures (x_{ClO_2} = 0.5) or for more diluted mixtures at higher initial pressures, the wave propagation becomes unstable : Large pressure oscillations are observed ahead and behind the shock front (Fig. 9) and velocity fluctuations between the observation points are appreciable. The mean shock velocity remains quite below CJ velocity and no cell structure is observed, as shown in Fig. 10 for a mixture containing 50 mol.% ClO_2 at 25 Torr initial pressure. In such conditions, the ignition delay is too large to ensure the onset of a detonation wave. A chemically enhanced shock wave is observed.
3) With pure ClO_2, the shock velocity rapidly accelerates just as the conditions of temperature and pressure exceed the coupling parameters (T_c, P_c) and tends to CJ velocity, as shown in Fig. 11 for an initial pressure of 25 Torr.

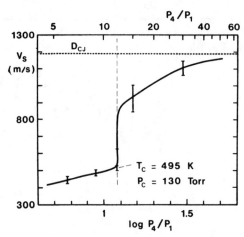

Fig. 10 Effect of shock strength on shock velocity. Mixture |0.50 ClO$_2$ + 0.50 Ar|. P$_1$ = 25 Torr, T$_1$ = 293 K. (T$_c$ and P$_c$ are the "coupling" temperature and pressure.)

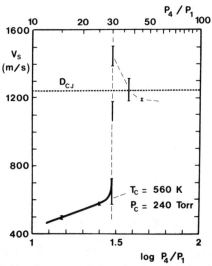

Fig. 11 Effect of shock strength on shock velocity for nondiluted ClO$_2$. P$_1$ = 25 Torr, T$_1$ = 293 K. (T$_c$ and P$_c$ are the "coupling" temperature and pressure.)

In any case, for initial pressure in ClO$_2$ equal or lower than 50 Torr, an unstable detonation takes place with large velocity fluctuations and pressure oscillations. The experimental characteristics of a shock wave passing through pure ClO$_2$ are summarized in Table 2.

The analysis of soot traces shows that when the explosive gas is submitted to pressure and temperature

Table 3 Detonation parameters for nondiluted ClO_2 (T_1 = 295 K)

P_1, Torr	P_{CJ}, Torr	T_{CJ}, K	P_{VN}, Torr	T_{VN}, K	τ_i, µs	τ_v, µs	D_{CJ}, m/s	D_{50}, m/s
200	4342	2121	7800	1285	11	0.01	1305	1280
50	1037	1974	1890	1202	44	0.06	1265	1230
25	508	1915	925	1160	91	0.13	1245	unstable
10	197	1820	360	1110	235	0.34	1220	unstable

conditions higher than T_c and P_c, the appearance of detonation cells, their decay to a single-head spin, and the
disappearance of any detonation are observed. With P_4/P_1 =
75 and P_1 = 10 Torr, for example, a cell structure appears
at the first location A of the smoked foil in the 50-mm-
i.d. tube and then is destroyed to give a unique trace
corresponding to a single-head spin detonation at the
second location B of the smoked foil. At this low initial
pressure, even with a strong shock strength, shock velocities measured between piezoelectric gages located in positions 1, 2, 3, 4 are generally lower than CJ detonation
velocity. At higher pressure, around 50 Torr, a spinning
detonation is found for a shock strength P_4/P_1 = 29. In
these conditions, the velocity is close to CJ value (about
3% lower). At initial pressure of 200 Torr, a regular cell
structure could be observed with a velocity deficit of 2%
(Ben Caïd 1966).

In table 3, detonation parameters for nondiluted ClO_2
at various initial pressures are summarized. The Chapman-
Jouguet and von Neumann characteristics (T_{VN}, P_{VN}) were
computed for an enthalpy of ClO_2 formation equal to 104.6
kJ/mole and the presence of O_2, Cl_2, and Cl at thermodynamic equilibrium. Concentrations of ClO, O, and other species were found to be negligible. The CJ detonation velocity D_{CJ} was compared to experimental velocities measured
in a 50-mm-i.d. tube, D_{50}.

Ignition delay times τ_i and vibrational relaxation
times τ_v have been evaluated from the results presented
above for diluted ClO_2, with the assumption of von Neumann
parameters behind the leading shock. In pure ClO_2, τ_i is
much greater than τ_v. Thermal equilibrium can be assumed
to be established when the reaction becomes explosive.
Vibrational relaxation may influence the decomposition
rate of chlorine dioxide when the explosive reaction is
initiated with a strong shock wave at low initial pressures of ClO_2 diluted in argon.

Conclusion

The present study on the explosive decomposition of chlorine dioxide has yielded measurements of ignition delay times as a function of temperature, pressure, and diluent concentration over the range corresponding to the limits of detonation. Data analysis indicates an apparent activation energy of 87 kJ/mole, and the relationship $\log \tau_i = f$ (T, P, x) provides useful information for calculation of the dynamic parameters of ClO_2 detonation.

Vibrational relaxation times have been determined for $|ClO_2 - Ar|$ mixtures, and it appears that near detonation limits, vibrational relaxation has no influence on the rate of ClO_2 decomposition. From the effects of reaction exothermicity and shock strength on shock velocities, the coupling conditions between the shock wave and the reaction zone were observed. Over a large range of initial pressures, the cellular detonation degenerates in a single head spin detonation. This is probably due to a small activation energy of the overall decomposition rate of chlorine dioxide.

Acknowledgments

This work was partly supported by the french ministry of Defence, General Delegation for Armament, Direction of Research, Studies, and Techniques, under Contract No 84075.

References

Ben Caïd, M. (1966) Etude de la flamme de décomposition du bioxyde de chlore gazeux, limites d' inflammabilité, détonation, déflagration. Ph. D. Thesis, University of Paris, France.

Bethune, D.S., Schell-Sorokin, A.J., Lankard, J.R., Loy, M.M.T., and Sorokin, P.P. (1983) Advances in Laser Spectroscopy (edited by B.A. Garetz and J.R. Lombardi), Vol. 2, pp. 1–43. John Wiley and Sons, New York.

Kiefer, J.H. (1981) Shock Waves in Chemistry (edited by A. Lifshitz), pp. 219–277. Marcel Dekker, Inc., New York.

Mc Hale, E.T. and von Elbe, G. (1968) The explosive decomposition of chlorine dioxide. J. Phys. Chem. 72(6), 1849–1856.

Paillard, C., Dupré, G., Charpentier, N., Lisbet, R., and Youssefi, S. (1985) A shock tube study of the decomposition of chlorine dioxide. 15th International Symposium on Shock Tubes and Waves, Berkeley, Calif. (to be published).

Schumacher, H.J., and Stieger, G. (1930) Thermal decomposition of chlorine dioxide. Z. phys. Chem. 7B, 363–386.

Semenov, N.N. (1959) Some Problems in Chemical Kinetics and Reactivity, Vol 2. Princeton University Press, Princeton, N.J.

Westbrook, C.K. (1984) Chemical kinetic factors in gaseous detonations. ACS Symp. Ser. 249, 175–192.

Measurement of Ignition Delay Time on Unsteady Hydrogen Jets

Fumio Higashino*

Tokyo-Noko University, Tokyo, Japan

and

Yoshimi Ishii†

Tokyo Metropolitan College of Aeronautical Engineering, Tokyo, Japan

and

Akira Sakurai‡

Tokyo Denki University, Tokyo, Japan

Abstract

In the present study attention is focused on the non-steady aspect of ignition of hydrogen jets into the compressed air. Two types of shock tubes, namely a conventional and a free piston shock tube, are utilized for the rapid compression of air. Experiments were made near the temperature of the ignition limit. In the case of shock compression, ignition is controlled by the mixing process between hydrogen and air for higher temperature region. Whereas it is limited by chemical kinetics for comparatively low temperature region. In the case of isentropic compression by a free piston, the ignition delay time is several times longer than those observed in shock compression, since the temperature in the test section is very low during the early stage of compression and this period does not contribute significantly to chemical reaction rates.

Introduction

In the last few years, hydrogen fuel has been recognized as advantageous for both direct-injection spark-ignition and diesel engines since the injection of hydrogen into the cylinder can simultaneously prevent backfire and increase the volumetric efficiency. In practice there have been

---

Presented at the 10th ICDERS, Berkeley, California, August 4-9, 1985. Copyright © 1986 by the American Institute of Aeronautics and Astronautics, Inc. All rights reserved.

*Professor, Department of Mechanical Engineering.

†Associate Professor.

‡Professor, College of Science and Engineering.

numerous attempts to burn hydrogen fuel in reciprocating engines. To predict the performance characteristics of these engines, one must understand the mixing process between the hydrogen and the high-temperature oxidizer and the ignition phenomena. While extensive theoretical and experimental studies have been performed on the steady jet issuing into an isothermal atmosphere, few studies on the mixing and combustion processes have been performed (Adachi and Suzuki, 1984; Higashino et al.,1984; Ishii et al.,1983; Sakurai and Takayama, 1983; Takayama and Sakurai,1983).

In the present study, attention is focused on the non-steady aspect of the ignition of hydrogen jets injected into the compressed air. Two types of shock tubes are utilized for rapid compression of the air. The experiments were made at temperatures near the ignition temperature limit. With shock compression, the ignition is found to be controlled by the mixing process between hydrogen and air at higher temperatures, whereas it is limited by the reaction rates at relatively low temperatures. With isentropic compression by a free piston, the ignition delay time is several times longer than that with shock compression, since the temperature in the test section is very low during the early stages of compression and a period that does not contribute significantly to the chemical reaction rates.

Shock Compression

Figure 1 is a schematic diagram of the experimental setup of a conventional nondiaphragm shock tube. The low-pressure section is 4.5 m long, has a square cross section of 40×40 mm^2, and has observation windows near the end plate. Hydrogen gas stored in a commercial cylinder was injected into the test section through an orifice on the end plate of the

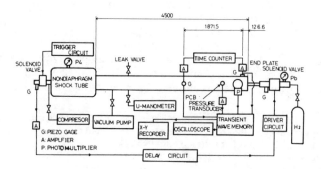

Fig.1 Schematic diagram of experimental setup of a conventional nondiaphragm shock tube.

shock tube. The hydrogen injection was controlled by a
solenoid valve in the pipe connecting the gas cylinder to
the shock tube. The valve was triggered by an electrical
pulse and remained open for 54 ms. The flow rate of the
hydrogen through the orifice of the end plate was about 2
cm^3/ms measured at standard temperature and pressure. The
fuel was injected into the test chamber for 8.5 - 11 ms
before the reflection of the incident shock waves at the
end plate. In the present work, the ignition phenomena was
investigated near the region of the ignition limits so that
the strength of the incident shock Mach numbers was 2 - 3.
Air was used as both the driving and driven gases. The
pressure change behind the incident and reflected shock
waves was measured with pressure transducers. They are
mounted on the sidewall of the chamber. Additional dimensions of the test chamber are shown in Fig. 2.

Piston Compression

The schematic diagram of the experimental setup of the
free piston shock tube is shown in Fig. 3. The stainless
steel driven section was 2 m long and had a circular cross
section of 26.7 mm i.d. An aluminum test section of 65 mm
long and 26 mm i.d. was attached by a flange at the end
of the driven tube. The test section has three measuring
ports at which either a pressure gage or a photocell was
mounted to analyze the ignition process in hydrogen jets.
Hydrogen fuel was injected into the test section through
a solenoid valve mounted on the end plate of the test section. To simulate compression process in an engine cylinder the following four pistons were tested: 0.01 and 0.022
kg Teflon, 0.099 kg Duralmine, and 0.29 kg steel. Heavier
pistons were found to be suitable for isentropic compression of the air. Triggered by an electric valve, the fuel

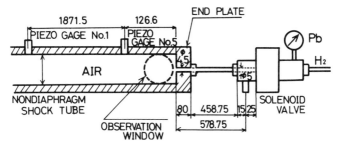

Fig.2 Test section of the nondiaphragm shock tube and hydrogen
injection equipment.

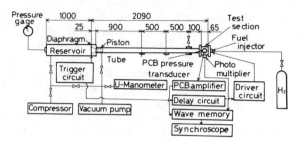

Fig.3   Schematic diagram of experimental setup of a free piston
shock tube.

injection valve opened for 10–13 ms. The mean flow rate of
the hydrogen jet in this period was about 1.5 cm^3/ms measured at standard temperature and pressure.   The pressure
change as well as the light emission due to combustion, was
observed on an oscilloscope.

Results and Discussion

   A uniform region of temperature and pressure is usually
obtained behind a reflected shock wave in the test section.
In the present work, the effect of the jet on the flowfield
behind the shock wave must be taken into account. A schlieren photograph and a schematic of the interaction between
the reflected shock wave and the jet are shown in Fig. 4.
In this figure, a strong interaction region exists near the
apex of a bow shock wave produced by the jet-shock interactions.   The propagation speed of the apex was determined
from successive schlieren photographs. Thus the pressure
and temperature change with time along the passage of the
apex was estimated from the Rankine-Hugoniot relations,
with the physical quantities ahead of the bow shock taken
as those behind the incident shock wave.   The typical
changes in the pressure and temperature of interaction
region, A, are shown in Fig. 5. The maximum temperature of
A is two times higher than it is in the absence of the jet.
The effect of the jet on the flowfield behind the shock is
limited to times of a few tenths of microsecond after the
shock reflection. Thereafter the values of both the pressure and temperature approach the uniform values behind the
reflected shock wave. This time interval is negligible in
comparison with the ignition delay time. The ignition delay
time, td, defined by the time interval between the instant
of shock reflection at the end wall and that of ignition,
is plotted against the reciplocal of temperature behind the

reflected shock wave, 1/T, in Fig. 6. The measured ignition
delay time was several times greater than that indicated in
the existing data for premixed gases. In this figure, the
regions A and B may be dominated by diffusion processes,
whereas region C may be dominated by chemical reaction
rates. This difference can be attributed to the time
needed for unsteady mixing of the hydrogen and air prior to
ignition.

In the case of piston compression, one of heavier pistons
was employed to compress the air isentropically. The experiments were made in the region near the autoignition temperature limit for a premixed gas. Typical traces on the
oscilloscope are shown in Figs. 7 and 8. The upper and
lower traces show the temporal historics of the pressure

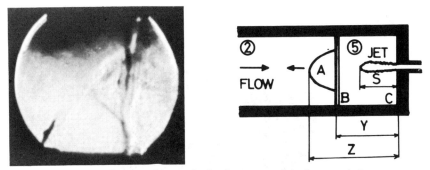

Fig.4   Interaction of shock wave and hydrogen jet.

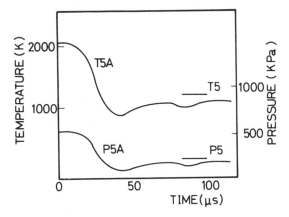

Fig.5   Change in pressure and temperature of A in Fig.4. (Hydrogen
pressure in a reservoir Pb=0.6 MPa, shock Mach number Ms=2.75,
initial pressure in the test section of the nondiaphragm shock tube
P1=533Pa)

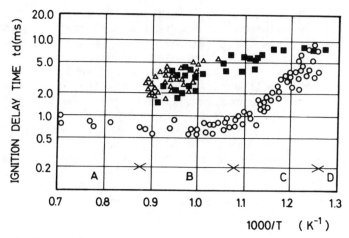

Fig.6 Ignition delay time td observed behind reflected shock waves in the nondiaphragm shock tube O and in the free piston shock tube, △ 0.099 kg and ■ 0.29 kg pistons.

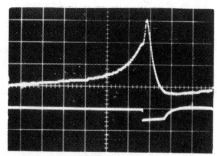

Fig.7 Typical oscilloscope traces of pressure and light emission in the test section of the free piston shock tube (ignition: 5 ms/div, upper trace denotes pressure change in time, 1.2 MPa/div).

jump and light emission, respectively. Both the pressure and light emission are apparent and are indicative of the ignition and combustion of the injected hydrogen. The scant light emission observable in Fig. 8 suggests the absence of ignition and hydrogen combustion.

The operational range of a regulator controlling the hydrogen pressure in a reservoir Pb is limited for safety reasons to comparatively low pressures of 2.5 MPa; thus the hydrogen was injected into the cylinder 30-50 ms before the compressed air pressure reached the maximum value. If we define the ignition delay time as the time interval between the instant of hydrogen injection and that of ignition, the ignition delays are some 10 times longer than the values of

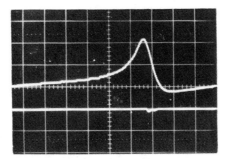

Fig.8 Typical oscilloscope traces of pressure and light emission in the test section of the free piston shock tube (no ignition: 5 ns/div, upper trace denotes pressure change in time 1.2 MPa/div).

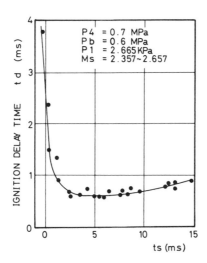

Fig.9 Relation between ignition delay time td observed behind reflected shock waves and hydrogen injection time ts (Ms: shock Mach number, P4: pressure in the driving section, P1: in the driven section, Pb: hydrogen reservoir pressure).

the shock compression. Thus the ignition delay time is defined here as the time interval between the instant at the autoignition temperature for premixed gas of 858 K and that of the ignition detected by the photocell.

In this case, the temperature was estimated from the measured pressure profiles under the assumption of isentropic compression. The ignition delay time for the piston compression is also plotted in Fig. 6 against the reciplocal of the temperature. From this figure, we found that the ignition delay times of unsteady hydrogen jets compressed by means of a free piston were much greater than those produced by the shock compression in the higher-temperature region B. In the early stage of the piston compression, the air temperature is low and the mixing process is dominant for the flowfield. In the lower temperature region C , however, the delay time for the piston compression approaches the values of the shock compressions in the near region of the ignition limit. This limit may exist in the vicinity of the boundary between regions C and D, since ignition was not observed in region D. In region C, the mixing process can continue until the temperature approaches the ignition limit, thereafter, the ignition process is controlled by the reaction rates.

To justify this explanation, the ignition delay time of the shock compression is plotted against various hydrogen injection times, in Fig. 9. We found that the ignition delays were very long when hydrogen was injected into the test section just before or after the initial shock reflection at the end wall. The results suggest that the temperature difference between the fuel and the oxidizer during the unsteady process of compression may play an important role for the ignition. In fact, the sudden compression resulting from the use of the lighter piston makes the ignition delay time greater (see Fig. 6).

Conclusions

In this work, the ignition delay times in unsteady hydrogen jets were investigated. The conclusions of the investigation were:

(1) The ignition delay times of unsteady hydrogen jets compressed by means of a free piston are greater than those caused by shock compressions in the higher-temperature region.

(2) The ignition delay times in both experiments approach each other in the near region of the ignition limit and are controlled by reaction rates.

(3) The temperature difference between the fuel and the oxidizer during an unsteady process of compression may play an important role for the ignition delays.

Acknowledgments

This work was partially suported by the Japan Ministry of Education under a grant in aid for special scientific research on energy and the SECOM Science and Technology Fund.

References

Adachi, T. and Suzuki, T., (1984) Ignition of a spurt fuel gas. Res. Rept., SECOM Science and Technology Fund., pp.55-61 (in Japanese).

Higashino, F., Shioneri, T. and Sakurai, A., (1984) Study on compression process in a free piston shock tube. Paper presented at Japanese Society of Mechanical Engineers Conf.(in Japanese).

Ishii, Y., Higashino, F. and Sakurai, A., (1983) Interaction between reflected shock wave and nonsteady jet in a shock tube. Paper presented at Japanese Society of Mechanical Engineers Conf. (in Japanese).

Sakurai, A. and Takayama, F., (1983) Spurt of gas from a hole into air. J. Phys. Soc. Japan, 52, 2963-2964,(1983).

Takayama,F. and Sakurai, A., (1983) Numerical calculation on combustion process of hydrogen jets in air. Proceedings of 15th Conference Japan Society of Fluid Mechanics, pp. 196-200 (in Japanese).

Deviations from the Boltzmann Distribution in Vibrationally Excited Gas Flows

Friedrich Offenhäuser* and Arnold Frohn†
University of Stuttgart, Stuttgart, West Germany

Abstract

A new model for the exchange of vibrational energy in one-dimensional flows of CO_2-H_2O-N_2-O_2-He gas mixtures is presented. In contrast to previous models, the assumption of local Boltzmann distributions for the vibrational degrees of freedom is not required. This generalization was achieved by the assumption that the molecules are harmonic oscillators with one or more degrees of freedom represented by finite numbers of energy levels. The population densities of these energy levels are coupled by a set of rate equations. It is shown that in some cases of molecular gas flow the Boltzmann distribution for the vibrational degrees of freedom may be disturbed.

Introduction

The energy transfer between translational, rotational and vibrational degrees of freedom in flows of molecular gas mixtures is of fundamental importance in molecular gas lasers. For many applications local equilibrium energy distribution can be assumed for all degrees of freedom. The internal energy can be described by a set of temperatures; the temperature for rotation is usually assumed to be the same as for translation. These assumptions were made by Harrach and Einwohner (1973) who presented a four-temperature kinetic model for the calculation of CO_2 TE-lasers, and by Frohn and Berg (1977), who presented theoretical calculations of the reaction kinetics in shock-heated CO_2-O_2-N_2 gas mixtures. There are, however, some processes in molecular gas

Presented at the 10th ICDERS, Berkeley, California, August 4-9, 1985. Copyright © 1986 by the American Institute of Aeronautics and Astronautics, Inc. All rights reserved.
*Dr.-Ing., Department of Aeronautical Engineering, University of Stuttgart. Present address: DFVLR, German Aerospace Research Establishment Stuttgart, West Germany.
† Prof.-Dr.rer.nat., Department of Aeronautical Engineering, University of Stuttgart.

flows that may lead to departures from the local equilibrium energy distribution for the vibrational degrees of freedom; such processes are strong vibrational-vibrational energy transfer and excitation by electron impact and radiation fields. These processes have beeen studied in CO_2-H_2O-N_2-O_2-He gas mixtures by theoretical calculations with a model that does not require local Boltzmann distributions for the vibrational degrees of freedom.

The System of Reactions

It is assumed that all vibrational degrees of freedom of a molecule can have the same maximum energy. The total vibrational energy of one molecule may not exceed the dissociation energy D. The number of energy levels, which have to be taken into account, can be estimated by the relation

$$N_j = D/JE_j$$

where J is the number of vibrational degrees of freedom of the molecule and E_j is the energy difference between energy levels for the vibrational degree j. The values for N_j and the characteristic temperatures for vibration are given in Table 1. For H_2O only the bending mode was taken into account. The other vibrational modes of H_2O have characteristic temperatures $\Theta > 5000$ K; they are therefore negligible. The reactions and energy levels, which have been taken into account, are given in Table 2. The reactions listed without an asterisk have been suggested by Taylor and Bitterman (1969) for the description of CO_2-N_2 laser systems. The system of reactions is completed by the reactions listed with an asterisk; these reactions are important for high-temperature processes.

In contrast to previous kinetic models, which were presented by Glowacki and Anderson (1971) and by Armandillo and Kaye (1980), the symmetric modes of CO_2 are not coupled by an equilibrium assumption, but by a fast Fermi-resonance reaction, because the first vibrational levels of these modes may be populated much more rapidly by radiation due to the laser transitions $00°1 \rightarrow 10°0$ or $00°1 \rightarrow 02°0$ than the other levels of these vibrational modes. The kinetic data given by Taylor and Bitterman (1969) and by Blauer and Nickerson (1974) were brought up to date as far as experimental results were available in the literature. The kinetic data, which are not known from literature, have been calculated by different theoretical methods. The best result was obtained by the theory of Widom (1957), which reproduces known kinetic data with high accuracy. It is therefore concluded that the unknown kinetic data can be calculated as well by this theory. In some cases, the method of Millikan and White (1963) could be applied. The rate coefficients for electron impact have been taken from Nighan (1970). The mean electron energy is given by Nighan (1970) as a function of the reduced electric field strength E/n, where n is the particle density for the gas mixture. The mean velocity v_m of the electron drift can be determined by an elec-

Table 1   Number of energy levels that have been taken into account and characteristic temperatures for vibration

Degree of freedom	Number of levels	Characteristic temperature for vibration, K
CO_2 (ν_1)	18	1998
CO_2 (ν_2)	35	960
CO_2 (ν_3)	10	3381
H_2O (ν_2)	25	2296
N_2	33	3355
O_2	26	2240

Table 2   System of vibrational-translational (VT) and vibrational-vibrational (VV) reactions

VT reactions

CO_2 (j-1 0°0) + M + 1388/cm → CO_2 (j 0°0) + M^a		$j = 1,\ldots,18$
CO_2 (0 j-1°0) + M + 667/cm → CO_2 (0 j°0) + M		$j = 1,\ldots,35$
CO_2 (0 0° j-1) + M + 2349/cm → CO_2 (0 0° j) + M		$j = 1,\ldots,10$
H_2O (0 j-1 0) + M + 1595/cm → H_2O (0 j 0) + M		$j = 1,\ldots,25$
N_2 (j-1) + M + 2331/cm → N_2 (j) + M		$j = 1,\ldots,33$
O_2 (j-1) + M + 1556/cm → O_2 (j) + M		$j = 1,\ldots,26$

Intermolecular VV reactions

CO_2 (00°0) + N_2 (j) + 18/cm → CO_2 (00°1) + N_2 (j-1)		$j = 1,\ldots,33$
CO_2 (00°2) + N_2 (j-1) + 7/cm → CO_2 (00°1) + N_2 (j)		$j = 1,\ldots,33*$
O_2 (0) + CO_2 (02°0) + 270/cm → O_2 (1) + CO_2 (00°0)*		
N_2 (0) + O_2 (1) + 775/cm → N_2 (1) + O_2 (0)		
N_2 (0) + H_2O (010) + 736/cm → N_2 (1) + H_2O (000)		
H_2O (000) + O_2 (1) + 39/cm → H_2O (010) + O_2 (0)		

Intramolecular VV reactions

CO_2 (03^{1}0) + M + 416/cm → CO_2 (00°1) + M
CO_2 (02°0) + M + 102/cm → CO_2 (10°0) + M^b

[a]The variable M may be each species of the mixture. [b]Fermi resonance.

tron energy balance. Neglecting vibrational deactivations
by electron impact, one obtains

$$v_m = n/E \sum_{j,1} \psi_j u_{j1} k_{j1}$$

where ψ_j, u_{j1}, and k_{j1} are the molar fractions, the excitation energies, and the rate coefficients for electron impact
respectively. The vibrational degrees of freedom are numbered by the index j, and the vibrational energy levels are
numbered by the index 1. From the measured electric power P,
the current density j can be determined by the equation

$$P/V = Ej$$

where V is the volume, which is pumped by the electric discharge. The electron density n_e therefore is given by

$$n_e = j/ev_m$$

The gas flow was assumed to be frictionless, and heat conduction has been neglected. Together with the conservation
equations, one obtains a system of ordinary differential
equations, which was solved numerically by the method of
Runge-Kutta.

Results of the Calculations

Calculations were performed to study the influences of
electric discharges, radiation fields, and gasdynamic shocks

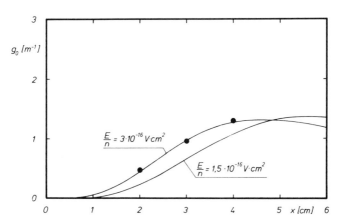

Fig. 1 Small signal gain for the 10.6-μm transition in the transverse flow CO_2 laser excited by radio frequency.

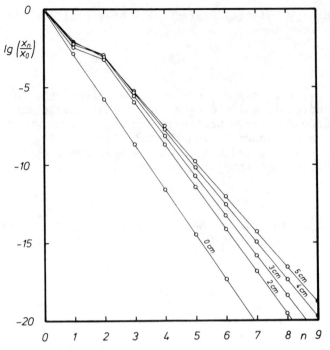

Fig. 2   Population density distribution for $CO_2(\nu_1)$ in the transverse flow CO_2 laser without radiation.

on the population density distributions of the vibrational
degrees of freedom. Very strong deviations from the Boltzmann distribution have been obtained for a transverse flow
CO_2 laser which is excited by radio frequency. A gas mixture of 4.4% CO_2, 18.6% N_2, and 77% He, and a reduced electric
field strength of 3×10^{-16} V.cm² was assumed. Figure 1 shows
the small signal gain for the 10.6-μm CO_2 laser transition;
these results are in good agreement with experimental results given by Jacoby (1983). In Figs. 2-5, profiles of the
population density distributions for all vibrational degrees
of freedom are shown as a function of the vibrational quantum number. The most important deviations are those of the
stretching modes, because the deviations are limited to the
first and second excited energy levels, which are the laser
levels. This gives as a result that the small signal gain
is increased by the deviations from the Boltzmann distribution.
     Now lasing is admitted. It is assumed that the inten-
sity inside the cavity is zero, if the small signal gain is
less than the threshold value given by the gain and loss

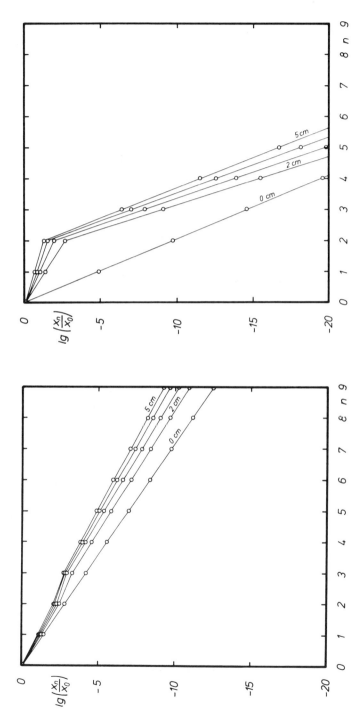

Fig. 3 Population density distribution for CO_2 (ν_2) in the transverse flow CO_2 laser without radiation.

Fig. 4 Population density distribution for CO_2 (ν_3) in in the transverse flow CO_2 laser without radiation.

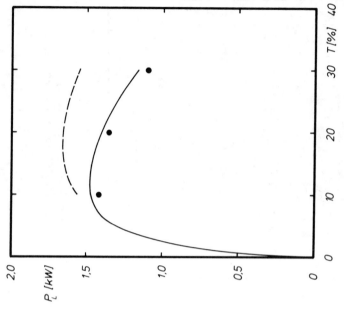

Fig. 6 Laser power at 10.6 µm for the excited CO_2 laser as a function of the transmission fraction.
●: Experimental results of Jacoby (1983).
---: Calculations of Jacoby (1983).
——: Results of the present calculations.

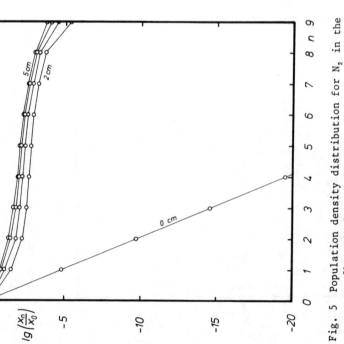

Fig. 5 Population density distribution for N_2 in the transverse flow CO_2 laser without radiation.

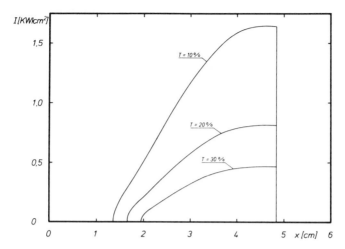

Fig. 7 Profiles for the intensity inside the cavity of the 10.6-µm
CO_2 laser excited by radio frequency.

balance

$$R_1 R_2 < \exp\ (-2gl)$$

where R_1 and R_2 are the reflectivities of the mirrors; the
gain length is $1 = 48$ cm. Figure 6 shows the laser power as
a function of the transmission; the transmitting mirror has
an absorption of about 2%, and the reflectivity of the other
mirror is about 98%. The results of these calculations show
good agreement with the theoretical and experimental results
reported by Jacoby (1983). Figure 7 shows the intensity profile inside the cavity. In Figs. 8-11, profiles for the population density distributions are shown for this case. The
departures from the Boltzmann distribution are decreased for
the symmetric stretching mode of CO_2 because the lower laser
level is populated by radiation transitions. The same happens to the bending mode of CO_2 by a strong energy transfer
due to the Fermi resonance. This result might have been expected. The asymmetric stretching mode, however, still
shows the same deviations from the Boltzmann distribution.
This is evoked by the strong vibration-vibration energy
transfer from nitrogen. The disturbances of the population
density distribution for N_2 are not reduced by this process,
because the molar fraction of N_2 is much higher than the
molar fraction of CO_2 and the N_2 molecules are still pumped
by radio frequency.
  The model is now applied to the relaxation zone behind
gasdynamic shocks in CO_2-H_2O-N_2 gas mixtures. The gas tem-

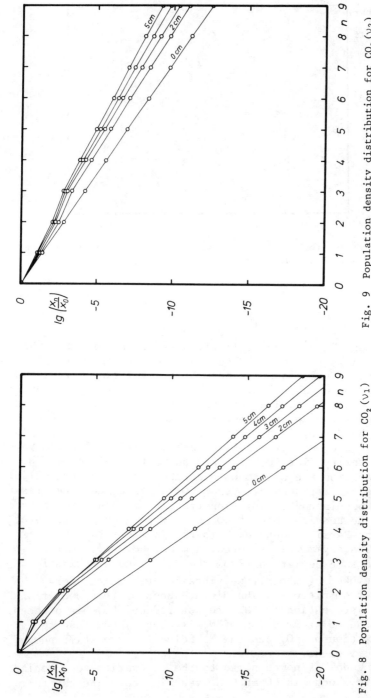

Fig. 9   Population density distribution for CO_2 (ν_2) in the transverse flow CO_2 laser with radiation.

Fig. 8   Population density distribution for CO_2 (ν_1) in the transverse flow CO_2 laser with radiation.

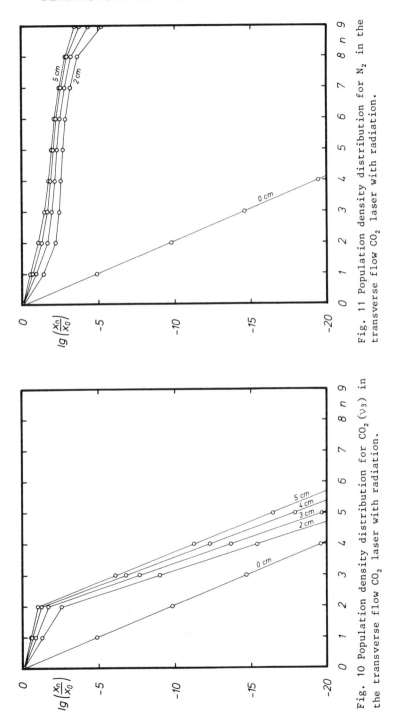

Fig. 10 Population density distribution for CO_2 (ν_3) in the transverse flow CO_2 laser with radiation.

Fig. 11 Population density distribution for N_2 in the transverse flow CO_2 laser with radiation.

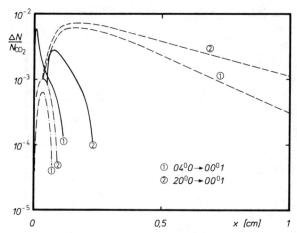

Fig. 12 Population density differences for the transitions
$04°0 \rightarrow 00°1$ - P32 and $20°0 \rightarrow 00°1$ - P32 with the wavelengths
50.194 μm and 22.321 μm behind a shock with Ma_s = 4 in a gas mixture of 10% CO_2, 2% H_2O, and 88% N_2 at an initial pressure of 8
mbars. ——: Results of calculations with the present model. ---: Results of calculations reported by Losev (1981). -.-.-: Experimental
results reported by Losev (1981).

perature and the pressure at the beginning of vibrational
relaxation are given by the equations of Rankine-Hugoniot.
In Fig. 12, profiles for the population density difference
are shown that belong to the transitions $04°0 \rightarrow 00°1$ and
$20°0 \rightarrow 00°1$ of CO_2. For these cases, theoretical and experimental results have been reported by Losev (1981) for a
gas mixture of 10% CO_2, 2% H_2O, and 88% N_2; the initial
pressure is 8 mbars and the shock Mach number Ma_s = 4. The
results obtained by the present kinetic model show a much
better agreement with the experimental results than the
theoretical results of Losev (1981). In this case, deviations from the Boltzmann distribution are obtained only for
the asymmetric stretching mode of CO_2; these deviations are
due to the strong vibrational energy transfer from nitrogen.
In Fig. 13, profiles for the difference between the vibrational temperatures of the energy levels and the mean vibrational temperature for CO_2 (ν_3) are shown. At the beginning of the relaxation, this temperature difference increases and reaches a maximum of about 130 K for the first
excited vibrational energy level; behind this maximum, the
system converges to thermodynamic equilibrium, where the
population density distribution can be described by a Boltzmann distribution.

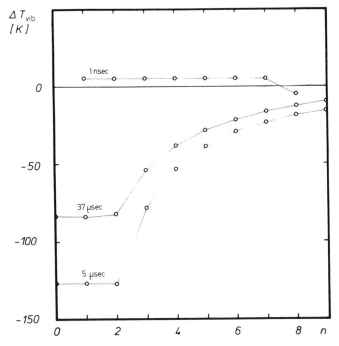

Fig. 13 Deviations of the temperatures for the vibrational energy levels from the mean vibrational temperature for $CO_2(\nu_3)$ behind a shock with Ma_s = 4 and an initial pressure of 8 mbars in a gas mixture of 10% CO_2, 2% H_2O, and 88% N_2.

Conclusions

The presented results show that there are some cases for which it is not justified to assume local equilibrium energy distributions for the vibrational degrees of freedom in gasdynamic flow problems. This is especially the case for gas flow lasers that are pumped by electric discharge.

References

Armandillo, E. and Kaye, A.S. (1980) Modelling of transverse-flow cw CO_2 lasers; Theory and experiment. J. Phys.: Appl. Phys. D 13, 321-338.

Blauer, J.A. and Nickerson, G.R. (1974) A survey of vibrational relaxation rate data for processes important to CO_2-N_2-H_2O infrared plume radiation. AIAA Paper No. 74-536, AIAA, New York.

Frohn, A. and Berg, G. (1977) Theoretical and experimental investigation of physico-chemical reaction processes in consideration of vibrational relaxation of CO_2-O_2-N_2 mixtures in a shock tube. For-schungsbericht des Landes Nordrhein-Westfalen, Nr. 2683, Fachgruppe Physik, Chemie, Biologie, 1. Aufl. Westdeutscher Verlag.

Glowacki, W.J. and Anderson, J.D. Jr. (1971) A computer program for CO_2-N_2-H_2O gasdynamic laser gain and maximum available power. NOLTR 71-210, Naval Ordnance Laboratory, White Oak, Md.

Harrach, R.J. and Einwohner, T.H. (1973) Four-temperature kinetic model for a CO_2 laser amplifier. UCRL-51399, Lawrence Livermore Laboratory, University of California, Livermore.

Jacoby, H. (1983) Calculation of performance of an RF-excited trans-verse flow CO_2 laser.DFVLR-FB 83-06, DFVLR Stuttgart, West Germany.

Losev, S.V. (1981) Gasdynamic Lasers,1st. ed., Springer, Berlin.

Millikan, R.C. and White, D.R. (1963) Systematics of vibrational relaxation. J. Chem. Phys. 39 (12), 3209-3213.

Nighan, W.L. (1970) Electron energy distributions and collision rates in electrically excited N_2, CO, and CO_2. Phys. Rev. 2 (5), 1989-2000.

Taylor, R.L. and Bitterman, S. (1969) Survey of vibrational relax-ation data for processes important in the CO_2-N_2 laser systems. Rev. Mod. Phys. 41 (1), 26-47.

Widom, B. (1957) Inelastic molecular collisions with a Maxwellian interaction energy. J. Chem. Phys. 27 (4), 940-952.

Nozzle and Cavity Flowfields in Gas Chemical Laser: Numerical Study

M. Imbert,* D. Zeitoun,† and R. Brun‡
Université de Provence, Marseille, France

Abstract

Experiments on chemical lasers are difficult and expensive due to the small dimensions of the apparatus, the gases that are used, and the large number of parameters. Thus, it is interesting to utilize calculation codes for simulating such flows. Furthermore, it is well known that the quality of the laser beam depends strongly on the structure of the flowfield in the cavity, which itself depends on the absolute values and on the different profiles in the laser nozzles. Thus, the Navier-Stokes equations are solved here in the unsteady, two-dimensional, laminar case, in both nozzles and in the cavity. The numerical method is adapted from an explicit MacCormack scheme, and the two-dimensional operator is split into a sequence of one-dimensional operators. Different physical cases are treated that take into account the presence of a step between the exit sections of the nozzles. The first two cases, with nitrogen as a test gas, show the importance of the dissipative effects in the nozzles and the modifications of the structure of the flow in the cavity induced by a change of the reservoir conditions. Thus, the influence of the pressures and the velocities, at the exit of the nozzles, on the wave structure in the cavity are studied. The third case takes into account the diffusion effects in the cavity, the jet being constituted by a mixture of F, F_2, HF, He for the primary nozzle and a mixture of D_2, He for the secondary one.

Presented at the 10th ICDERS, Berkeley, California, August 4-9, 1985. Copyright © 1986 by the American Institute of Aeronautics and Astronautics, Inc. All rights reserved.
*Assistant, Dept. of Mathematics, Fluid Dynamics and Thermophysics Lab.
† Research Engineer, Dept. of Mathematics, Fluid Dynamics and Thermophysics Lab.
‡ Research Director, Dept. of Mathematics, Fluid Dynamics and Thermophysics Lab.

Nomenclature

D_{km}	= species diffusion coefficient
e	= specific internal energy
E	= specific total energy
h_k	= specific enthalpy of species k
k	= thermal conductivity
$\hat{M}_m$	= mixture molecular weight
P	= pressure
q_x, q_y	= components of energy flux (heat conduction and species diffusion)
R	= universal gas constant
u, v	= velocity components
Y_k	= mass fraction of species k
ν_{kx}, ν_{ky}	= components of diffusion velocity of species k
ρ	= mass density of mixture
ρ_k	= mass density of species k
τ_{xx}, τ_{xy}, τ_{yy}	= components of viscous stress tensor

Introduction

It is well understood that the very small dimensions, the gases that are used, the complexity of the phenomena, and the great number of parameters contribute to create difficult and costly conditions for experiments on chemical lasers. Thus, the numerical approach is of great interest for the simulation of the corresponding reactive flows, including subsonic, transonic, and supersonic regions.

In past years, the numerical determinations of these flows have been the subject of many works. Among them, those of Parthasarathy et al. (1979), Fomin and Soloukhin (1979), and Rapagnani (1983) can be cited. From an experimental point of view, Cenker and Driscoll (1982) have presented a visualization of the flow in the cavity of a supersonic trip nozzle flow.

Before analyzing the flow in the cavity, it is necessary to study the flows in the nozzles and to determine the influence of the reservoir conditions on these flows. Therefore, the numerical integration of the complete two-dimensional and unsteady Navier-Stokes equations has been recently performed in the nozzles and in the cavity (Zeitoun et al. 1984). The numerical method (Imbert and Zeitoun 1982) is based on the explicit MacCormack scheme (MacCormack and Baldwin 1975).

Some specific cases are treated which take into account the presence of a step separating the exit of the nozzles. The first two cases (test gas: nitrogen) point out the importance of the dissipative effects in the nozzles and the

modifications of the flowfield in the cavity due to a
variation of reservoir conditions. The third case takes into
account the diffusive effects in the cavity, the jets being
constituted by a mixture of F, F_2, HF, He in the primary
nozzle and of mixture of D_2, He in the secondary nozzle.

Governing Equations

In the case of a two-dimensional and laminar flow, the
Navier-Stokes equations can be written in the following
conservative form :

$$\frac{\partial \vec{U}}{\partial t} + \frac{\partial \vec{F}}{\partial x} + \frac{\partial \vec{G}}{\partial y} = 0 \qquad (1)$$

where

$$\vec{U} = [\rho, (\rho_k, k = 1, KM-1), \rho u, \rho v, \rho E]^t$$

$$F = \{\rho u, [\rho_k(u + v_{kx}), k = 1, KM-1], \rho u^2 + P$$

$$- \tau_{xx}, \rho uv - \tau_{xy}, (\rho E + P)u - u\tau_{xx} - v\tau_{xy} + q_x\}^t$$

$$\vec{G} = \{\rho v, [\rho_k(v + v_{ky}), k = 1, KM-1], \rho uv$$

$$- \tau_{xy}, \rho v^2 + P - \tau_{yy}, (\rho E + P)v - u\tau_{xy}$$

$$- v\tau_{yy} + q_y\}^t$$

with

$$\rho_k \vec{v}_k = -\rho D_{km} \vec{\nabla} Y_k$$

$$\vec{q} = -k_m \vec{\nabla}T - \sum_{k=1}^{KM} \rho_K h_K \vec{v}_k$$

$$E = e + (u^2 + v^2)/2$$

$$P = \rho r T \text{ with } r = R/\hat{M}_m$$

Computational Domain and Boundary Conditions

The integration of the equation system (1) is performed
in the domain represented on Fig. 1. Given the finite
distance between the exit sections of the nozzles, this
integration is made in two steps. The first one enables us
to calculate the flow in every nozzle and therefore to know

the profiles of the quantities in these exit sections. The
second one gives the flow in the cavity up to a distance of
ten times the distance between the nozzle axes.

The boundary conditions are the following ones. On the
symmetry axes :

$$\frac{\partial f}{\partial y} = 0 \ (\forall f = \rho_k, \ u, \ P, \ T) \ \text{and} \ v = 0$$

On the walls : $u = v = 0$; $\partial e/\partial n = 0$ or $e(T_w)$ imposed ;
$\partial P/\partial n$ is given from the second momentum equation, taking
into account the no-slip condition at the wall of the
nozzles ; and $\partial P/\partial n = 0$ at the wall separating the nozzle
exits.

For each domain, the outflow conditions are determined
by extrapolation.

The flows in the nozzles are obtained from a particular
case which takes into account the flow-rate conservation at

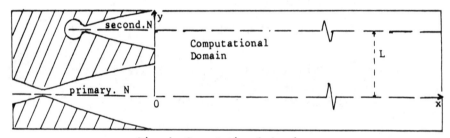

Fig. 1 Computational domain.

Table 1 Reservoir conditions for the three studied cases

	Cases	P_o,atm	T_o,K	D*,mm	D_E,mm	Test gases	Re_o
A	Primary nozzle	1	1500	.12	2.78	N_2	410
	Secondary nozzle	1.2	300	.10	1.45	N_2	2660
B	Primary nozzle	3	1500	.12	2.78	N_2	1230
	Secondary nozzle	1.2	300	.10	1.45	N_2	2660
C	Primary nozzle	3	1500	.12	2.74	He:.75 F :.05 F_2:.005 HF:.195	630
	Secondary nozzle	.8	300	.10	1.50	He:.5 D_2:.5	790

each instant, and those in the inflow cavity are kept fixed during the solving procedure.

In every part of the integration domain, which is rendered rectangular by a change of coordinates, the step Δx is constant and the step Δy variable following an exponential law. The number of grid points is (97×32) for the nozzles and (101×47) for the cavity.

Numerical Scheme

The conservative form of the two-dimensional time-dependent Navier-Stokes equations is solved by using a predictor-corrector finite difference scheme (MacCormack and Baldwin 1975). The two-dimensional operator is separated into two one-dimensional sweep operators in the x and y directions (time-splitting technique). The dependent variable vector $\vec{U}$ is advanced in time as

$$U(x, y, t + \Delta t) = L_Y(\Delta t/2)\, L_X(\Delta t)\, L_Y(\Delta t/2)\, U(x, y, t)$$

Every one-dimensional operator is developed with an explicit predictor-corrector scheme.

The important variations of the time step Δt, coming from the stability criteria, in the integration domain impose a technique of local temporal step. This method becomes nonconsistent with the unsteady equations, but this does not affect the flow quantities in the steady state, which is alone of interest here. The evolution of dependent variables up to the steady state is controlled from maximum residuals for the velocity and pressure. This state is considered as reached when these values are smaller than 10^{-5} for each point of the computational domain.

Results and Discussion

Three cases are computed. The corresponding reservoir conditions (ρ_o, T_o) throat dimensions, nozzle exit sections (D^*, D_E) and Reynolds numbers are presented in Table 1. The first two cases (A and B) have been treated with nitrogen as a test gas for evaluating the influence of the reservoir pressure on the flowfield. The third case (C) considers a homogeneous mixture in each nozzle, and the composition is also indicated in Table 1.

Nozzle Flows

It is well known (Rapagnani 1983) that the solutions derived from boundary-layer-type equations are not entirely

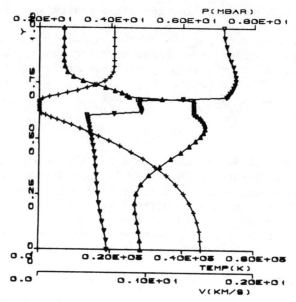

Fig. 2   Pressure ∇, temperature Δ, and velocity + profiles at the nozzle exit sections (case A).

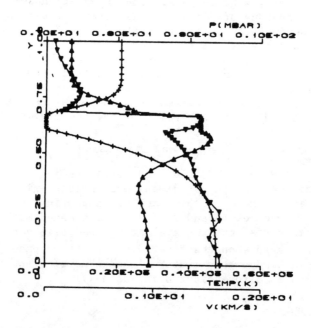

Fig. 3   Presure ∇, temperature Δ, and velocity + profiles at the nozzle exit sections (case B).

satisfying. It is therefore necessary to determine the
profiles of the quantities in each nozzle by solving the
Navier-Stokes equations.

It should be pointed out that the wall temperature
conditions differ in the nozzles. For the primary nozzle,
the wall temperature law is given, and for the secondary
one, no wall heat transfer is assumed.

The profiles of pressure, temperature, and velocity at
the nozzle exit sections are represented in Figs. 2-4,
for the three cases. Thus the importance of the dissipative
effects, the influence of the wall conditions on the
temperature profiles, and the nonuniform pressure distribution can be observed. The nozzle exit values on the
symmetry axes are also given in Table 2.

The increase of the reservoir pressure of the primary
nozzle (case B) can give at the exit section a pressure
higher than the exit pressure of the secondary nozzle; the
pressure and velocity gradients are then in the same sense,
contrary to case A. Case C is similar to case B.

The influence of the profiles on the structure of the
flow in the cavity is presented in the following section.

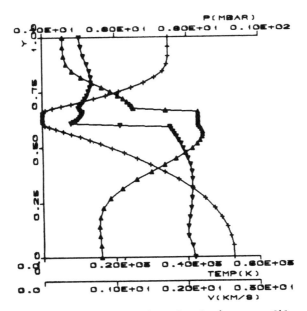

Fig. 4 Pressure ∇, temperature Δ, and velocity + profiles at the
nozzle exit sections (case C).

Cavity Flow
      In all cases, a .15-mm distance separates the extreme
points of the exit sections of each nozzle, which
represents 7% of the distance between the nozzle axes. The
differences concerning the flowfield structures and the
intensity of phenomena are clearly evidenced on Figs. 5-8,
which represent respectively the isobars and isomachs for
cases A and B. On the symmetry axes, the first compression,
which is also the more intense compression corresponds to
the axis on which the initial pressure was the weakest one.
The velocities of the jets being almost the same for both
cases, the pressure at the primary nozzle exit, smaller for
case A, gives a flowfield much less disturbed in this case.
Futhermore, the impact and penetration of the primary jet
in the secondary one is very important in case B.
      In case C , the flow is computed taking into account
the diffusion of species. As shown in Table 2, the pressure
gradient is in the same sense as the velocity gradient.
Thus the isobar curves are similar to case B (Fig. 9);
only one reflection is visible at the level of the primary
nozzle axis, and the flow in the cental part of the cavity
is less perturbed. The complexity of the flow can also be

Table 2 Nozzle exit values on symmetry axes

	Cases	P,mb	T,°K	V,m/s	M_s
A	Primary nozzle	3.9	285	1.5×10^3	4.36
	Secondary nozzle	7.0	69.5	6.9×10^2	4.07
B	Primary nozzle	9.2	288	1.58×10^3	4.58
	Secondary nozzle	7.0	69.5	6.9×10^2	4.07
C	Primary nozzle	6.6	150	2.7×10^3	5.4
	Secondary nozzle	5.0	55	1.74×10^3	4.18

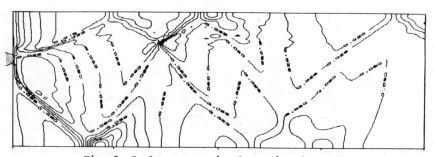

Fig. 5 Isobar curves in the cavity (case A).

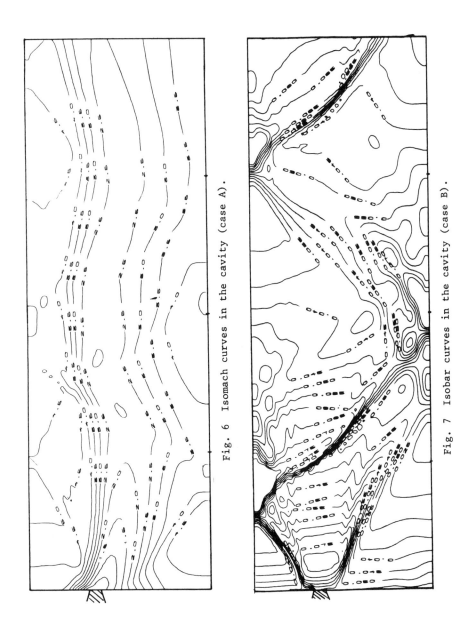

Fig. 6   Isomach curves in the cavity (case A).

Fig. 7   Isobar curves in the cavity (case B).

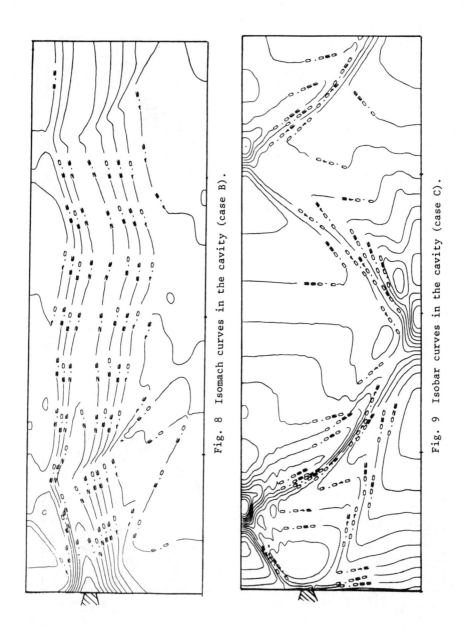

Fig. 8   Isomach curves in the cavity (case B).

Fig. 9   Isobar curves in the cavity (case C).

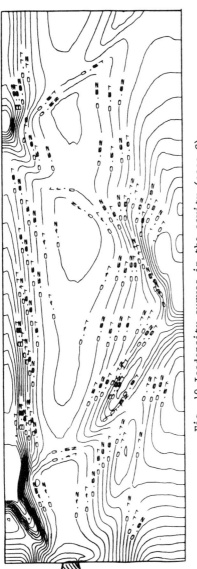

Fig. 10 Isodensity curves in the cavity (case C).

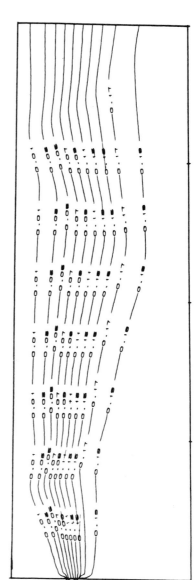

Fig. 11 Isoconcentration curves (HF) in the cavity (case C).

seen in Fig. 10, where isochore curves are represented. In
the same way, isoconcentration curves (mole fraction of HF)
are represented in Fig. 11, the inflection points of these
curves corresponding to the previously mentioned large
gradients.

Conclusions

For the study of the flows in laser nozzles and the
cavity of a chemical laser, the Navier-Stokes equations
have been solved with realistic conditions of geometry and
reservoir conditions. The results show the important
influence of the dissipative effect on the nozzle flows
and of the reservoir conditions on the flow structure in
the cavity. The conjugated effects of the pressure and the
velocity on the wave positions and intensities in the
cavity have been clearly shown. The species diffusion terms
having been taken into account, the next step will be the
study of the coupling between the flow and chemical
reactions.

Acknowledgments

This work was partially supported by ONERA and CCVR.

References

Cenkner, A.A., and Dirscoll, J.R. "Laser-Induced Fluorescence
    Visualization on Supersonic Mixing Nozzles that Employ Gas-
    Trips," AIAA Journal, Vol. 20, No 6, 1981, pp. 812-819.

Fomin, N.A., and Soloukhin, R.I. "Modeling of gas dynamic and
    relaxation phenomena in a mixed flow laser," AIAA Progress in
    Astronautics and Aeronautics, (edited by M. Summerfield),
    Combustion in Reactive Systems, Vol. 76, 1979, pp. 46-75.
    AIAA, New York.

Imbert, M., and Zeitoun, D., "Etude numérique d'un écoulement
    à nombre de Reynolds modéré dans une tuyère," Journal de
    Mécanique théorique et appliquée, Vol. 1, No 4, 1982,
    pp. 595-609.

MacCormack, R.W., and Baldwin, B.S., "A numerical method for
    solving the Navier-Stokes equations with application to
    shock-boundary layer interactions," AIAA Paper 75-1, 1975.

Parthasarathy, M.N., Anderson, J.D., and Jones, E., "Downstream
    Mixing Gasdynamic Lasers. A numerical Solution," AIAA Journal,
    Vol. 17, No 11, 1979, pp. 1208-1215.

Rapagnani, N.L., "Unsteady analysis of chemical laser cavity
    flowfields," Paper presented at the Ninth International
    Colloquium on Gasdynamics of Explosions and Reactive Systems,
    Poitiers, France, July 4-8, 1983.

Zeitoun, D., Imbert, M., and Brun, R. "Numerical study of a
    two-dimensional laser cavity flow," Paper presented at the
    Fifth International Symposium on Gas Flow and Chemical Lasers,
    Oxford, England, 20-24 August 1984.

CARS Spectroscopy of the Reaction Zone CH_4/N_2O and Nitramine Propellant Flames

L.E. Harris*

*U.S. Army Armament Research and Development Center
Dover, New Jersey*

Abstract

Coherent anti-Stokes Raman scattering spectra were obtained for a $\phi = 3.2$ CH_4/N_2O model and nitramine propellant flames throughout the reaction and postflame zones in order to assess kinetic mechanisms occurring in these flames. Spectra were obtained in the regions 4200-3900, 2400-2050, and 1900-1200 cm^{-1}. The reaction zone of the rich CH_4/N_2O flame was studied primarily to provide a stationary flame analog to the transient propellant flame. In the CH_4/N_2O flame, the decay of the initial products was observed through the Q branch of the ν_1 and ν_3 modes of N_2O and Q, O, and S branches of the ν_2 and $2\nu_2$ modes of CH_4. The formation of the products N_2, H_2 [Q(v" = 0 and v" = 1) and S(5)-S(9)], CO, and CO_2 (ν_1) were also observed. Temperatures were obtained from both the H_2 and N_2 Q branches. In the nitramine propellant flame, the upper bound of the gas-surface interface temperature was measured as 900 ± 100 K from the H_2 Q branch. Near the surface of the propellant, reactant RDX (1599 cm^{-1} tentatively assigned as asymetric NO_2 stretch) and transients HCN (ν_1) and NO are observed at moderate concentration (>1%).[1] The final product N_2 is observed at low concentration (~1%); H_2 (Q and S branches) and CO are observed at higher concentration (>10%). RDX and HCN decay within 2 mm of the propellant surface, while NO remains constant until 4 mm, where it decays with a concomitant rise in N_2 concentration and temperature. H_2 and CO also increase in temperature and concentration. The nitramine reaction zone is seen to consist of two regions characterized by

Presented at the 10th ICDERS, Berkeley, California, August 4-9, 1985. This paper is declared a work of the U.S. Government and therefore is in the public domain.

*Armament Engineering Directorate, Energetics & Warheads Division.

the reactions of RDX and HCN near the surface, consistent
with the high-temperature mechanism of RDX decomposition,
and the conversion of NO to N_2 to generate the luminous
flame further above the surface.

Introduction

    Obtaining direct experimental information on the
dynamics of the deflagration and detonation of energetic
materials has been difficult with the use of conventional
optical techniques. The advent of nonlinear optical techniques such as coherent anti-Stokes Raman scattering
(CARS) provides an opportunity to extend the information
obtainable on energetic systems. Propellant flames are
often transient, particle-laden, incandescent, and under
some conditions turbulent. However, single-shot CARS
spectra were obtained from the postflame region of double-
base propellant flames (Harris and McIlwain 1981, 1982).
This demonstration of the direct applicability of CARS to
obtaining temperature and concentration from propellant
flames led to further investigations in the reaction zone
and postflame region of CH_4/N_2O model propellant flames
(Harris 1982, 1983; Aron et al. 1983). These studies were
then extended to the reaction zone of nitramine propellant
flames through measurements at the surface and 6 mm above
the propellant surface (Aron and Harris 1984; Harris
1984). In these studies, both the spectral and spatial
ranges are extended for the reaction zone of CH_4/N_2O and
nitramine propellant flames. Rich CH_4/N_2O flames are used
as stationary model flames of the transient propellant
flames.
    The combustion of nitramine propellants has been
recently reviewed (Boggs 1984; Schroeder 1981). Nitramine
propellants contain a substantial percentage of nitramines
(~75%) along with some percentage of energetic binder
(i.e., nitrocellulose) and/or nonenergetic binders (i.e.,
organic esters). Current models of nitramine propellant
combustion are essentially models of HMX, cyclotetramethyl-
ene tetranitramine, and RDX, hexahydro-1,3,5-trinitro-s-
triazine deflagration. The burning rate expression for
nitramine propellants (Ben-Reuven and Caveny 1981)

$$r = ap^{1/2} (1 + p/b)^{1/2}$$

is such that at low pressure,

$$p \ll b, \quad r \sim p^{1/2}$$

whereas at high pressure,

$$p \gg b, \quad r \sim p^1$$

Much of the modeling of nitramine propellant has been to explain this complex burning rate behavior. Ben-Reuven and Summerfield (1983) have recently reviewed nitramine propellant modeling and derived improvements to the comprehensive (Ben-Reuven and Caveny 1981) model of nitramine deflagration . The Ben-Reuven and Caveny model consists of the following mechanism. First, partial decomposition of RDX in the liquid phase:

$$RDX(L) \rightarrow 3CH_2O + 9/4 \ N_2O + 3/4 \ NO + 3/8 \ N_2$$

followed by gas phase decomposition of RDX in the near field (close to the propellant surface):

$$RDX(G) \rightarrow NO_2 + N_2O + 3CH_2O + 3/2 \ N_2$$

and finally, oxidation of formaldehyde by NO_2:

$$CH_2O + NO_2 \rightarrow H_2O + CO + N_2$$

in the far field (relatively far from the propellant surface).

Recently, Kubota (1981) has made thermocouple measurements of temperature profiles in nitramine/polyurethane composite propellants. He obtained surface temperatures between 690 K and 730 K in the pressure range between 10 and 30 atm. Kubota (1981) found from flame quenching studies that both RDX and the polymeric binder melt at the surface to produce an homogeneous liquid layer that produces an homogeneous flame. The flame was found to consist of dark and luminous flame zones, with the luminous zone approaching the propellant surface with increasing pressure. The measured dark zone thickness at 20 atm for a 75% RDX formulation was 2 mm. The dark zone thickness was observed to vary inversely with the square of the pressure. At 1,000 psi, the dark zone would be compressed to 200 μm, spatially limiting accessibility for optical diagnostics. The luminous flame, however, would still be accessible to optical diagnostics. Kubota (1981) attributed the dark zone reaction to NO reduction to N_2 to produce the luminous flame. The burning rate was found to be controlled not by the luminous flame but rather by exothermic reactions of RDX at the propellant surface.

Ben-Reuven and Summerfield (1983) have added to the Ben-Reuven and Caveny model a nonequilibrium evaporation

law at the melt/gas interface, an improved melt phase
model including a decomposition-gas bubble, and an improved model for far-field processes including several
simultaneous secondary reactions.

Schroeder (1981) has reviewed nitramine decomposition
chemistry. At low temperature (500 to 600 K), the gas
phase reaction mechanism by which RDX initially decomposes
to CH_2O and N_2O is postulated as

$$(CH_2\ NNO_2)_3 \rightarrow 3CH_2O + 3N_2O$$

This is thought to occur through HONO elimination
and/or cyclic decomposition via the intermediate N-
nitroformimine, CH_2NNO_2. Crossover to a high-temperature
reaction mechanism in the gas phase is thought to occur
above about 600 K. This mechanism is thought to occur via
breakage of an initial NN bond followed by fragmentation
to CH_2NNO_2, which decomposes to HCN and NO_2, leading to
the overall initial reaction

$$(CH_2NNO_2)_3 \rightarrow 3HCN + 6NO_2 + 3H_2$$

The liquid phase reaction is also thought to occur by
a mechanism similar to the high-temperature gas phase
reaction mechanism.

Price et al. (1982) have recently modified the
Beckstead-Derr-Price (BDP) framework for HMX decomposition
to incorporate both the high- (endothermic) and low- (exothermic) temperature nitramine decomposition mechanisms in
the solid and exothermic and endothermic second-order
reactions in the gas phase. Cohen et al. (1985) have explicitly included reactions of HCN and NO in the BDP
model.

Miller (1982) and Miller and Coffee (1983[a,b]) have
given a detailed comparisons of the various methods used
to model propellant combustion. Their assessment was that
the methods place too much emphasis on matching experimental burning rate data that are relatively insensitive
to mechanistic details. They conclude that mechanisms
used in propellant modeling should be independently vali-
dated. Propellant surface temperature T_s enters many of
these models as a parameter used to match experimental
burning rates. Independent measurement of T_s will allow
validation and further development of these models.

CARS spectroscopy, which was reviewed (Druet and
Taran 1981; Ecbreth and Schreiber 1981), provides an ideal
tool for the further elucidation of nitramine propellant
kinetic mechanisms. CARS has the necessary spatial (100
μm), temporal (10 ns single shot), and spectral resolution

to provide the detailed temperature, concentration, and
rovibrational state distribution profiles necessary to
successfully model propellant flames from independently
measured elementary kinetic reactions (Gardner 1984).

Experimental

CARS spectra were generated using folded BOXCARS to
achieve phase matching. The output of a Quanta-Ray DCR-2A
Nd/YAG laser at 1.06 μm (700 mJ) is doubled to generate
the pump beam at 5320 A (250 mJ) with a bandwidth of near
1 cm^{-1}. The pump beam is separated from the primary beam
with prisms. The pump beam is split to generate ω_{1s} and
ω_{1p}. ω_{1s} is used to pump a dye laser to generate the
Stokes beam ω_2. The dye laser consists of a flowing dye
cell in a planar Fabry-Perot oscillator cavity pumped
slightly off-axis by 20% of ω_{1s}, with the output amplified
by an additional flowing dye cell pumped by the remainder
of ω_{1s}. The dye laser is operated broadband with the
laser dyes given in Table 1. To achieve BOXCARS geometry,
ω_{1p} is reflected onto the front surface of a dichroic
mounted at a 45 deg angle. The dichroic reflects 50% of
ω_{1p} from half of its front surface and 100% of the remainder of ω_{1p} from its back surface while transmitting
ω_2, which is introduced from the rear below ω_{1p}. ω_{1p} is
split into two beams, ω_1 and ω_1', such that at the focusing lens (200 mm focal length), ω_1, ω_1', and ω_2 are parallel and situated on a circle of 12.5 mm diameter with ω_1
and ω_1' on the central horizontal plane of the lens with
ω_2 in the central vertical plane. Telescopes are inserted
in the ω_{1p} and ω_2 beams to allow the focal spots to be
equalized and intersecting. This was achieved with 0.85X
and 2X Galilean telescopes in ω_{1p} and ω_2, respectively. A
5-mm-diam iris centered on ω_{1p} prior to splitting further
restricts intersection to the central portion of ω_{1p}. To
optimize phase matching, a 12.5-mm-thick optical flat
rotatable about its horizontal axis was inserted into ω_2
before the focusing lens. After passage through the sample, the beams are recollimated with a 200-mm focal length
lens, after which ω_3 is located below the plane of
ω_1 and ω_1'. ω_1, ω_1', and ω_2 are terminated using a neu-
tral density filter. ω_3 is then focused with a 100-mm
focal length lens onto the slits of a 1/3-m monochromator
equipped with a 2400-line/mm holographic grating and 100-
μm slits. The signal was detected by a PAR SIT detector
and processed by a PAR OMA2 system. The full width at
half-height (FWHH) of calibration lines near the center of
the SIT detector is near 3.0 cm^{-1}, giving approximately 1
cm^{-1} per channel over the spectral range investigated.

Table 1　Laser dyes

Dye	Solvent	Concentration (x 10^5 M)		Nonresonant background		
		Oscillator	Amplifier	I_{max} (cm^{-1})	FWHH (cm^{-1})	Species observed
DCM LDS698	EtOH	5.8 2.0	0.6 2.0	4210	300	$H_2(Q)$
DCM LDS698	EtOH	25 16	8.4 5.4	3930	420	$H_2(Q)$ H_2O
Rh640 Ox725	MeOH (NaOH)	44.5 7.8	4.1 1.5	3920 2270	150 110	$H_2(Q)$ N_2
LD690	MeOH	25	8.3	3390 3225	80 70	CH_4 overtones
LD690	MeOH	15	8.3	3190	95	CH_4 overtones
Rh640	MeOH	24	3.2	2300	110	N_2, N_2O
Kiton red Rh640	MeOH	10 3	10 3	2180	150	N_2O, CO, HCN, $H_2(S)$
Kiton red Rh640	MeOH	21 4.5	2.8 0.3	1860	100	NO, $H_2(S)$
Kiton red Rh640	MeOH	21 2.8	2.8 0	1820	110	NO, $H_2(S)$, O_2
Kiton red	MeOH	23	2.8	1730	110	$H_2(S)$, O_2, $CH_4(S)$
Kiton red	EtOH	23	2.8	1660		O_2, $CH_4(S)$
Rh610	MeOH with 4 drops saturated NaOH	21	3.7	1540	110	CH_4, $H_2(S)$ CO_2, O_2
	EtOh with aqueous NaOH			1335	100	CO_2, $H_2(S)$, N_2O

Calculated from nonresonant background.

Stationary flame measurements were made on a premixed CH_4/N_2O flame maintained on a circular burner with a 2.0-cm-diam head. The burner surface was constructed of a matrix of steel syringe needles of 0.09 cm outer diameter, so that a flat flame is achievable under suitable flow conditions. Matheson technical grade methane and chemically pure nitrous oxide were separately flowed through 603 Matheson rotameters prior to premixing. The flow through the burner was adjusted to 13 cm/s to maintain a 3.2 equivalence ratio (ϕ) flame, where ϕ is defined here as the fuel/oxidant ratio divided by the stoichiometric fuel/oxidant ratio. At this flow, there is a dark zone extending about 5 mm above the burner surface followed by a dark yellow reaction zone extending to about 13 mm above the burner. A bright yellow postflame region surrounded by a light blue afterburning diffusion flame was situated above the reaction zone. To obtain CARS spectra in the reaction zone, the center of the burner surface was displaced vertically from the focus of ω_1, $\omega_1{}'$, and ω_2 to 2 cm below the focus. Spectra were obtained at intervals of 0.5 mm (0.25 mm in the vicinity of the reaction zone). The flame exhibited macroscopic structure in that the position of the reaction zone fluctuated with respect to the burner surface. Spectral scans through the reaction zone were obtained in periods of stability between these large-scale fluctuations. Spectra obtained at the same positions relative to the position of initial attainment of full flame temperature were reproducible within the precision of the flowmeters.

The nitramine propellant grains were 14×14-mm^2 cylinders of mass 3.2 g. The propellant grains were burned in a′r with spectra taken along the centerline above the burning propellant surface during the approximatley 1-min burn time. The propellant flame can be characterized as consisting of an inner dark zone approximately 4 mm thick, a bright yellow postflame region, and an outer blue diffusion flame. The calculated approximate gas velocity from the burning cylinder is 50 cm/s.

Results

Theory

The observed CARS spectrum is proprotional to the square of the modulus of the third-order susceptibility $\chi^{(3)}$, which is the sum of a resonant term χ_r related to a nuclear displacement, and χ_{nr} related to electronic displacement:

$$\chi^{(3)} = \chi_r + \chi_{nr} \qquad (1)$$

The resonant term can be considered as a sum of Lorentzian line shapes of each Q(J), O(J), or S(J) transition (Tolles et al. 1977; Hall 1979).

$$\chi_r = \Sigma_j \; \frac{kj \; \Gamma j}{2\Delta\omega_j - i \; \Gamma j} \tag{2}$$

where

$$kj = (N/m\omega_o) \; |Mj|^2 \; (\Delta j) \; \Gamma j^{-1} \tag{3}$$

where Mj, Δj, and Γj are the polarizability matrix element, normalized population difference, and line width, respectively; $\Delta\omega_j = \omega_1 - \omega_2 - \omega_j$; m is the reduced mass; and ω is the resonant Raman frequency. $M = \alpha^2 \; (\nu + 1)$ and 7745 $b_J \pm 2^J \; \gamma^2 (\nu + 1)$ for the Q and O,S branches, respectively, where α, γ, and b_J1^J are the derivatives of the mean isotropic and anisotropic molecular polarizability, and b_J1^J are the Placzek-Teller coefficients, ν is the vibrational quantum number, and ($\nu + 1$) is contributed by the vibrational matrix element. The observed spectrum is convoluted over the laser linewidths and instrumental slit function.

χ_r is the sum of real and imaginary components χ' and χ'' respectively, such that

$$| \chi^{(3)} |^2 = \chi'^2 + 2 \; \chi' \; \chi_{nr} + \chi''^2 + \chi_{nr}^2 \tag{4}$$

χ' and χ'' display dispersive and resonant behavior, respectively. Normalizing eq. (4) with respect to the observed χ^2nr at the resonance gives

$$\left(\frac{| \chi^{(3)} |^2}{\chi_{nr}^2} \right)_o = \left| Xi \frac{(\overline{ikj})}{\overline{\chi}_{nr}} + 1 \right|^2 \tag{5}$$

where $\overline{\chi}$ and $\overline{k}$ are defined as χ/N and k/N, respectively, and Xi = N/N_T with N and N_T the number density of the resonant species and the total number density, respectively.

Summing of eq. (5) over the populated levels gives

$$\Sigma I^2 corr = \Sigma \; (\frac{1}{g_j})^2 \; (\frac{| \chi^{(3)} |^2}{\overline{\chi}_{nr}^2})_o = \left| Xi \; \Sigma \frac{\overline{ikj}}{\overline{\chi}_{nr}} + 1 \right|^2 \tag{6}$$

where $\overline{kj} = 1\overline{M}1^2/m \; \omega_o \; \Gamma j$, since $\Sigma\Delta_i = 1$ (Q branch) and $\Sigma\Delta_i/(2J + 1) \; n_J = 1$ (O,S branches) when gi = ($\nu + 1$) for Q

Table 2 Summary of species identified in a $\phi = 3.2$ CH_4/N_2O flame

Observed I_{max} (cm^{-1})	Species	Comment
4155–4075	H_2 Q branch	Temperature calculations indicate a temperature of 500 K in preheat zone.
3240–3100	CH_4 $(2\ \nu_2)$	Rapid decrease in reaction zone
2325	N_2	Gradual increase in reaction zone
2222	N_2O (ν_3)	Rapid decrease in reaction zone
2136	CO	Increase in reaction zone
2129	H_2 S(9)	Rapid increase of intensity in reaction zone; large compared to CO
1813	H_2 S(7)	Signal intensity increases up the reaction zone
1636	H_2 S(6)	Weak signal seen in post flame
1531	CH_4 (ν_2)	Gradual decrease in reaction zone
1447	H_2 S(5)	Signal intensity increases up the reaction zone
1294	N_2O (ν_1)	Intense signal that decreases rapidly in reaction zone

branches without rotational structure and $J(J + 1)$ nj $(\nu + 1)$ for O and S branches where n_J is the nuclear spin degeneracy. Γj is taken as the experimentally observed spectral line width, Γexp, since the observed CARS intensity $|\chi^{(3)}|^2$ is convoluted by the instrument function. If the peak height is measured from the maximum to the minimum of the resonant peak modulation of the nonresonant background spectra, only the imaginary term is measured (Tolles et al. 1977) so that

$$X_i = \frac{\left(\Sigma\ I_{corr}^2\right)^{1/2}}{\left(\Gamma_{exp} m\ \omega_o\right)\ \bar{\chi}_{nr}/\bar{M}^2}\ (7)$$

with $\bar{M} = \alpha^2$ and $7/45$ γ^2 for Q branches (where anisotropic contributions are neglected) and O,S branches, respectively. Equation (7) forms the basis for a qualitative interpretation of the spectra. Concentration and temperature

are separable such that concentration is related to the
sum of I_{corr} while temperature is related to the ratio of
I_{corr} of populated levels. Equation (7) was used to obtain relative concentrations through the reaction zone,
which are discussed. In addition, for N_2 and H_2, the
spectra were synthesized using the method of Hall (1979)
and fitted to the experimental spectra using a least-
square procedure to obtain temperature and concentration.

CH_4/N_2O Flames

Thermochemical calculations (Gordon and McBride 1976)
were performed for $\phi = 3.2$ CH_4/N_2O. The calculated flame
temperature was 1745 K with 23% CO, 1% CO_2, 42% H_2, 5%
H_2O, and 29% N_2. CARS spectra were obtained in the regions 4200–3900, 2400–2050, and 1900–1200 cm^{-1} as a function of distance above the center of the surface of the
burner. The spectra are given in Figs. 1–6 and summarized
in Table 2.

Reaction occurs over a region extending from 2 to 12
mm with the steepest concentration gradients occurring
between 6 and 10 mm. The flame may be roughly characterized as consisting of a dark or preheating zone extending
from the surface to 6 mm, a reaction zone extending from 6
to 10 mm, and a postflame region above 10 mm. The spectra
shown in Figs. 1–6 show one representative spectra from
each of these regions.

The spectra shown in Fig. 1 show the decay of reactant N_2O and the increase of the products N_2, CO, and H_2.
The N_2O ν_3 at 2222 cm^{-1} and associated hot bands has been
previously observed and assigned from spectra occurring in
the reaction zone of lean CH_4/N_2O flames (Harris 1982).
The N_2 and CO vibrational and the H_2 pure rotational S(7)
[the notation used is S(ν'')] spectra have been previously
discussed for spectra arising from the postflame region of
rich CH_4/N_2O flames (Aron et al. 1983). The simultaneous
observation of the decay of N_2O and the increase of the
three principal products provides an opportunity to obtain
concentration gradients to test kinetic mechanisms.

The spectra shown in fig. 2 show the decay of the
other reactant together with the increase of product H_2
S(7) and S(6) pure rotational lines. The CH_4 structure
shown in fig. 2 is assigned to ν_2 transitions. The prominent features of the methane transitions are observed at
(some lines were observed using other dyes) 1531, 1562,
1583, 1606, 1628, 1653, 1676, 1722, 1745, 1769, 1792 cm^{-1},
in agreement with previously observed (Champion and Berger
1975) and calculated (Gray and Labrette 1976) Raman lines

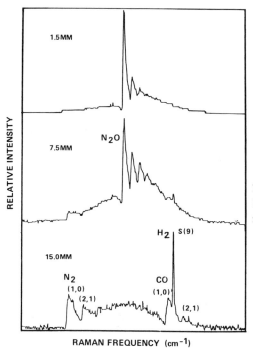

Fig. 1 CARS spectra of N_2O, CO, N_2, and H_2 S(9) from a ϕ = 3.2 CH_4/N_2O flame.

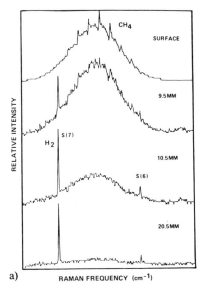

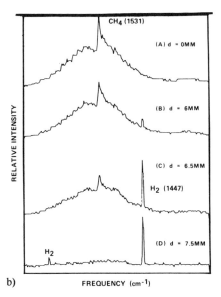

Fig. 2a) CARS spectra of CH_4 (ν_2) and H_2 S(7) and S(6) from a ϕ = 3.2 CH_4/N_2O flame. b) Fig. 2b CARS spectra of CH_4 (ν_2) and H_2 S(5) and S(6) from a ϕ = 3.2 CH_4/N_2O flame.

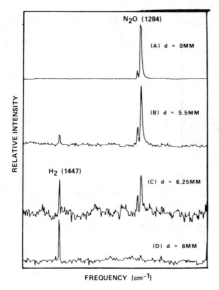

Fig. 3   CARS spectra of N_2O and H_2 S(5) from a ϕ = 3.2 CH_4/N_2O flame.

at 1534 (Q), 1566 [S(0)], 1587 [S(1)], 1610 [S(2)], 1632 [S(3)], 1654 [S(4)], 1677 [S(5)], 1724 [S(7)], 1746 [S(9)], and 1770 cm^{-1} [S(10)]. The observed position of 1792 cm^{-1} is consistent with S(11), assuming a separation of 4B for a ground state B value of 5.24 (Gray and Robiette 1976).

N_2O and H_2 S(5) spectra are shown in fig. 3. The N_2O CARS at 1284-, 1290-, and 1295-cm^{-1} transitions, which have not been previously reported, are assigned to the ν_1 and $\nu_1 + \nu_2 - \nu_2$ and $\nu_1 + 2 \nu_2 - 2 \nu_2$, in agreement with transitions observed in the Raman at 1285 and calculated at 1289.7 cm^{-1} (Herzberg 1945). The ν_1 transition has a larger Raman cross section than the ν_3 together with hot bands populated at low temperature, providing an attractive option for quantitative temperature and concentration profiles in the reaction zone.

Figure 4 shows the decay of the CH_4 2 ν_2 band. The observed bands are 3101 (Q), 3129 [S(0)], 3149 [S(1)], 3168 [S(2)], 3182 [S(3)], 3205 [S(4)], 3223 [S(5)], 3235 [S(6)], and 3258 cm^{-1} [S(7)], in agreement with assignments given by Hunt et al. (1982). Hydrogen Q band structure as it increases through the reaction zone is given in fig. 5. Hydrogen is seen at a concentration less than 1% at 4 mm. The line positions as shown for $\nu'' = 0$ in fig. 5 and $\nu'' = 0$ J $\leqslant$ 11 and $\nu'' = 1$ J $\leqslant$ 9 in fig. 6 and the S bands J = 5–9 have been shown to be in excellent agreement with the results of constants derived from ab initio calculations (Fendell et al. 1983) and constants (Dabrowski 1984) derived from the $B´\Sigma u^{+} \longleftarrow X´ \Sigma g+$ and $C´ \pi u \longleftarrow X´ \Sigma g+$ band of H_2 (Haw et al. 1985).

Temperature was calculated from the $v´´ = 0$ Q branch throughout the reaction zone. From 4 to 8 mm, there is a gradual increase in temperature from 500 K to 900 K, and the distribution appears Boltzmann. Above 8 mm, there is an apparent bimodal distribution in which approximatley half of the observed spectra have Boltzmann distributions consistent with the random experimental error, while the other spectra show much larger deviations from Boltzmann distribution much larger (greater than 2 σ) than the random experimental error. In these non-Boltzmann distributions, the odd levels are preferentially populated over the even levels, with the higher J value of both even and odd levels showing excess population. At present, it is not certain whether these apparently non-Boltzmann distributions reflect the actual hydrogen rotational distributions, instrument errors, or flame instability for hydrogen above the reaction zone. Flame stability lessens with distance above the burner. The outer blue diffusion flame is noticeably floppy. Further work is being done in more stable, rich CH_4/N_2O flames to clarify the interpretation of the hydrogen Q branch spectra. The H_2 spectra that have Boltzmann distributions give results consistent with temperatures obtained from N_2 CARS spectra and thermochemical calculations.

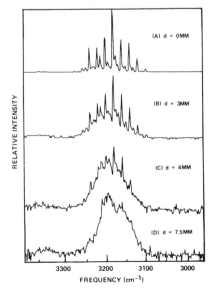

Fig. 4   CARS spectra of CH_4 ($2v_2$) from a $\phi = 3.2$ CH_4/N_2O flame.

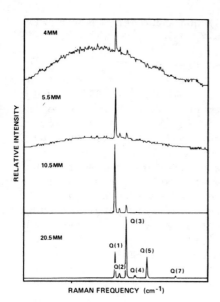

Fig. 5   CARS spectra of the
H_2 Q branch ($v'' = 0$) in a
$\phi = 3.2$ CH_4/N_2O flame.

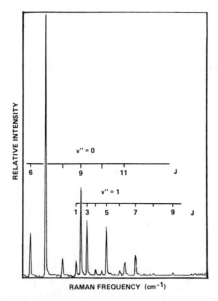

Fig. 6   CARS spectra of the
H_2 Q branch ($v'' = 0$ and $v'' = 1$)
in a $\phi = 1.8$ CH_4/N_2O flame.

    N_2 CARS spectra in the reaction zone are difficult to
interpret because of the unknown nonresonant
susceptibility. However, in the postflame region least-
squares-fits to the N_2 spectra give a temperature of 1890
± 100 K and concentration of 0.26 ± 0.05 that are close,
if slightly high for temperature, to the results of the
thermochemical calculations.

Table 3   Summary of species identified in nitramine propellant flame

Observed I_{max} (cm^{-1})	Species	Comment
4155–4075	H_2 Q branch	Temperature calculations indicate a temperature of 900 K at the surface of the propellant
2325	N_2	Slow increase until near the end of reaction zone, a large increase occurs
2136	CO	Signal increases up reaction zone
2129	H_2 S(9)	Observed similar intensity to CO
2086	HCN (ν_1)*	Strong signal initially which diminishes rapidly
1872	NO	Low concentration modulation which remains constant through-out reaction zone; decreases rapidly at end of reaction zone
1814	H_2 S(7)	Signal intensity increases up the reaction zone
1599	RDX (NO_2 asymetric stretch)*	Moderate concentration early in the reaction zone
1447	H_2 S(5)	Signal intensity increases up the reaction zone
1387	CO_2 (ν_1)	Moderate signal early in reaction

* Tentative.

products calculated as 38% CO, 27% H_2, 22% N_2, 10% H_2O, and 3% CO_2. CARS spectra were obtained in the regions 4200–3900, 2400–2050, and 1900–1200 cm^{-1} both as a function of distance above the center of the propellant surface and time at a given initial distance. The spectra are given in figs. 7–14 and summarized in Table 3. The 14-mm-high grain burns in approximately 60 s for an average burning velocity of 0.2 mm/s. The average flow calculated is 50 cm/s assuming that the cylindrical geometry is retained throughout the burn (a residual residue retaining approximately the original cylindrical geometry is retained throughout much of the burn). In the middle of the burn, the propellant flame consists of a dark region ex-

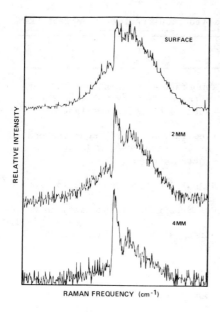

Fig. 7 Time-averaged CARS spectra of N_2 at various distances above the surface of a nitramine propellant flame.

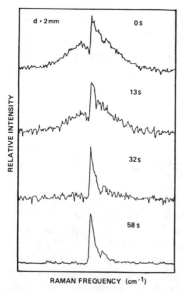

Fig. 8 Time-resolved CARS spectra of N_2 at various times after ignition at 2 mm above the surface of a nitramine propellant.

Nitramine Propellant Flames

Thermochemical calculations (Gordon and McBride 1976) were performed for the nitramine propellant with the result that the adiabatic, constant-pressure, flame temperature was calculated as 2076 K, with the equilibrium final

tending 4 mm above the propellant surface, above which a
conical dark yellow postflame region extending to 3 cm
above the propellant surface, followed by conical blue
afterburning diffusion flame of CO, H_2, and air extending
to 8 cm above the propellant surface. The initial region
of steepest formation of N_2 occurs from 3 to 6 mm above
the propellant surface. The extent of the flame was
smaller during ignition and extinguishment of the flame
(the initial and final 10 s, approximately).

Average spectra (100 scans, 10 s) were taken as a
function of distance from the propellant surface to 6 mm
above the surface at intervals of 1 mm. Each spectrum was
taken nominally 10 s after ignition. In addition, time
sequences of ten-scans (1 s) spectra were taken approximately every 6 s from ignition to extinguishment. For H_2,
the intensity of the signal permitted the acquisition of
single-shot spectra.

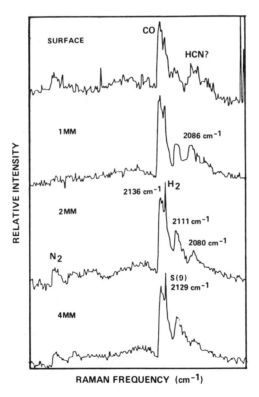

Fig. 9   Time-averaged CARS spectra of HCN, CO, N_2, and H_2 S(9)
above the surface of a nitramine propellant flame.

The N_2 spectrum as a function of distance above the propellant surface is shown in fig. 7. N_2 appears at low concentration (<1%) at the propellant surface and gradually increases to final concentration near 6 mm. Time-resolved N_2 spectra at 2 mm above the propellant surface are shown in fig. 8. N_2 is seen initially at low concentration and takes more than half the burn time to approach equilibrium concentration.

The CO region is shown in fig. 9. CO is seen at higher concentration than N_2 at the surface, with an increase in temperature most evident with increasing distance above the surface. A peak at the frequency of HCN at 2086 cm^{-1} is moderately strong at the surface and 1 mm, but has decayed by 2 mm to reveal the CO hot band at 2080 cm^{-1}. In addition, N_2 and H_2 S(9) at 2129 cm^{-1} are seen to be in approximately constant ratio to CO above 2 mm. In time-resolved spectra of this region (not shown), near the end of the burn (55 s after ignition), CO and H_2 are no longer present and only the N_2 resonance is seen.

NO and H_2 S(7) spectra are shown in fig. 10 as a function of distance above the propellant surface. NO is seen as a modulation of the nonresonant background. However, since NO has a Raman cross section only half that of N_2, the concentration of NO may be greater than 1% at the surface. In time-resolved spectra, NO is seen to persist until the steep increase in N_2 which occurs near 4 mm above the propellant surface. The H_2 S(7) band is seen to undergo a continuous increase with distance above the propellant surface up to 6 mm.

Prominent spectra at 1599 cm^{-1} tentatively associated with the RDX NO_2 asymmetric stretch transition reported at 1596 cm^{-1} in the Raman spectrum of RDX powder (Iqbal et al. 1972), H_2 S(5) at 1446 cm^{-1}, and CO_2 at 1387 cm^{-1} are shown in fig. 11. The 1599-cm^{-1} transition is present at the surface and at 1.0 mm but absent at 3.0 mm. The H_2 S(5) transition is seen to increase relative to CO_2. This is consistent with the final equilibrium concentrations of these species. At 1 mm, the moderate intensity feature between H_2 and CO_2 near 1408 cm^{-1} is seen to be complex and tentatively associable with CO_2 and HCN bending modes (Harris 1984).

Time-resolved spectra of the H_2 Q branch at the propellant surface are given in figs. 12-14. The H_2 Q branch has a cross section twice as large as nitrogen, and hydrogen is present at high concentration throughout the propellant flame so that the hydrogen Q branch spectra is not perturbed appreciably by the nonresonant susceptibility. This simplified interpretation of the spectra to obtain temperature. As shown in fig. 12, single-shot spectra are

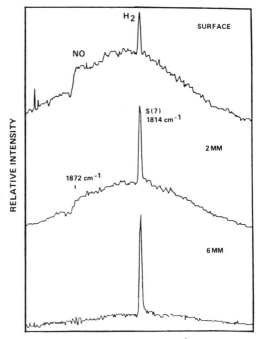

Fig. 10 Time-averaged CARS spectra of NO and H_2 S(7) above the surface of a nitramine propellant.

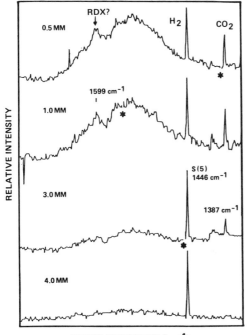

Fig. 11   Time-averaged CARS spectra of RDX (tentative), H_2 S(5), and CO_2 above the surface of a nitramine propellant.

*FOR CLARITY, SPIKES DUE TO ARCING NEAR THE PROPELLANT SUR-FACE HAVE BEEN RE-MOVED.

L.E. HARRIS

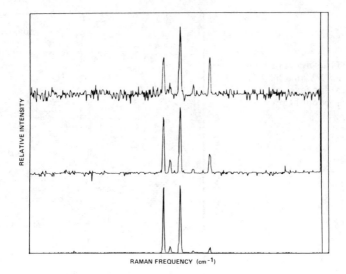

Fig. 12 Time-resolved (single-shot) CARS spectra of the H_2 Q branch taken at nominal 6-s intervals after ignition, and shown sequentially from bottom to top.

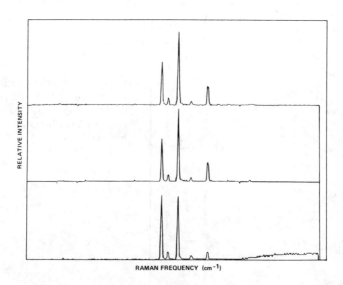

Fig. 13   Time-resolved (10-shot averaged) CARS spectra of the H_2 Q branch taken at nominal 6-s intervals after ignition and shown sequentially from bottom to top.

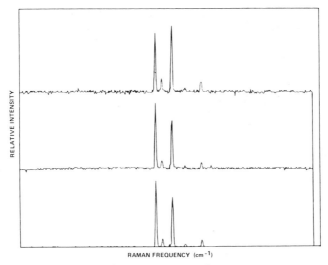

RAMAN FREQUENCY (cm⁻¹)

Fig. 14   Time-resolved (3-shot averaged) CARS spectra of the H_2 Q branch taken at nominal 6-s intervals after ignition and shown sequentially from bottom to top.

obtained at good signal-to-noise and are in substantial agreement with 10-shot averaged spectra. The reduction of the signal-to-noise from bottom to top is a reflection of the increasing temperature as a function of distance above the propellant surface. Temperatures have been obtained from the H_2 Q branch spectra. The average temperature of the spectra taken at the surface is 1000 ± 200 K. The dispersion reflects not only the noise in the individual spectra but also the variation of the distance of the surface with respect to the CARS sampling volume, since spectra are taken as close to the surface as possible. A propellant surface/gas interface temperature is more properly associated with the lower bound of the propellant surface temperature measured with good signal-to-noise rather than the average surface temperature. The spectra shown in fig. 14 are three-shot averages and are near the lower bound of measured surface/gas interface temperatures. The lower two spectra were least-squares fit to a temperature of 900 ± 100 K, which may then be associated with an upper bound of the surface/gas interface temperature. Time-resolved H_2 Q branch CARS spectra taken at the surface during the duration of the propellant burn were used to obtain temperature. (The spectra used were Boltzmann within experimental accuracy. The non-Boltzmann distributions were not obtained with the frequency en-

countered in CH_4/N_2O flames.) These spectra show an initial temperature of 1000 ± 200 K, with a gradual rise to
1500 ± 200 K over the first 40 s of the burn and a sharp
increase within a second to 2000 ± 200 K which is maintained for approximately 10 s before a rapid rise to 2600
± 200 K at 55 s of the burn. The initial gradual temperature rise to 1500 K may be associated with the dark zone,
while the steep temperature rise to 2000 K may be associated with the reaction zone of the adiabatic flame, which
culminates in flame temperature near the calculated adiabatic flame temperature of 2076 K. The subsequent steep
temperature rise to 2600 K encountered at 55 s into the
burn is associated with the afterburning or diffusion
flame of CO, H_2 with air. This is consistent with the
absence of CO and H_2 at this time in the time-resolved
spectra discussed above. As in the CH_4/N_2O flame, flame
stability decreases with distance above the surface of the
propellant. Temperature and concentration were also obtained from N_2 spectra in the adiabatic postflame region
(the spectra have not been analyzed in the dark and reaction zones because the composition of these regions is not
as yet sufficiently characterized to permit estimation of
the nonresonant susceptibility) as 2010 ± 115 K and 0.26 ±
0.04, which are in good agreement with the calculated values (2076 K and 0.22) and the temperature estimated from
hydrogen spectra in this region, 2000 ± 200 K.

Discussion

The reaction zone of the rich CH_4/N_2O flame was studied primarily to provide a stationary flame analog to the
transient nitramine propellant flames. Since the rich
CH_4/N_2O flame has similar atomic composition, adiabatic
flame temperature, and final products as the propellant
flame, it is to be expected that some features of the
kinetic mechanisms will be similar in both systems. In
the CH_4/N_2O flame, the decay of the initial products was
observed through the Q branch of the ν_1 NN and ν_3 NO
stretching modes of N_2O and the Q, O, and S branches of
the ν_2 and 2 ν_2 modes of CH_4. The formation of the products N_2, H_2, CO, and CO_2 was also observed.
No intermediate species were detected in the spectral
range investigated. Initial decomposition of the reactants was observed to occur near 500 K. The first-order
rate constants for N_2O (Balakhnine et al. 1977) and CH_4
(Tabayaski and Bauer 1979)

$$N_2O + M \rightarrow N_2 + O + M \quad k = 1.3\text{x}10^{15} \exp(-56,500/RT)$$

$$CH_4 + M \rightarrow CH_3 + H + M \quad k = 1.0\text{x}10^{17} \exp(-85,800/RT)$$

preclude observable reaction at 500 K. However, secondary reactions for N_2O given by Balakhnine et al.(1979)

$$H + N_2O \rightarrow N_2 + OH \quad k = 6 \times 10^{13} \exp (-13,100/RT)$$

and Wagel et al. (1982)

$$CH + N_2O \rightarrow N_2 + CHO \quad k = 4.7 \times 10^{13} \text{ cm}^3/\text{mole s at 300 K}$$

and CH_4 given by Tabayaski and Bauer (1979)

$$CH_4 + H \rightarrow CH_3 + H_2 \quad k = 7.23 \times 10^{14} \exp (-15,600/RT)$$

explain reactivity at 500 K.

The conversion of methane to final product is thought to proceed by the following global mechanism (Westbrook and Dryer 1984):

$$CH_4 \rightarrow CH_3 \rightarrow CH_2O, \ CHO \rightarrow H_2, \ CO$$

The conversion of fuel-bound nitrogen is an area of active current research. Recently, $CH_4/O_2/AR$ flames doped with HCN, NO, and NH_3; and $H_2/O_2/AR$ flames doped with HCN have been studied by Zabielski (1984) and Miller et al. (1984), respectively. These and previous studies have shown that the conversion of fuel nitrogen to HCN is almost quantitative and independent of the chemical nature of the initial fuel nitrogen. Since conversion to HCN is not rate limiting, research has focused on conversion to NO and N_2. These processes are thought to occur by the global mechanism.

$$\text{fuel N} \rightarrow \text{HCN} \rightarrow \text{NHi} \rightarrow \text{NO} \qquad i = 0, \ 1, \ 2$$

with NO pathway leading to N_2

Very good agreement between the theoretical and experimental temperatures and the concentration profiles in these studies has been obtained in terms of the following key reactions:

$$HCN + O \rightarrow NCO + H \quad R1 \quad k_1 = 1.26 \times 10^7 \ T^{1.87} \exp (-6800/RT)$$

$$NCO + H \rightarrow NH + CO \quad R2 \quad k_2 = 5 \times 10^{13}$$

$$NH + H \rightarrow N + H_2 \quad R3 \quad k_3 = 3 \times 10^{13}$$

$$N + OH \rightarrow NO + H \quad R4 \quad k_4 = 2.22 \times 10^{14} \exp (-50,500/RT)$$

$$N + NO \rightarrow N_2 + O \quad R5 \quad k_6 = 1.84 \times 10^{14} \exp{(-76250/RT)}$$

The reaction constant k_1 is based on a measurement by Perry and Melius (1984); k_2 is an estimate (Miller et al. 1984); and k_3-k_5 are taken from a recent compilation (Miller et al. 1983). These reactions, with the addition of the hydrocarbon combustion cycle given globally above, account qualitatively for the formation of the principal products H_2, N_2, and CO in the rich CH_4/N_2O flame studied here. Further reduction of the data obtained from the CH_4/N_2O flame to quantitative temperature and concentration profiles will enable a quantitative test of these kinetic mechanisms. The results for CH_4/N_2O, as presented, provide a comparison for results obtained in nitramine propellants.

In nitramine propellant, the upper bound of the gas-surface interface temperature was measured at 900 ± 100 K. To our knowledge, this is the first direct measurement of the gas-surface interface temperature in the gas phase. The measured temperature is in good agreement with a propellant surface temperature of 700 K measured in the solid phase using thermocouples (Kubota 1981). Near the surface of the propellant, reactant RDX and transient species HCN and NO were observed at moderate concentration (>1%). The final product N_2 was observed at low concentration ($\sim$1%) while H_2 and CO are observed at higher concentration ($\geq$10%). A gradual temperature rise from the surface to 4 mm above the surface to 1500 K is observed, with a steep rise to 2000 K occurring near 4 mm. RDX and HCN are observed to decay within 2 mm of the surface, with NO remaining constant to 4 mm. Near 4 mm, NO decays rapidly, with a concomitant increase in N_2 concentration and temperature to adiabatic flame temperature. H_2, CO, and CO_2 increased in concentration throughout this region.

The physical structure of the flame is similar to that modeled by Ben-Reuven and Summerfield (1983) in terms of near-field and far-field reactions occurring in the dark zone. However, these results suggest that the chemistry in this model (Ben-Reuven and Caveny 1981) must be modified. The observation of HCN and lack of observation of N_2O (<0.1%) is consistent with the high-temperature (T>600 K) nitramine decomposition mechanism

$$(CH_2NNO_2)_3 \rightarrow 3HCN + 6NO_2 + 3H_2$$

NO_2, although observable at low pressure in CARS, is not observable at atmospheric pressure, perhaps due to the adsorption by NO_2 in the region of the laser beams used in these experiments. Additional experiments are needed (laser fluorescence is a possibility) to determine the concentration of NO_2 in the reaction zone. NO_2 is presumably converted to NO by fast radical recombination reactions of NO_2 with H, N, and O (Baulch et al. 1973).

Thus, RDX decomposition in the flame differs from processes occurring in the CH_4/N_2O flame in that RDX decomposes directly into products that are thought to be the principal intermediates in the conversion of fuel-bound nitrogen to final products. This allows, in contrast to CH_4/N_2O processes, a substantial buildup of the intermediates HCN and NO near the propellant surface so that they are directly observable in CARS.

At these pressures, it is the decomposition processes of the species at the surface, RDX and HCN, which supply the heat that determines the burning rate of the nitramine propellant. NO conversion to N_2, which provides the heat for the luminous flame, occurs too far upsteam to affect the surface. HCN decomposition processes, which are given above, are initiated by Rl. Since Rl depends on the oxygen concentration, the ignition of nitramine propellants depends critically on oxygen concentration. The steep rise in temperature to 2600 K indicates the presence of the afterburning reaction of hot CO ahd H_2 (which constitutes 65% of the products) with air. This reaction serves to shield the inner flame from the influence of atmospheric oxygen and serves as a flameholder for the inner flame.

The reaction zone of nitramine propellant is thus seen to consist of two characteristic areas: (1) an inner flame area near the solid gas-interface which is at a temperature of 900 ± 100 K, characterized by the gas-phase reactions of RDX and HCN that provide the heat that determines the burning rate, and (2) an outer flame area farther upstream where NO is converted to N_2 to generate the luminous flame.

References

Aron, K., Harris, L. E., Fendell, J. (1983) N_2 and CO vibrational CARS and H_2 rotational CARS spectroscopy of CH_4/N_2O flames. Appl. Opt., 22(22), 3604-3611.

Aron, K. and Harris, L. E. (1984) CARS probe of RDX decomposition. Chem. Phys. Lett., 103(5), 413-417.

Balakhnine, V. P., Vandooren, J., Van Tiggelen, P.J. (1977) Reaction-mechanism and rate constants in lean hydrogen nitrous oxide flames. Combust. Flame, 28(2), 165-173.

Baulch, D. L., Drydale, D. D., and Horne, D. G. (1973) Evaluated Kinetic Data for High Temperature Reactions, Vol 2. Butterworths, London.

Ben-Reuven, M. and Summerfield, M. (1983) Combustion of nitramine propellants, ABRL-CR-00507, U.S. Army ARRADCOM, Dover, N.J.

Ben-Reuven, M. and Caveny, L. H. (1981) Nitramine flame and deflagration interpreted in terms of a flame model, AIAA J., 19(10), 1276-1285.

Boggs, T. L. (1984) Fundamentals of Solid-Propellant Combustion, (edited by Kenneth K. Kuo and Martin Summerfield), Chap. 5, AIAA, New York.

Champion, J. P. and Berger, J. (1975) High-resolution Raman-spectrum of V2-band of $CH_{12}(4)$. J. Phys., Paris, 36(2), 135-139.

Cohen, N. S., Lo, G. A., and Crowley, J. C. (1985) Model and chemistry of HMX combustion. AIAA J., 23(2), 276-282.

Dabrowski, I. (1984) The Lyman and Werner bands of H_2. Can. J. Phys., 62(12), 1639-1664.

Druet, S. A. J. and Taran, J. P. E. (1981) CARS spectroscopy. Prog. Quantum Electron., 7(1), 1-72.

Ecbreth, A. C. and Schreiber, P. (1981) Coherent anti-Stokes Raman spectroscopy (CARS). Chemical Application of Nonlinear Raman Spectroscopy (edited by A. B. Harvey), p. 27, Academic Press, New York.

Fendell, J., Harris, L. E., and Aron, K. (1983) Theoretical calcu-lation of H_2 CARS spectra for propellant flames. ARLCD-TR-83048, U.S. Army Armament Research and Development Center, Dover, N.J.

Gardner, W. C. (1984) Combustion Chemistry. Springer-Verlag, New York.

Gordon S. and McBride, B. J. (1976) Computer program for calcu-lation of complex chemical equilibrium compositions, rocket performance, incident and reflected shocks, and Chapman-Jousuet detonations. NASA SP-273.

Gray, D. L. and Robiette, A. G. (1976) Simultaneous analysis of NU2 and NU4 bands of methane. Mol. Phys., 32(6), 1609-1625.

Hall, R. J. (1979) CARS spectra of combustion gases. Combust. Flame, 35(1), 47-60.

Harris, L. E. and McIlwain, M. E. (1982) Coherent anti-Stokes Raman (CARS) temperature-measurements in a propellant flame. Combust. Flame, 48(1), 97-100.

Harris, L. E. (1982) Broad-band N_2 and N_2O CARS spectra from a CH_4/N_2O flame. Chem. Phys. Lett., 93(4), 335-340.

Harris, L. E. (1983) CARS spectra from lean and stoichiometric CH_4/N_2O flames. Combust. Flame, 53(1), 103-121.

Harris, L. E. (1984) CARS probe of RDX decomposition. Chem. Phys. Lett., 109(1), 112-113.

Harris, L. E. and McIlwain, M. E. (1981) Coherent anti-Stokes Raman spectroscopy in propellant flame. Fast Reactions in Energetic Systems (edited by C. Capellos and R. F. Walker), pp. 473-484, Reidel, Boston.

Haw, T., Cheung, W. Y., Baumann, G. C., Chiu, D., Harris, L. E. (1985) A study of flame species using CARS. 40th Symposium on Molecular Spectroscopy, Ohio State University Abstract WH10, p. 106.

Herzberg, G. (1945) Infrared and Raman Spectra p. 278, Van Nostrand Reinhold, New York.

Hunt, R. H., Lolck, J. E., Robiette, A. G., Brown, L. R., and Toth, R. A. (1982) Measurement and analysis of the infrared-absorption spectrum of the 2-NU-2 band of $12CH_4$. Mol. Spectrosc., 92(1), 246-256.

Iqbal, Z., Suryanarayana, K., Bulusu, S., and Autera, J. R. (1972) Infrared and Raman spectra of 1,3,5-Trinitro-1,3,5-Triazacyclohexane (RDX). Technical Report 4401, Picatinny Arsenal, Dover, N.J.

Kubota, N. (1981) Combustion mechanisms of nitramine composite propellants. 18th Symposium on Combustion, pp. 187-194, The Combustion Institute, Pittsburgh, Pa.

Miller, J. A., Branch, M. C., McLean, W. J., et al. (1984) On the conversion of HCN to NO and N_2 in H_2-O_2-HCN-AR flames at low pressure. WSS/CT 84-36, Sandia National Laboratories, Livermore, Calif.

Miller, J. A., Smooke, M. D., Green, R. M., and Kee, R. J. (1983) Kinetic modeling of the oxidation of ammonia in flames. Combust. Sci. Technol., 34(1-6), 149-176.

Miller, M. S. (1982) In search of an idealized model of homogeneous solid-propellant combustion. Combust. Flame, 46(1), 51-73.

Miller, M. S. and Coffee, T. P. (1983a) A fresh look at the classical approach to homogeneous solid-propellant combustion modeling. Combust. Flame, 50(1), 65-74.

Miller, M. S. and Coffee, T. P. (1983b) On the numerical accuracy of homogeneous solid-propellant combustion models. Combust. Flame, 50(1), 75-88.

Perry, R. A. and Melius, C. F. (1984) The rate of the reaction of HCN with oxygen atoms over the temperature range 549-900 K. 20th Symposium (International) on Combustion, University of Michigan, p. 2209.

Price, C. F., Boggs, T. L., Parr, T. P., and Parr, D. M. (1982) A modified BDP model applied to the self-deflagration of HMX. Proceedings of the 19th JANNAF Combustion Meeting, CPIA Publication No. 366, pp. 299–310, The Johns Hopkins University, Laurel, MD.

Schroeder, M. A. (1981) Critical analysis of nitramine decomposition data product distribution from HMX and RDX decomposition. Proceedings of the 18th JANNAF Combustion Meeting, CPIA Publication No. 347, 395–414, The Johns Hopkins University, Laurel, MD.

Tabayaski, K. and Bauer, S. H. (1979) Early stages of pyrolysis and oxidation of methane. Combust. Flame, 34(1), 63–83.

Tolles, W. M. Nibler, J. W., McDonald, J. R., and Harvey, A. B. (1977) Review of theory and application of coherent anti-Stokes Raman-spectroscopy (CARS). Appl. Spectrosc., 31(4), 253–271.

Wagel, S. S., Carrington, T., Filseth, S. V., and Sadowski, C. M. (1982) Absolute rate constants for the reactions of $CH(X^2 pi)$ with NO, N_2O, NO_2, and N_2 at room temperature. Chem. Phys. Lett. 69(1–2), 61–70.

Westbrook, C. K. and Dryer, F. L. (1984) Chemical kinetic modeling of hydrocarbon combustion. Proj. Energy Combust. Sci., 10(1), 1–57.

Zabielski, M. F. (1984) Mechanism and reaction dynamics related to methane combustion. UTRC Report 956114-24, United Technology Research Center, East Hartford, Conn.

Author Index

PROGRESS IN ASTRONAUTICS AND AERONAUTICS
SERIES VOLUMES

VOLUME TITLE/EDITORS

*1. Solid Propellant Rocket Research (1960)
Martin Summerfield
Princeton University

*2. Liquid Rockets and Propellants (1960)
Loren E. Bollinger
The Ohio State University
Martin Goldsmith
The Rand Corporation
Alexis W. Lemmon Jr.
Battelle Memorial Institute

*3. Energy Conversion for Space Power (1961)
Nathan W. Snyder
Institute for Defense Analyses

*4. Space Power Systems (1961)
Nathan W. Snyder
Institute for Defense Analyses

*5. Electrostatic Propulsion (1961)
David B. Langmuir
Space Technology Laboratories, Inc.
Ernst Stuhlinger
NASA George C. Marshall Space Flight Center
J.M. Sellen Jr.
Space Technology Laboratories, Inc.

*6. Detonation and Two-Phase Flow (1962)
S.S. Penner
California Institute of Technology
F.A. Williams
Harvard University

*7. Hypersonic Flow Research (1962)
Frederick R. Riddell
AVCO Corporation

*8. Guidance and Control (1962)
Robert E. Roberson
Consultant
James S. Farrior
Lockheed Missiles and Space Company

*9. Electric Propulsion Development (1963)
Ernst Stuhlinger
NASA George C. Marshall Space Flight Center

*10. Technology of Lunar Exploration (1963)
Clifford I. Cummings and
Harold R. Lawrence
Jet Propulsion Laboratory

*11. Power Systems for Space Flight (1963)
Morris A. Zipkin and
Russell N. Edwards
General Electric Company

*12. Ionization in High-Temperature Gases (1963)
Kurt E. Shuler, Editor
National Bureau of Standards
John B. Fenn, Associate
Editor
Princeton University

*13. Guidance and Control—II (1964)
Robert C. Langford
General Precision Inc.
Charles J. Mundo
Institute of Naval Studies

*14. Celestial Mechanics and Astrodynamics (1964)
Victor G. Szebehely
Yale University Observatory

*15. Heterogeneous Combustion (1964)
Hans G. Wolfhard
Institute for Defense Analyses
Irvin Glassman
Princeton University
Leon Green Jr.
Air Force Systems Command

*16. Space Power Systems Engineering (1966)
George C. Szego
Institute for Defense Analyses
J. Edward Taylor
TRW Inc.

*17. Methods in Astrodynamics and Celestial Mechanics (1966)
Raynor L. Duncombe
U.S. Naval Observatory
Victor G. Szebehely
Yale University Observatory

*18. Thermophysics and Temperature Control of Spacecraft and Entry Vehicles (1966)
Gerhard B. Heller
NASA George C. Marshall Space Flight Center

*19. Communication Satellite Systems Technology (1966)
Richard B. Marsten
Radio Corporation of America

---

*Out of print.

470

79. **Electric Propulsion and Its Applications to Space Missions** (1981)
Robert C. Finke
NASA Lewis Research Center

80. **Aero-Optical Phenomena** (1982)
Keith G. Gilbert and Leonard J. Otten
Air Force Weapons Laboratory

81. **Transonic Aerodynamics** (1982)
David Nixon
Nielsen Engineering & Research, Inc.

82. **Thermophysics of Atmospheric Entry** (1982)
T.E. Horton
The University of Mississippi

83. **Spacecraft Radiative Transfer and Temperature Control** (1982)
T.E. Horton
The University of Mississippi

84. **Liquid-Metal Flows and Magnetohydrodynamics** (1983)
H. Branover
Ben-Gurion University of the Negev
P.S. Lykoudis
Purdue University
A. Yakhot
Ben-Gurion University of the Negev

85. **Entry Vehicle Heating and Thermal Protection Systems: Space Shuttle, Solar Starprobe, Jupiter Galileo Probe** (1983)
Paul E. Bauer
McDonnell Douglas Astronautics Company
Howard E. Collicott
The Boeing Company

86. **Spacecraft Thermal Control, Design, and Operation** (1983)
Howard E. Collicott
The Boeing Company
Paul E. Bauer
McDonnell Douglas Astronautics Company

87. **Shock Waves, Explosions, and Detonations** (1983)
J.R. Bowen
University of Washington
N. Manson
Université de Poitiers
A.K. Oppenheim
University of California at Berkeley
R.I. Soloukhin
Institute of Heat and Mass Transfer, BSSR Academy of Sciences

88. **Flames, Lasers, and Reactive Systems** (1983)
J.R. Bowen
University of Washington
N. Manson
Université de Poitiers
A.K. Oppenheim
University of California at Berkeley
R.I. Soloukhin
Institute of Heat and Mass Transfer, BSSR Academy of Sciences

89. **Orbit-Raising and Maneuvering Propulsion: Research Status and Needs** (1984)
Leonard H. Caveny
Air Force Office of Scientific Research

90. **Fundamentals of Solid-Propellant Combustion** (1984)
Kenneth K. Kuo
The Pennsylvania State University
Martin Summerfield
Princeton Combustion Research Laboratories, Inc.

91. **Spacecraft Contamination: Sources and Prevention** (1984)
J.A. Roux
The University of Mississippi
T.D. McCay
NASA Marshall Space Flight Center

92. **Combustion Diagnostics by Nonintrusive Methods** (1984)
T.D. McCay
NASA Marshall Space Flight Center
J.A. Roux
The University of Mississippi

93. **The INTELSAT Global Satellite System** (1984)
Joel Alper
COMSAT Corporation
Joseph Pelton
INTELSAT

94. **Dynamics of Shock Waves, Explosions, and Detonations** (1984)
J.R. Bowen
University of Washington
N. Manson
Universite de Poitiers
A.K. Oppenheim
University of California
R.I. Soloukhin
Institute of Heat and Mass Transfer, BSSR Academy of Sciences

95. **Dynamics of Flames and Reactive Systems** (1984)
J.R. Bowen
University of Washington
N. Manson
Universite de Poitiers
A.K. Oppenheim
University of California
R.I. Soloukhin
Institute of Heat and Mass Transfer, BSSR Academy of Sciences